Recovery of Bioactives from Food Wastes

This book is a concise presentation of important aspects involved in dealing with extraction and utilization of bioactive compounds from the food industry waste. It starts with a focus on the global scenario of food waste generation and potential of food waste in various industries. Thereafter, the various conventional and advanced extraction techniques are discussed to highlight on how to separate bioactive compounds from the food waste. The application of industrial food waste-derived bioactive compounds in various emerging sectors are highlighted.

FEATURES

- Presents critical discussion on advancement in various extraction processes including future trends.
- Provides elaborative description of food waste sources and challenges associated with it.
- Highlights potential of bioactive compounds in various industries.
- Quantitatively discusses existing as well as new technologies/methodologies.
- Includes a separate chapter on pertinent policies of various countries.

This book is aimed at researchers and graduate students in chemical and food engineering, separation technology and bioactive compounds.

Recovery of Bioactives from Food Wastes

Mihir Kumar Purkait, Prangan Duarah
and Pranjal Pratim Das

CRC Press
Taylor & Francis Group
Boca Raton London New York

CRC Press is an imprint of the
Taylor & Francis Group, an **informa** business

First edition published 2023
by CRC Press
6000 Broken Sound Parkway NW, Suite 300, Boca Raton, FL 33487-2742

and by CRC Press
4 Park Square, Milton Park, Abingdon, Oxon, OX14 4RN

CRC Press is an imprint of Taylor & Francis Group, LLC

Library of Congress Cataloging-in-Publication Data

Names: Purkait, Mihir K., author. | Duarah, Prangan, author. | Das, Pranjal Pratim, author.
Title: Recovery of bioactives from food wastes / Mihir Kumar Purkait, Prangan Duarah, Pranjal Pratim Das.
Description: Boca Raton : CRC Press, 2023. | Includes bibliographical references and index. |
Identifiers: LCCN 2022050158 (print) | LCCN 2022050159 (ebook) | ISBN 9781032325255 (hardback) | ISBN 9781032325262 (paperback) | ISBN 9781003315469 (ebook)
Subjects: LCSH: Food industry and trade--Waste minimization. | Food waste--Recycling. | Bioactive compounds.
Classification: LCC TD899.F585 P87 2023 (print) | LCC TD899.F585 (ebook) | DDC 664--dc23/eng/20230119
LC record available at https://lccn.loc.gov/2022050158
LC ebook record available at https://lccn.loc.gov/2022050159

ISBN: 9781032325255 (hbk)
ISBN: 9781032325262 (pbk)
ISBN: 9781003315469 (ebk)

DOI: 10.1201/9781003315469

Typeset in Times
by Deanta Global Publishing Services, Chennai, India

Contents

About the Authors

Mihir Kumar Purkait is a Professor in the Department of Chemical Engineering, Dean of Alumni and External Affairs and ex-Head of Centre for the Environment at Indian Institute of Technology Guwahati (IITG). Prior to joining as faculty in IITG (2004), he received his PhD and MTech in Chemical Engineering from Indian Institute of Technology Kharagpur (IITKGP) after completing his BTech and BSc (Hons) in Chemistry from University of Calcutta. He has received several awards, such as Dr. A.V. Rama Rao Foundation's Best PhD Thesis and Research Award in Chemical Engineering from IIChE (2007), BOYSCAST Fellow award (2009–10) from the DST, Young Engineers Award in the field of Chemical Engineering from the Institute of Engineers (India, 2009) and Young Scientist Medal award from the Indian National Science Academy (INSA, 2009). Purkait is a Fellow of Royal Society of Chemistry (FRSC) UK and Fellow of Institute of Engineers (FIE) India. He is the director of two incubated companies (viz. RD Grow Green India Pvt. Ltd. and Vixudha Bio Products Ltd.). He is also technical advisor of Gammon India Ltd and Indian Oil Corporation, Bethkuchi, for their water treatment plant.

His current research activities are focused in four distinct areas viz. (i) advanced separation technologies, (ii) waste to energy, (iii) smart materials for various applications and (iv) process intensification. In each of the area, his goal is to synthesize stimuli-responsive materials and to develop a more fundamental understanding of the factors governing the performance of the chemical and biochemical processes. He has more than 20 years of experience in academics and research and has published more than 250 papers in different reputed journals (citation >12,000, h-index = 64, 10 index = 152). He has 10 patents and has completed 35 sponsored and consultancy projects from various funding agencies. Purkait has guided 20 PhD students so far. He is the author of ten books and several book chapters published in reputed international journals.

Prangan Duarah is a doctoral research fellow at Centre for the Environment at Indian Institute of Technology Guwahati (IITG). He worked on a research project at the Indian Institute of Technology, Bombay, before joining IITG as a research fellow. He graduated from North Eastern Hill University in Shillong, India, with a B.Tech in Energy Engineering. His current field of study focuses on naturally derived bioactive compounds and green nanomaterial synthesis for various applications. He is a recipient of the prestigious Prime Minister's Research Fellowship (PMRF), which is given to the country's most outstanding research scholars as a token of encouragement to pursue their research. He has published several peer-reviewed articles in reputed international journals, along with several book chapters and one book.

Pranjal Pratim Das is currently a research scientist at the Department of Chemical Engineering, Indian Institute of Technology Guwahati, Assam, India. He received his M.Tech (2017) and B.Tech (2014) degrees in Food Engineering and Technology from Tezpur (Central) University, Assam, India. His research work is purely dedicated to industrial wastewater treatment via electrochemical and advanced oxidation treatment techniques. His research focus is also related to the application of integrated (electrochemical and oxidation) water treatment processes for the remediation of specific unit operations of steel industry effluents. He is currently working in the treatment of cyanide- and phenol-contaminated wastewater of Tata Steel Industry, Jamshedpur India. He has published several peer-reviewed articles in reputed international journals, along with patents and book chapters. He is also a potential reviewer of the Journal of Water Process Engineering. Furthermore, he has fabricated and demonstrated many pilot plants for green energy generations from wastewater. He has worked extensively in various iron and steel making industries and has also delivered many pilot-scale set-ups to several water treatment facilities across the state of Assam, India for the treatment of toxic contaminated wastewaters.

Preface

Food is essential not just to our existence as a species but also to our culture. However, in order to fulfil our requirements, we generate a large amount of waste. According to the UN Food and Agricultural Organization (FAO), one-third of all food produced for human consumption is wasted, amounting to 1.3 billion tonnes of food waste per year. Improper food waste disposal is presently causing major concerns, including global emissions and climate impact. The enormous carbon footprint of food waste stems not only from the energy required to transport, prepare and manufacture the food that ends up as waste but also from the potent methane gas produced when the food waste decomposes in landfills. However, in recent years various research has indicated that these waste-considered materials can be a valuable source of bioactive compounds, which can be extracted and implemented in new processes and products and replace costly raw materials for various other products.

Bioactive compounds are primarily secondary metabolites found in plants, although certain primary metabolites have recently been identified as bioactive molecules as well. They have been used to reduce the risk of sickness and to treat a wide range of illnesses. Given the fact that bioactive substances exist in varying quantities, it is vital that their production is developed in order to obtain the maximum quantity and identify new, cheaper sources of alternatives. Several studies have shown that various types of food waste derived from diverse sources can be used as a potential source to extract bioactive compounds with considerable applicability in a variety of applications.

This book is a concise presentation of a variety of important aspects involved in dealing with the extraction and utilization of bioactive compounds from food industry waste. Each and every fundamental concept with respect to the utilization of such waste for bioactive compound extraction will be explained elaborately, and importance will be given to the minute details. The first few chapters of the book are concerned with the worldwide scenario of food waste generation and the potential of food waste in various sectors. Following that, numerous conventional and advanced extraction strategies are addressed in order to provide a mechanistic understanding of the bioactive component separation processes from food waste. Furthermore, advancements in the extraction of bioactive compounds from various forms of food waste are highlighted. The advanced utilization of industrial food waste-derived bioactive compounds in several developing areas is emphasized in the next chapter. Finally, government policies for industrial food waste management of several countries are presented. Moreover, readers of this book will learn everything there is to know about industrial food waste valorization, from the fundamentals to the most recent research achievements.

Mihir Kumar Purkait
Prangan Duarah
Pranjal Pratim Das

Acknowledgements

Writing this book has been a lifelong academic endeavour infused with the love, trust and support of some of our most important individuals and communities. Today, at the acme of our publication, we fondly remember our parents, family, friends and well-wishers. Without their enthusiastic support and active participation, this book would not have taken shape. The honour is elevated manifold because of the very opportunity bestowed upon us by our esteemed institution—the Indian Institute of Technology, Guwahati. We will be forever grateful for that. We would also like to express our gratitude to the Indian National Academy of Engineering, Gurgaon, India, for the support.

1 Food waste as a potential source of bioactive compounds

1.1 OVERVIEW OF FOOD WASTE

1.1.1 ORIGIN OF FOOD WASTE

Food waste happens mostly, but not solely, during the last consumer step of the food supply chain, i.e., the household level. Processing, distribution, retail, ultimate consumption and the processes after consumption all result in food loss. A general model indicating the various steps of the food supply chain along with instances of food waste and loss is shown in Figure 1.1. The majority of mechanical loss and/or leakage during harvest activities are the primary reason of food losses in the first phase viz. crop sorting, fruit picking and threshing. The predominant causes of the aforementioned food losses are natural forces (such as weather conditions and temperature) and economic aspects (such as rules/regulations and private/public standards for quality) [1]. Food losses are any reduction in edible food mass, which was lost accidentally or unintentionally throughout the post-harvest, storage, processing and distribution phases. For instance, spillage, deterioration, unavailability of storage provisions and freightage between the farm and distribution unit all contribute to food loss during the post-harvest handling stage. Pests and microbes cause a significant proportion of food loss during storage.

Food waste during industrial processing, however, results from spills or deterioration such as canning and juice production. Food waste at this step can also result from slicing, washing, boiling, peeling, process stoppage or crops that have been sorted [2, 3]. Food waste is produced at the distribution stage as a result of ineffective transportation techniques, poor packing, time constraints, supplier/buyer interactions and inadequate infrastructure. Food loss is typically referred to as food waste at the store level since it mostly results from a deliberate choice to discard food. Still healthy and safe for ingestion by people, this food. The Department for Environment, Food & Rural Affairs asserts that retailers' poor demand prediction, catalogue management errors, weather conditions and temperature susceptibilities throughout freightage, dumping of unsold food, improper packaging and misinterpretation of food regulations are to blame for the creation of food waste [4]. Retailers often discard large amounts of food that has passed the "use by," "sell by," or "best before" dates. In order to share unsold food or provide customers advice on the utilization of food, which may be in danger of going to waste, some merchants work with charitable organisations along with food redistribution and recovery institutions. Other merchants refrain from giving free food to reduce their liability in the event of food contamination. Additionally, lack of information and awareness about how to utilize food efficiently, cultural challenges, along with the absence of proper planning and packaging all contribute to food waste at the consumer level. Food waste can be broken down into three categories at the post-consumer level: 1) waste that can be strictly avoided (thrown away foods which had been expired), 2) waste that can be possibly avoided (food that can be consumed when prepared one way but not another, or that some people will eat but others won't) and 3) waste that cannot be avoided (inedible foods that are left out after food preparation) [5]. It is estimated that the food lost through the producer to distributor stage could feed one billion people. Additionally, food losses squander human labour, agricultural inputs, livelihoods, capital and limited natural resources like water. Currently, the upstream supply chain (from production

DOI: 10.1201/9781003315469-1

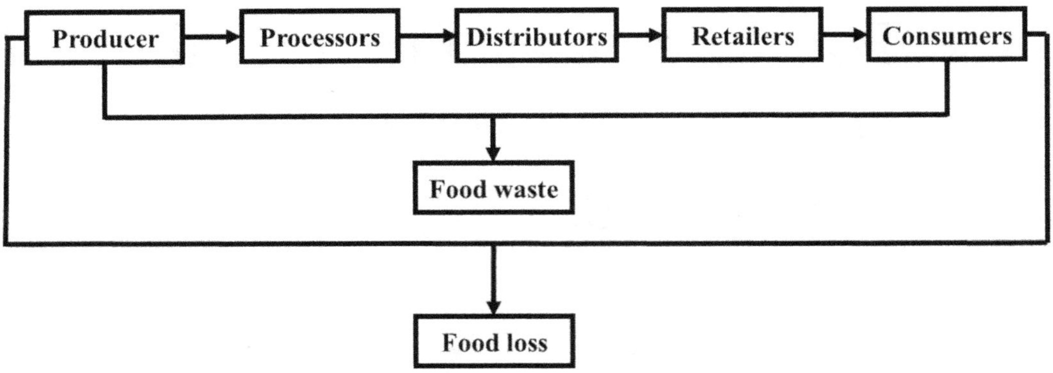

FIGURE 1.1 Stages of food waste and food loss occurrence in a food supply chain [Reproduced with permission from Otles et al. (2015) [2]].

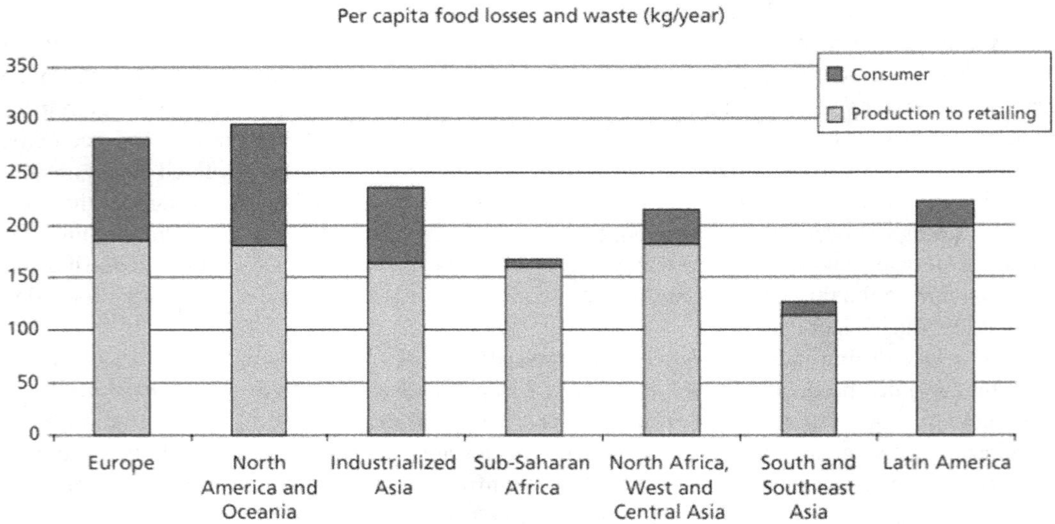

FIGURE 1.2 Per capita food waste and food loss in different continents [Reproduced with permission from Otles et al. (2015) [2]].

to distribution) accounts for the majority of food losses in underdeveloped nations, whereas the downstream supply chain is the focus of losses and waste in developed nations. Decreasing food losses in Africa is crucial owing to the way the continent's food supply networks are set up. Losses in this area are caused by a wide range of technological and administrative restrictions in harvesting methods, infrastructure, packaging, marketing systems, storage, transportation, processing and cooling facilities. There have been some attempts to estimate worldwide food loss and waste [6]. To quantify the real loss amounts, meanwhile, is challenging because methodology, evaluation levels and food goods vary widely. Figure 1.2 displays the per capita food waste and food loss for industrialized Asia, Europe, North Africa, South and Southeast Asia, North America and Oceania, West and Central Asia and Latin America. As can be observed, even in Europe's industrialized nations, the bulk of food loss and waste occurs from the manufacturing to the retail phases. Consumer food loss and waste is primarily a challenge in Europe, Oceania, North America and industrialized Asia. The bulk of food is lost from manufacturing to retail phases, whereas South and Southeast Asia and Sub-Saharan Africa appear to have the least amount of food loss at the consumer stage. Food losses and waste are produced throughout supply chains in industrialized nations and can be attributed to managerial choices, market signals, insufficient access to the right techniques, regulatory structures and their erroneous interpretation, social norms and ineffective waste management techniques [7]. Rapid transitions to a highly sustainable food supply chain, addressing tangible activities from initial manufacturing to the consumption level, might be made possible by a legislative enabling environment, focused business sector investments and civil society engagement. It is also crucial to remember that data on food loss and waste varies depending on the product. Seven subcategories and two groupings (plant and animal) can be used to classify food items (e.g., fruit and vegetables, root and tubers, fish and seafood, oil crops and pulses, diary, cereals and meat products). The highest amounts of waste are seen in fruits, vegetables, roots and tubers. Globally, 40–50% of grains, 30% of fish, 30% of root, fruit and vegetable production and 20% of meat, dairy and oil seed production are lost or squandered [8].

1.1.2 Global food waste generation

All across the world, large amounts of food are wasted. One of the main conclusions is that food is dispersed in about the same amounts in industrialized and underdeveloped countries. More than 40% of food losses take place at the processing and post-harvest stages in poor countries, compared to around 45% at the retail and consumer levels in developed nations. Nearly as much food is wasted at the consumer level in developed nations (0.222 billion tonnes) as is produced in sub-Saharan Africa overall (0.230 billion tonnes). According to the survey, most of the food waste is comprised of fruits, vegetables, roots and tubers. Of all the food that is purchased each year in UK households 25% is wasted (by weight). For two primary reasons, unnecessary food and beverage waste is thrown away: Too much food is cooked, prepared or served, resulting in the waste of 2.2 million tonnes; and an additional 2.9 million tonnes as it was not utilized in time. Food waste is expensive: in the UK, consumers spend £12 billion annually on good food that is ultimately squandered. This equates to £480 for the typical UK family, rising to £680 for those with children—an average of little more than £50 each month [9]. In Japan, over 20 million tonnes of food waste are produced annually, similar to the UK. This implies that up to $10 trillion worth of food is wasted each year. Food waste in Japan was recycled to the tune of 70% in 2008; of this, 50% was used as animal feed, 30% as fertilizer and 5% as methane. Most of the remaining food waste was either burned or dumped in landfills [10]. Every year, Taiwan produces over 16.5 million tonnes of food waste. Food waste is said to make up more than 22% of municipal solid trash in the Republic of Korea, and it is produced at a rate of approximately 0.25 kg per person each day. This equates to 4.3 million tonnes of food waste per year. It is astounding how much food is wasted in the United States. The US Environmental Protection Agency estimates that the US produces around 35 million tonnes of food waste annually. More than 14% of the entire municipal solid waste stream is made up of food waste. In 2009, 34 million tonnes of food waste were generated; less than 3% of that was collected and repurposed. Of the remainder, 33 million tonnes were wasted. Food waste currently makes up the majority of municipal solid trash that is transferred to landfills and incinerators. Over 96% of food waste in the United States (US) is now thought to be buried in landfills [11]. Food waste is considered to be a substantial methane source, a powerful greenhouse gas having more than 20 times the global warming potential of CO_2, which quickly rots when it is disposed of in a landfill. More than 20% of all methane emissions in the US come from landfills, which is a significant supply of human-related methane that may be exploited as a source of energy. Nevertheless, there are methods to keep this garbage out of landfills, as shown by a variety of municipal and state-level initiatives and legislation, such as collection schemes for source-separated organic waste. The food waste from households accounts for US$ 48.3 billion of the US$ 90–100 billion in annual food losses in the US. Norway generates 71 kilogrammes of garbage per person per year, which is comparable to the UK. Moreover, each family in Holland wastes between 43 and 60 kilogrammes of food each year. Each year, the Netherlands wastes € 2.4 billion ($2.4 billion) worth of food, or more than 20% of all the food sold. The estimates range widely from 25% to 50% in the US, while an average Australian household disposes of roughly AUD 239 (US$ 222) per person yearly [5]. The unexpectedly huge discrepancies across cultures that appear to share comparable socio-economic and cultural traits may be partially attributed to methodological issues and challenges in defining the measurement's bounds. An estimated 50% of the food produced in Europe is lost or squandered. In the best situation, about 20% of our food is wasted; however, this varies from nation to country and from industry to industry. Over 50 million people in Europe are at the danger of relative poverty. At the same moment, 17% of all manufacturing waste in the European Union (EU27) and more than 40% of waste in Ireland, Cyprus, Hungary and the Netherlands comes from the production of food. In Sweden, it is estimated that 25% of the food purchased is wasted by an average family [8]. The quantity of food waste and food loss along the entire food supply system is shown in Figure 1.3.

The amount of food wasted greatly relies on the society in which it was produced and eaten. The majority of food wasted in poor nations is lost at the point of production, whether it be in the fields, owing to inadequate cooling and storage facilities, or during transit. Lack of infrastructure and related management along with technical abilities in the production of food and post-harvest processing have been highlighted as major factors in the formation of food waste in the developing nations, both today and in the near future [12]. The loss of fresh vegetables, for instance, is said to be between 35 and 40% in India since neither retail nor wholesale stores have cold storage facilities. The cheap cost of food in relation to disposable income, the high expectations of consumers for food aesthetic grades, as well as the growing separation between customers and the production of food are the main causes of post-consumer food waste in industrialized nations. Similar to how growing urbanisation in developing nations may cut off residents from food supplies, this trend is expected to result in more food waste being produced. Poor harvesting and growing practises and a lack of infrastructure in many developing nations are expected to continue to be key contributors to the production of food waste. Postharvest systems get less than 6% of the funding for agricultural projects, despite the fact that cutting these losses is acknowledged as a key factor in enhancing food security. Regardless of the global area, the food supply chain has to successfully use culture-specific innovations and technology to cut losses [13, 14].

1.1.3 Sustainability of the food supply chain

On a global scale, it looks like there is enough food to feed everyone on the planet. On the other hand, this conceals a broad range of food consumption, from severe food shortages to excessive eating and significant amounts of food waste. For more than one billion people, food insecurity is an unpleasant reality. On the opposite end of the scale, individuals also frequently overeat and waste

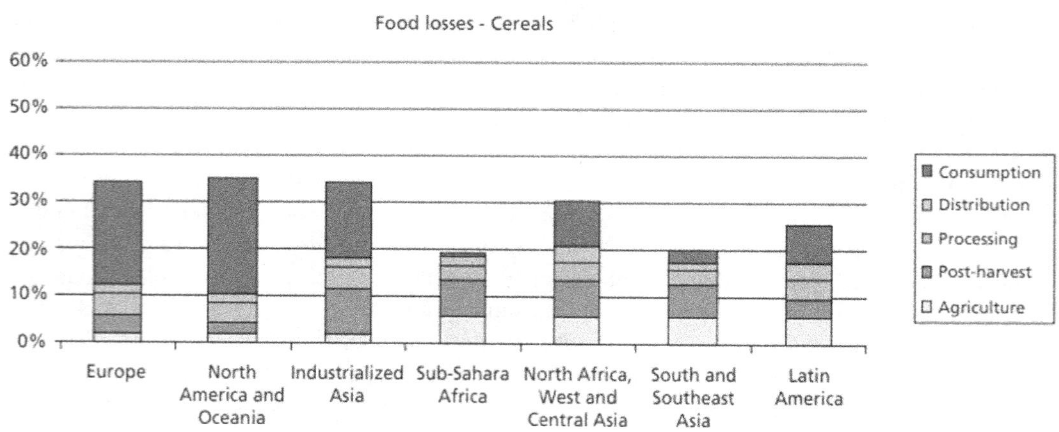

FIGURE 1.3 The amount of food loss and food waste along the food supply chains [Reproduced with permission from Otles et al. (2015) [2]].

food. The issue of obesity and overweight in adolescents and children is mostly a result of increasing calorie consumption. The incidence of childhood obesity may have been impacted by long-term trends showing considerable increases in the availability of fruits and vegetables, added oils, frozen dairy products, meat, fruit juices and cheese. For the food category of meat, estimated losses from spoilage, waste and cooking procedures can reach 57%. Analysis of poor food chain efficiency is hampered by a lack of empirical evidence [15]. The food supply in many nations is far higher than both the recommended daily consumption of food and the international average. Since there are significant losses and conversions, the variation between the amount of food consumed and the amount of food produced in the field is significantly greater. The danger of a twin problem—a worsening public health condition and an increase in waste—increases as the availability of food rises and is regarded as being relatively inexpensive and widely available. This has detrimental effects on the demand on resources, the environment and societal productivity. The whole food supply chain (FSC), from initial agricultural production to ultimate home consumption, is plagued by food waste. There is significant food waste and loss even in the early stages of the FSC. Between 25 and 50% of food output is thought to be lost along the supply chain. The early and middle phases of the FSC contribute to the majority of food loss in low-income nations; consumer waste accounts for a substantially smaller portion of food loss. Food production that will not be consumed leads to unwarranted emissions of CO_2 along with a decline in the food's economic worth [16, 17].

For reasons related to the economy, ecology and food security, food waste is a major issue on a worldwide scale. According to the experts, if more intelligent and sustainable management of food consumption and production is not implemented, food prices in a world of 7 billion people—which is expected to be more than 9 billion by 2050—may in fact increase and become more volatile. Because of climate change, land degradation and water shortages, more than 20% of the global food supply may be "lost" in this century [18]. These are the negative effects on the ecosystem brought on by the untenable utilization of natural resources including land, energy and water [19, 20]. The 21st century saw a considerable improvement in the amount of food produced throughout the world, mostly as a consequence of increased yields brought on by irrigation, fertilizer usage and agricultural expansion into new territories, with little thought given to the energy efficiency of the food being produced. In addition, it is predicted that the global price of food would increase by 30–50% and become more volatile over the next few decades. The lives and livelihoods of those who are already malnourished or poor who spend 75–85% of their daily wages on food have been drastically affected by rising food costs. The combined consequences of food stock speculation, harsh weather occurrences, low grain stockpiles, expansion in biofuels vying for cropland and high oil prices are major contributors to the present food crisis. In order to increase food-energy effectiveness by 30–50% at present production levels, the United Nations Environmental Programme advised capturing and recycling post-harvest losses/waste and developing innovative technologies [21]. The daunting issues of water scarcity, climate change adaptation and mitigation, biodiversity loss, collapse of petroleum-derived energy and continuing food insecurity in expanding populations require new approaches. Food waste also has an economic impact, which eventually has an influence on all the businesses and people engaged in the food supply system, including the final consumer. By improving supply via optimization of food-energy efficacy, we may boost food security without just concentrating on raising output. Our capacity to reduce energy loss in food from potential harvest by processing to actual recycling and consumption is referred to as food energy efficiency. In a way comparable to the performance increment in the conventional energy sector, this chain may be optimized to allow for an increase in food supply while causing far less environmental harm [22].

1.2 ENVIRONMENTAL EFFECTS AND MANAGEMENT APPROACH

Evaluation of the carbon footprint of the food and beverage retail supply system has received a lot of attention, as carbon is a popular source to measure environmental impact. The greenhouse gas (GHG) emissions linked to food and packaging waste were evaluated by Waste & Resources Action

Programme (WRAP). The GHG emissions related to food and drink waste were calculated using WRAP's assessment of the carbon effect of the food and drink supply system and domestic waste. At each level of the supply chain, conversion factors were created to translate waste volumes into CO_2 equivalent quantities (3.8 tonnes of CO_2 eq. per tonne of waste is taken as the overall conversion factor). The GHG emissions of Finnish families' unwarranted food waste almost equal the emissions of 100,000 vehicles. Food and packaging waste in the supply system have a combined carbon effect on the UK of 10 million tonnes of CO_2 equivalent, while home garbage has a carbon impact of 26 million tonnes. Additionally, the by-product that is used as animal feed has a 3.7 million tonnes CO_2 equivalent greenhouse gas effect [23, 24]. Food waste has the greatest influence on the home and manufacturing sectors, whereas packaging waste has the greatest impact on distribution and retail. Additionally, a lot of food waste that is dumped in landfills releases methane, which is a stronger greenhouse gas than even CO_2. Uninitiated individuals may not be aware that excessive levels of greenhouse gases, such as methane, CO_2 and chlorofluorocarbons, heat up the earth's atmosphere by absorbing infrared radiation. This process results in climate change and global warming. Of course, methane is a GHG that many scientists think has a negative impact on the earth's climate and temperature (i.e., global warming/climate change). Compared to CO_2, methane is better at retaining heat in the atmosphere. Twenty percent of greenhouse gas emissions are made up of methane [25]. We generate less methane gas when we waste less food, which is much healthier for the environment. The biodiversity is also impacted by food waste in a variety of ways. The term "biodiversity" refers to the whole range of life found among various species and types of organisms in an ecosystem. Deforestation is carried out in order to raise or increase cattle. This has an impact on our native flora and animals. More natural area is converted into pastures as the number of animals increases. The more sheep graze on a piece of land, the less arable it becomes. Land loses its natural diversity and variety. We catch a lot of marine fish without considering how their population decline may impact our biodiversity. Some of these fish are rejected by retailers because they don't meet quality requirements, thrown out by customers, or otherwise allowed to perish in the stores. Food suppliers and manufacturers have a variety of options for lowering food waste. Food producers might spend money on technologies for preservation and storage. If food is not stored correctly after harvest, it may go bad. Refrigeration is not practicable in some locations—particularly so in many emerging nations. In these locations, scientists are looking for alternatives to refrigeration. Dehydrating the food and putting it in moisture-resistant storage containers is one possibility [26]. Additionally, stores may lower the cost of food that is deemed "imperfect" or "ugly." The "best before" dates on their meals can be carefully observed. In this manner, perfectly good foods won't be wasted. Additionally, supermarkets can give the unsold food to the poor and needy. Some of the food can be utilized as animal feed rather than being wasted. Currently, cultivating crops for livestock feed occupies a sizable portion of agricultural land. Alternatives are always worth considering, as clearing land for farming harms the habitat of local animals and plants. Finally, composting our food waste rather than tossing it away would benefit the earth [27].

1.3 VALORIZATION OF FOOD WASTE IN VARIOUS APPLICATIONS

1.3.1 Production of organic acids

Production of lactic acid: Lactic acid is utilized in a variety of product sectors, including chemical, medicine and food (as an acidulant, preservative and flavouring agent). Lactic acid is now produced at roughly 67 million kg per year, and it is anticipated that the global demand for lactic acid will increase by around 13% every year. As a source of raw materials for the formation of lactic acid, apple pomace exhibits a number of benefits, such as: (a) high concentration of free fructose and glucose, two efficient sources of carbon for the formation of lactic acid, (b) high concentration of monosaccharides that can be produced by the enzymatic hydrolysis of polysaccharides (hemicelluloses, cellulose and starch), (c) presence of substances that lactic bacteria can metabolise, such as malic acid, citric acid and monosaccharides apart from fructose and glucose and (d) metal ions, viz. Fe, Mn and Mg, are present, which may lower the price of the nutrient addition for fermentation medium [28]. Polysaccharides must be hydrolyzed, resulting in solutions with high amounts of sugars and other fermentable components, to improve the lactic acid output from apple pomace. Gullon et al. (2008) assessed: (a) apple pomace's analytical characterization, (b) cellulase–cellobiase combinations' enzymatic saccharification of the apple pomace and (c) fermentation of enzymatic hydrolysates collected under certain operating circumstances to produce lactic acid [29]. The effects of the liquor-to-solid ratio (LSR) and the cellulase-to-solid ratio (CSR) were studied in connection to the kinetics of glucose and total monosaccharide generation. Additionally, they used the Response Surface Methodology to create a collection of statistical regression models to replicate and forecast how the operating circumstances affected the composition of the hydrolysate. The glucose and fructose, which were present as free monosaccharides, were removed at the start of the enzymatic hydrolysis process when samples of apple pomace were used. After 12 hours, 79% of the total glucan was saccharified using low cellulase (8.5 FPU/g solid) and cellobiase charges (8.5 IU/g solid). This resulted in a solution with up to 42.6 g of monosaccharides per litre (rhamnose: 2.8g/L, galactose: 2.5g/L, fructose: 14.8 g/L, glucose: 22.8 g/L). These results translate to a volumetric productivity of 3.55 g of monosaccharides per litre hour and a monosaccharide/cellulase ratio of 0.05 g/FPU. Using liquors acquired under these circumstances, Lactobacillus rhamnosus CECT-288 was utilized to produce fermentative lactic acid, resulting in a media with around 31.5 g/L of L-lactic acid after 6 hrs. After 6 hrs, lactic acid was produced at a high volumetric productivity (Qp 55.41 g/(Lh)) due to the quick metabolism of glucose and flaxseed oil (FXO) [22].

Production of citric acid: Citric acid is an economically useful substance that is widely utilized as an acidifying and flavour-enhancing ingredient in the food, pharmaceutical and beverage sectors. It is mostly manufactured commercially by submerged fermentation (SmF) from a medium based on sucrose or molasses [30]. Shojaosadati and Babaeipour (2002) utilized apple pomace as a substrate for the SSF in a column reactor method of producing citric acid using Aspergillus niger [31]. They assessed a number of culture factors, including the aeration rate (0.8, 1.4 and 2.0L/min), bed height (4, 7 and 10 cm), particle size (0.6–1.18, 1.18–1.70 and 1.70–2.36mm) and moisture content (70%, 74% and 78%). The aeration rate and particle size were the most crucial variables for citric acid output. It was discovered that citric acid production was not considerably impacted by either bed height or moisture level. Low aeration rate (0.8L/min), high bed height (0.10m), big particle size (1.70–2.36 mm) and high moisture content were the operational parameters that enhanced citric acid production (78%). The synthesis of citric acid by A. niger under SSF conditions utilizing pineapple, mixed fruit and masoumi waste as substrates was studied by Kumar et al. (2003) [32]. Based on the amount of sugar ingested, they each achieved a maximum citric acid output of 51.4%, 46.5% and 50%. In a different investigation, wheat bran and sugarcane bagasse were used as substrates. After nine days of fermentation, they discovered that the latter produced the most (9.4 g/100 g of dried substrate) [33]. Imandi et al. (2007) investigated pineapple waste as the only substrate for Yarrowia lipolytica's synthesis of citric acid under SSF conditions [34]. A citric acid output of 202.35 g/kg ds (g citric acid generated per kilogramme of dried pineapple waste as a substrate) was achieved after maximizing the composition of the growth medium.

1.3.2 Extraction of bioactive compounds

Phenolic compounds: The health benefits of food generated from plants are mostly attributable to dietary fibre with a high molecular weight and low-molecular-weight secondary plant metabolites with a lower molecular weight. The latter components include carotenoids, polyphenols, glucosinolates, saponins, alkaloids and other compounds that are chemically quite diverse. The polyphenols are a very varied family of secondary metabolites that are generically categorized as phenolic acids (hydroxycinnamic and hydroxybenzoic acids), flavonoids, xanthones and stilbenes. Numerous health-promoting qualities, including antithrombotic, antidiabetic, anti-inflammatory, anticarcinogenic, antioxidant and vaso-protective actions, have been connected to them [35]. Numerous studies on the extraction of secondary plant metabolites have been conducted as a result of the tendency to develop functional meals by adding bioactive substances, which in turn begs the question of where these ingredients originate. Fruit and vegetable processing wastes, viz. stems, seeds and peels have drawn a lot of attention in this area during the past ten years. They may be produced in vast numbers and can provide a significant disposal challenge for the food sector, depending on the raw materials used and the technology employed. Even larger percentages of by-products result from the preparation of a few fruits, viz. mangoes, where the seeds and peels can make up to around 62% of its total weight. Secondary plant metabolites like polyphenols are crucial to the plant's defence mechanism and help to shield it from both abiotic and biotic stress [22]. Also, phenolic compounds exhibit antibacterial properties against plant diseases. The outer layers of fruits and vegetables as well as their seeds primarily consist of the secondary metabolites because of their biological function in plants. Such plant portions are often separated via peeling during processing or are kept in the press leftovers such as seeds and skins in grape pomace. As such, these by-products are interesting sources of bioactive chemicals that might be used in functional meals because they are highly concentrated sources of secondary plant metabolites. The by-products' high water concentrations make them susceptible to microbial deterioration and require prompt drying following processing, which is an economically prohibitive step [36]. The principal phenolic fractions and a few sources of fruit and vegetable (FV) processing residues are presented in Table 1.1. The most prevalent class of polyphenolics in the average person's diet is a group of plant by-products called flavonoids. They are further split into the flavone, flavonol, flavanone and isoflavone categories. In contrast to other flavonoids, isoflavonoids have ring B connected to the C-3 position of ring C such as genistein, daidzein and puerarin. They play a crucial role in protecting plants from oxidative damage and giving them the colour that draws pollinators and deters assaults from insects and microorganisms. A significant portion of the population finds most of these medications appealing due to their potential effectiveness and relatively low toxicity. Although soy is the main source of dietary flavonoids, many are

TABLE 1.1
Fruit and vegetable processing waste as a source of phenolic compounds

Fruit and vegetable derived by-products	Phenolic compounds
Fruit-derived	
Apple pomace	Chlorogenic acid, quercetin, glycosides, dihydrochalcones
Black current residues	Anthocyanins
Blueberry processing waste	Anthocyanins, hydroxycinnamates, flavanols, glycosides
Cranberry pomace	Caffeic acid, ellagic acid
Grape pomace	Anthocyanins, flavanols, glycosides, stilbenes, phenolic acids
Mango peels/kernels	Flavanol glycosides, xanthone glycosides, hydrolysable tannins, alk(en) ylresorcinols
Star fruit residues	Procyanidins
Vegetable-derived	
Artichoke pomace	Hydroxycinnamates, flavonoids
Cauliflower by-products	Kaempferol glycosides, hydroxycinnamates
Olive mill waste	Oleuropein, hydroxytyrosol, verbascoside, dihydroxyphenylglycol
Onion peels,	Quercetin glycosides
Potato peels	Phenolic acids, total flavonoids

Reproduced with permission from Kosseva and Webb (2013) [22].

also present in fruits, nuts and other unusual sources. The impact of flavonoids on elements of the metabolic syndrome may provide the greatest support for their advantages in aging-related disorders. In hypertensive animal models and in a few human studies, flavonoids from grape seed, soy, kudzu (Puerarialobata) roots and other sources all reduce arterial pressure. Additionally, they buffer plasma glucose and lower fat concentrations in the blood [37]. Antioxidant activities, effects on the central nervous system, changes in gastrointestinal transport, the sequestration and processing of fatty acids, activation of the peroxisome proliferators-activated receptor (PPAR) and improvements in insulin sensitivity seem to be the underlying processes. Dietary flavonoids also have a protective impact against cognitive decline, cancer and metabolic illness in animal disease models. Research has shown that flavonoids are not always beneficial and can even be harmful in some circumstances. Determining the effect of these drugs on disease prevention, including optimal exposure and potential contraindications, is crucial as the population ages. With potential uses in the prevention of atherosclerosis, the citrus flavonoids naringin and naringenin were discovered to drastically reduce the expression levels of vascular cell adhesion molecule-1 (VCAM-1) and monocyte chemotactic protein-1 (MCP-1). They are usually referred to be vitamin P factors because bioflavonoids like hesperidin (from orange peel), naringin (from grape fruit peel), or rutin can correct capillary permeability and vascular brittleness. Neohesperidin is over 2000 times sweeter than saccharose, whereas hydrated naringin is around 300 times sweeter [38]. Hesperidin is used in vein treatment, functions as an antiviral in flu therapy and has artificial sweetening qualities. Figure 1.4 depicts the chemical composition of different dietary flavonoids.

In addition, a number of studies have focused on the phenolic components of wine, notably the flavanols (such as catechins and proanthocyanidins), because of their connection to the health benefits associated with moderate wine intake. Only a portion of these substances, which originate in grape, are transmitted to the must. Their capacity to be extracted mostly depends on the vinification process's technological setup. Because of this, significant amounts of phenolic compounds are

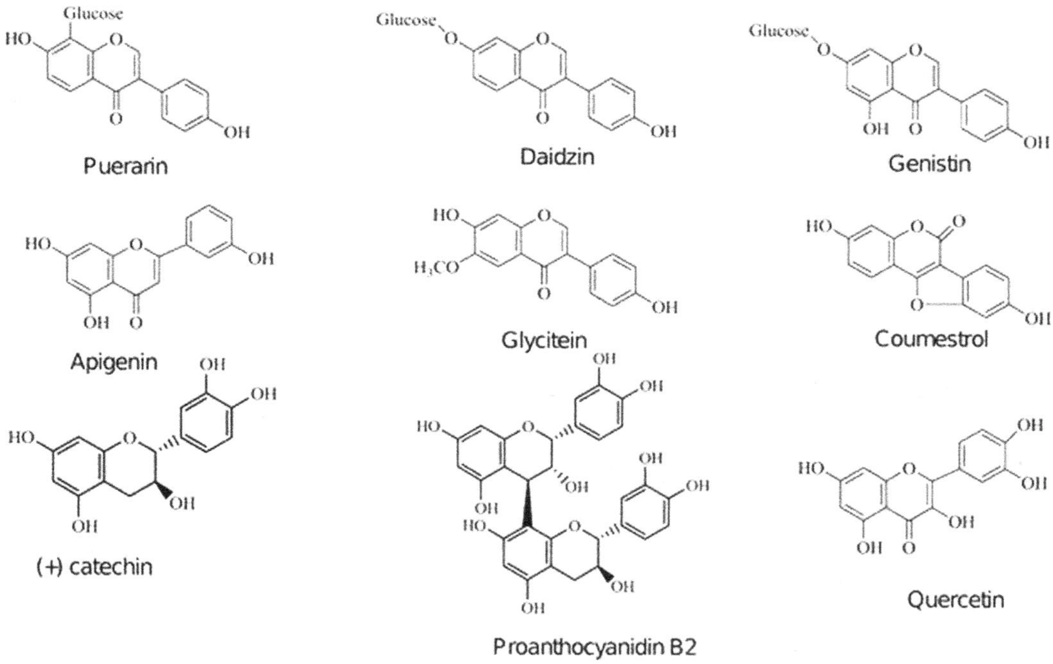

FIGURE 1.4 Chemical structures of different dietary flavonoids [Reproduced with permission from Prasain et al. (2010) [37]].

still present in wine by-products, and the exploitation of this class of grape by-product to acquire potentially beneficial phenolic compounds is of great interest [5]. Pro anthocyanidins with variable levels of polymerization are plentiful in grape seeds and are used as nutraceuticals in a wide range of goods. Grape seed extracts are generally acknowledged to be safe (GRAS) in the United States. Anthocyanins, the red and blue pigments also present in cranberries, elderberries, blueberries, blackberries, blackcurrants, strawberries, red cabbage and purple carrots, have long been extracted from red wine pomace.

The skin, stems and seeds of grapes have the highest quantities of polyphenols. Proanthocyanidins are a group of substances that may be isolated from grape seed and are present in a wide variety of plants. Catechin is their primary structural component. Catechin monomer, dimer and trimer, all of which are water-soluble molecules that include a number of phenolic hydroxyls, are components of proanthocyanidins. By removing free radicals from the body and lowering membrane lipid peroxidation, polyphenolic substances might prevent the onset of disorders linked to free radicals and perhaps slow down the ageing process. A variety of diseases brought on by free radicals, including myocardial infarction, atherosclerosis and drug-induced liver and kidney damage, can be prevented by using grape seed proanthocyanidins extract, according to recent studies. It can also neutralize free radicals, protect against overoxidative damage caused by free radicals and reduce the incidence of these diseases. They also have actions that are anti-fatigue, anti-radiation damage, anti-mutagenic, anti-tumour and anti-antithrombotic [8]. An exploratory investigation on the impact of grape seed proanthocyanidins extract (GSPE) on markers of free radical and energy metabolism during movement was conducted by Shan et al. (2010) [39]. According to Feng and Chen (2003), the extract rate of proanthocyanidins was 6.17% [40]. They looked at the anti-fatigue mechanism of GSPE in relation to the energy metabolism and antioxidation system in an effort to offer theoretical direction and experimental support for employing GSPE in athletic practise. According to their findings, GSPE may dramatically boost the activity of antioxidant enzymes in mice and eliminate free radicals from the body, protecting the organism from the harm that free radicals can cause. GSPE can also alter mouse glucose metabolism, boost their liver and muscle glycogen stores, decrease their need for glucose and maintain a steady blood glucose level. Additionally, GSPE can influence mouse fat metabolism and encourage fat use. The class of flavonoids includes phenolic chemicals known as anthocyanins. They are in charge of giving the fruit and petals of many different plants their distinctive colours, such as orange, rose, red, violet and blue. The use of anthocyanins as food colours in beverages, marmalades, ice creams and medicinal items is now permitted by the European Union [41].

Dietary fibres: According to several research studies, dietary fibres (DF) can prevent obesity, heart disease and type 2 diabetes. DF are crucial in the prevention and management of gastrointestinal problems as well as large intestine cancer. DF are made up of a variety of substances (pectin, lignin and cellulose) with various physical and chemical characteristics. FV by-products are employed as sources of DF supplements in refined food, including the pomace of apples, pears, oranges, peaches, black currants, cherries, artichokes, asparagus, onions and carrots [42]. Vegetable and fruit dietary fibre concentrates may have significant levels of dietary fibre and a better ratio of soluble (SDF) to insoluble (IDF) dietary fibre than cereal brans. The SDF:IDF ratio is significant for both technical and health-related qualities; a well-balanced ratio of 30–50% SDF and 70–50% IDF is thought to be necessary to achieve the physiological benefits of both fractions. Onions have a significant amount of DF and a favourable SDF:IDF ratio, which will have a variety of physiological and metabolic consequences. It's crucial that the fibre matrix retain its physical characteristics following processing. The SDF:IDF ratio is improved by sterilization, which results in a drop in IDF and a rise in SDF. Uronic acids from IDF often become solubilized after being heated, changing to SDF [43, 44]. Because insoluble fibres may absorb water like a sponge, swelling capacity (SWC), which is correlated with fibre structure, decreases as a result of IDF losses during sterilizing. However, sterilizing does not significantly alter the free hydroxyl groups that adsorb water through hydrogen bonds, hence the water-holding capacity (WHC) does not change in a way that is important. The decline in cation exchange capacity (CEC) may be connected to the modifications that pectic compounds go through following sterilization. Sterilization would be a useful way to stabilize onion by-products despite the alterations brought on by heat treatment because the physicochemical characteristics of sterilized by-products are often greater than those of cellulose. These by-products might thus be used as low-calorie bulk components in DF enrichment and could be useful in food items that need to retain moisture and oil. Waste potato peel has been suggested as DF for baked goods. Despite having a very high waste index, cauliflower is a fantastic source of protein (16.1%), cellulose (16%) and hemicellulose (8%). It is regarded as a rich source of DF and has anticarcinogenic and antioxidant qualities [45]. Stojceska et al. (2008) investigated the impact of cauliflower trimmings on the textural and functional characteristics of extrudates in ready-to-eat expanded products (snacks) [46]. It was discovered that adding cauliflower up to 10% enhanced the amounts of proteins and dietary fibre substantially. The vegetable pomace's high crude fibre content (20–65% DM overall) points to its potential use as a crude fibre bread improver. The researched carrot pomace functions as a stabilizer, acidifying agent, preservative or antioxidant in bread and bakery items, as well as in pastries, cereals and dairy products.

1.3.3 Energy generation

Organic waste and by-products are produced in enormous quantities by agricultural practises and the food industry, and these materials can be employed as substrates for bio-transformations achieved by the microbes or their enzymes. The end products might be sold as high-value commodities utilized in other businesses, or they can be used as sources of energy (such as biofuels and biogas) or as compost (pharmacy, cosmetics, food, dietetics, etc.) [22].

Bioethanol: The growing significance of biomass as a renewable energy resource is well acknowledged in order to execute regulations on the reduction of GHG emissions and reduce the dependency on fossil fuel supply. Fossil fuels now account for virtually all of the energy used in the transportation industry. Therefore, the potential utilization of biomass for the creation of vehicle-friendly biofuel has drawn particular interest. The usage of biofuels appears to have positive effects on the environment and the economy, but these sustainability claims need to be rigorously supported by life cycle assessment (LCA). Although biomass energy has a nearly complete carbon cycle, there are certain GHG emissions along its life cycle, such as N_2O and CH_4 from the use of fossil fuels, applying fertilizer and decomposing organic waste. The most widely used biofuels for transportation are bioethanol and biodiesel, both of which may be made from a variety of materials [47]. Most bioethanol is now manufactured on a big scale from sugar or starch (mainly dedicated crops of sugarcane and corn). However, there are disagreements between the usage of these goods for food, feed and industry. Straw, corn stover, forest thinning, sawdust, cotton stalks, along with food sector by-products such as sugarcane bagasse, rice husks, wheat bran and barley hull might be an intriguing, less divisive option. These lignocellulosic feedstocks can produce up to 0.4 L of bioethanol per kilogramme of dry mass when used as raw materials for fermentation. This additional hydrolysis stage is often required before fermentation. However, the benefits such as easy availability, cost-effectiveness, and their surplus presence can exceed this technological challenge. LCA has been used by several writers to examine the synthesis of bioethanol from various source materials [48]. Based on the management and nature of raw materials, its transformation and end-use techniques, varied outcomes were produced even for systems that were comparable. The majority of research have found that bioethanol can reduce the use of non-renewable energy and enhance environmental performance in key areas like global warming. However, the use of ethanol frequently has negative impacts viz. the generation of photochemical oxidants, eutrophication, acidification and ecological toxicity. In general, it can be said that using lignocellulosic wastes to produce energy results in greater reductions in fossil fuel use and greenhouse gas emissions. If residues are used as the feedstock for biofuels, it has been anticipated that these GHG reductions (in comparison to fossil fuels) will be more than 80%. GHG emissions for bioethanol made from lignocellulose and special crops range from 25 to 50 and 50 to 195, respectively, and for gasoline from 210 to 220.

Bioethanol made from lignocellulose has a "non-renewable energy input/energy output" ratio of 0.15 to 0.45, compared to 1.2 for gasoline or diesel [49]. Additionally, the production of bioethanol from waste bagasse has been shown to offer advantages in terms of acidification, eutrophication and human toxicity. When lignocellulosic biomass is converted or condensed into ethanol, a variety of by-products, including lignin and pentose sugars, may result. These by-products have a wide range of applications. The use of lignin wastes for energy generation to power biomass conversion facilities promotes the reduction of greenhouse gas emissions while using less fossil fuel. Additionally, a variety of other products (such as organic acids, alcohols and 1,2-propandiol) may be produced from residual sugars in addition to bioethanol, enhancing the plant's economic and environmental performance [50]. Lignocellulosic material might therefore be used to produce energy, heat, chemical compounds and valuable biofuels all at once. Some food wastes or by-products containing sugar can also be used to make bioethanol (e.g., fruit wastes, whey, or molasses). This is a smart choice since it lessens the demand for specialized crops while also valorising an industrial waste. According to a study conducted on the manufacture of molasses-derived ethanol, utilization of such fuel instead of regular gasoline reduces emissions by 76.6%. However, as there are so few studies that specifically address the environmental effects of producing bioethanol from these materials, further research from an LCA viewpoint needs to be conducted in order to draw reliable findings [51].

Biodiesel: Fatty acid alkyl esters are used in biodiesel, a diesel fuel made from vegetable or animal fat that may be manufactured from certain crops and food sources viz. rapeseed, olive, palm, peanut, corn, sunflower and soybean. Nevertheless, the "Energy vs Food" problem has spurred the development of different methods for the generation of biodiesel from waste materials, microalgae and non-edible sources (such as cotton, karanja, jatropha, neem and nagchampa) (e.g., used cooking oil, waste engine oil, animal fats, sewage sludges). Used cooking oil (UCO) is used as a raw material, which lowers manufacturing costs and aids in resolving environmental issues related to the disposal of this waste. The drawback is the scarcity of these UCOs. Although there are more waste animal fats available, processing them is more difficult. Although the procedure is still fairly expensive, the sludge from municipal sewage is another possible lipid source for biodiesel generation [52]. Both the process and the raw material have a significant impact on the outcomes of LCA studies, which might vary. The majority of LCA research concur that biodiesel made from virgin oils emits between 45% and 65% fewer greenhouse gases than regular fuel. According to the European Union Renewable Energy Directive (Annex V), certain biodiesel systems may reduce GHG emissions by an average of 88% for UCO biodiesel, 46% for rapeseed biodiesel and 63% for palm oil biodiesel. According to an investigation carried out for biodiesel generation in Ireland, UCO biodiesel has a GHG emission reduction value of 69% and animal fat biodiesel has a value of 54% [53]. In different research, GHG emission reductions for waste vegetable oil, beef tallow, poultry fat, sewage sludge, soybean and rapeseed biodiesels were determined to be 80, 72, 72, 75, 69 and 25%, respectively. According to reports, conventional low-sulphur diesel has substantially greater negative effects on energy consumption, global warming and ozone layer depletion than biodiesel fuels made from waste. However, somewhat higher levels for eutrophication and acidification were discovered. UCO-based procedures often have less of an adverse effect on the environment than those that employ sludge, animal fats or virgin oils. The alkali-catalyzed method with an free fatty acid (FFA) pretreatment, the acid-catalyzed process and the supercritical methanol process with propane as a co-solvent have all been taken into consideration in a recent study for the synthesis of biodiesel from UCO. The final choice is the one that is most advantageous for the environment, according to the results. The situations of marine aquatic eco-toxicity and the loss of abiotic resources have the most pertinent effect categories. The alkali-catalyzed method had a reduced overall environmental effect as compared to the acid-catalyzed process. From an environmental standpoint, first generation biodiesel and waste-derived biodiesel are both superior fuels to traditional diesel. UCO in particular emerged as the ideal lipid feedstock for the synthesis of biodiesel. Although UCO biodiesel satisfies the requirements, it is advised that animal fat-derived biodiesel be treated with UCO in order to achieve the required efficiency [54].

Biogas: Biogas produced via anaerobic digestion of renewable resources can be utilized as a fuel for transportation as well as for the generation of heat and power. Reducing the quantity of biodegradable trash dumped in landfills is only one of the many environmental advantages that may be realized. Agricultural harvesting residues, sewage sludges, ley crops, municipal organic wastes, manure, as well as organic wastes from the food sector are some of the basic materials that are readily accessible. Environmental gains may often be made when biogas systems take the place of fossil fuels. Overall, 20–40% of the energy included in the generated biogas is provided by the energy input into the biogas system. According to a research, the distance that raw materials may be delivered before the energy balance becomes negative varies depending on the biogas system [55]. Changes in land use and how organic waste is handled frequently provide indirect environmental advantages in addition to the direct ones. According to a study conducted on different biogas systems, such as the one that uses food industry waste as its raw product, application of biogas-based fuel to generate heat results in a reduction of GHG emissions of between 75% and 90%. About 60–75% of these GHG emissions are caused by CO_2, while 25–40% are by CH_4. Additionally, it was predicted that the emission of particles decreased by 30–70% and the potentials for eutrophication and acidification were reduced by up to 95%. The photochemical oxidant production potential, on the other hand, often rises by 20–70%. When biogas was employed for the co-generation of heat and power, similar ecological consequences were seen. Utilization of biogas instead of typical transportation fuels leads to a decrease in the global warming potential by 50–75% as well as a decrease in the generation of photochemical oxidants by 20–70% [56]. Similar reductions in acidity and eutrophication potentials were shown during the utilization of biogas for heating or both heat and power generation. The potential for anaerobic digestion to be used to remediate some packaging materials is another environmental consideration. Starch-polyvinyl alcohol (PVOH) combination have quickly evolved over the past few decades and are now frequently used as packaging or agricultural mulch film. The biodegradation of starch-PVOH-based biopolymers was obtained at 58–62% under anaerobic digesting conditions. An LCA assessment of a starch-PVOH biopolymer packaging system revealed that fuel combustion and atmospheric emissions discharged throughout the anaerobic digestion method are the primary causes of photochemical oxidation potentials, global warming, eutrophication and acidification, whereas ozone depletion, abiotic depletion and toxic effects are primarily due to energy consumption and infrastructure needs [57].

1.3.4 Composting

LCA was utilized by Finnvedan et al. (2005) to examine several methods for treating solid wastes in Sweden, including various fractions of municipal solid waste (MSW) [58]. Composting, digestion, incineration and landfilling were contrasted for the portion of food waste. Regarding energy utilization, greenhouse gas emissions and other aspects, anaerobic digestion was generally superior to composting and landfilling in the study. Composting was shown to be an intriguing alternative provided transport distances were kept to a minimum, whereas they were greater for the other treatment options. Because of this, large-scale composting was of little interest, but as a result, composting's benefits were also few. Remember that in this study it was believed that digested and composted wastes may be utilized as fertilizers, even though the authors note that this is not definite owing to the possibility of contaminants in the residues. Using LCA, Lundie and Peters (2005) demonstrated that home composting was the optimal method for managing food waste for the effect categories taken into account (among others analyzed, such as centralized composting and landfilling food waste with municipal waste as co-disposal) [59]. Policies seeking to lessen the landfilling of organic waste are primarily motivated by the emission of methane from the breakdown of organic materials. The authors emphasized that home composting, however, might significantly increase GHG emissions as a result of anaerobic methanogenesis if operated without the necessary regulated aerobic conditions. Composting has been suggested as a way to close the organic matter cycle by recovering stalk and dewatered wastewater sludge to create a sanitized organic supplement for use in the vineyard. Traditionally, landfills or incinerators have been used to dispose of these two organic wastes. The investigation used LCA to examine the environmental effects of composting and several alternatives for handling the substantial volume of organic wastes produced during wine production. While the energy balance revealed that composting systems used less energy than systems based on the consumption of mineral fertilizer, in situ composting had the highest performance in the majority of the impact areas considered [60]. The wide variety of oil-based polymers used for packaging is a significant drawback in the food business and food distribution sectors. Due to the complicated composites blends, they are typically nonbiodegradable and very challenging to recycle or reuse. The creation of packaging materials made from renewable natural resources has drawn more and more attention in recent years. The use of biopolymers, such as polylactide (PLA), thermoplastic starch (TPS), or others currently in use—as alternatives for packaging—is possible in the food industry, despite their present higher prices when compared to traditional plastics. Materials made of bio-based packaging may be appropriate for anaerobic digestion or the composting process. According to reports, home and/or municipal composting, which exhibits the same behaviour as organic matter in aerobic composting systems, is the most alluring method for treating bio-based packaging waste. Even still, it was noted that many LCAs of bio-based and biodegradable products overlook the post-consumer waste treatment phase due to a lack of reliable data, despite the fact that this part of the life cycle may significantly affect the results. It has been stated that the fast-food industry produces mixed heterogeneous waste as a result of using disposable cutlery (containing food waste and non-compostable plastic cutlery) [61]. Since this material cannot be recycled, it is now either burnt with or without energy recovery or dumped in inland landfills. With an LCA research that had as its functional unit "serving 1000 meals," Razza et al. (2009) came to the conclusion that employing biodegradable and compostable plastic cutlery allowed for an alternative management scenario by valorization through composting [62]. Significant improvements were observed when the alternate scenario was substituted for the existing one (an overall 10-fold energy saving and 3-fold GHG saving).

1.3.5 Nanoparticles

A promising method for waste valorization is the extracellular green production of metal nanoparticles utilizing plant-based food waste extracts. In recent years, investigations reporting the production of metallic nanoparticles using plant extracts have been widely documented in the field's literature. Numerous studies examine the existence of phytochemicals that are thought to have a role in the creation of nanoparticles, including polysaccharides, alkaloids, tannins, amino acids, saponins, phenolics, terpenoids, enzymes and vitamins. Metal ions combine to produce precursor building blocks, which then self-assemble into nanoparticles during the bottom-up synthesis process. Numerous noble metal nanoparticles, including Au, Ag, Pd and Pt, have been created using this method. However, relatively few research has examined the production of metal nanoparticles from food waste. However, a number of studies have noted the use of bioactive extracts from various fruits and vegetables, such as Ananas comosus (pineapple), Tanacetum vulgare (tansy fruit), Citrus sinensis (orange peel), Mangifera indica (mango peel), and Pyrus sp. (pear fruit) [63]. Table 1.2 depicts the structure of different metallic nanoparticles produced from plant-based sources. Green chemistry-based methods are used in a simple room-temperature procedure for the production of metal nanoparticles from plant-based food wastes. Getting an aqueous-based extract from the plant-based food waste is the first step in the technique. As shown in the schematic illustration in Figure 1.5, the extract is subsequently combined with an aqueous metal salt solution.

Bio-reduction starts right away, but the type of extract utilized affects how quickly it reacts. The reaction mixture changes colour in a distinctive way as the reaction proceeds, signalling the creation of nanoparticles. For instance, when $AgNO_3$ was bio-reduced using C. sinensis (orange) peel extract, the initially colourless liquid quickly took on a yellowish brown hue. Particle analysis performed after the fact showed that reactions occurring at 25°C resulted in spherical particles

TABLE 1.2
Summary of various metallic nanoparticles synthesized using plant-based sources Reproduced with permission from Poinern and Fawcett (2019) [63]

Nanoparticles	Size and shape	Extract source
Ag	20–50 nm, spherical	Potato (Solanum tuberosum)
Ag and Fe	10 nm, spherical	Bran powder (Sorghum spp.)
Au	6–18 nm, spherical	Mango (Mangifera indica) peel
Ag	5–35 nm, spherical	Pineapple (Ananas comosus)
Ag	3–12 nm, spherical	Orange (Citrus Sinensis) peel
Au	200–500 nm, triangular	Pear (Pyrus spp.)
Fe_3O_4	5–25 nm, cubes	Tea (Camellia Sinensis)
Mn_3O_4	20–50 nm, spherical	Banana (Musa paradisiaca) peel
Ag	20–80 nm, spherical	Elderberry (Sambucus nigra)
Pd	50 nm, crystalline	Banana (Musa paradisiaca) peel
MgO	29 nm, spherical	Orange (Citrus Sinensis) peel
Au	432 nm, triangular	Carrot (Daucus carota)
Au	50–100 nm, spherical	Rice bran waste
CeO_2	5–10 nm, irregular	Aloe vera

with a mean particle size of 35±2 nm. The spherical particles were found to be smaller when the synthesis procedure was conducted at 60°C, with a mean particle size of 10±1 nm [64]. The nucleation, development and production of stabilized metal nanoparticles can be influenced by a variety of operational factors during the synthesis process. Operating temperature, reaction time, precursor concentrations and the solution pH are some examples of these operational parameters. Parameters such as morphology, size distribution and particle size can all be significantly influenced by changes in any one of the factors. For instance, the physical characteristics of the resultant nanoparticles can

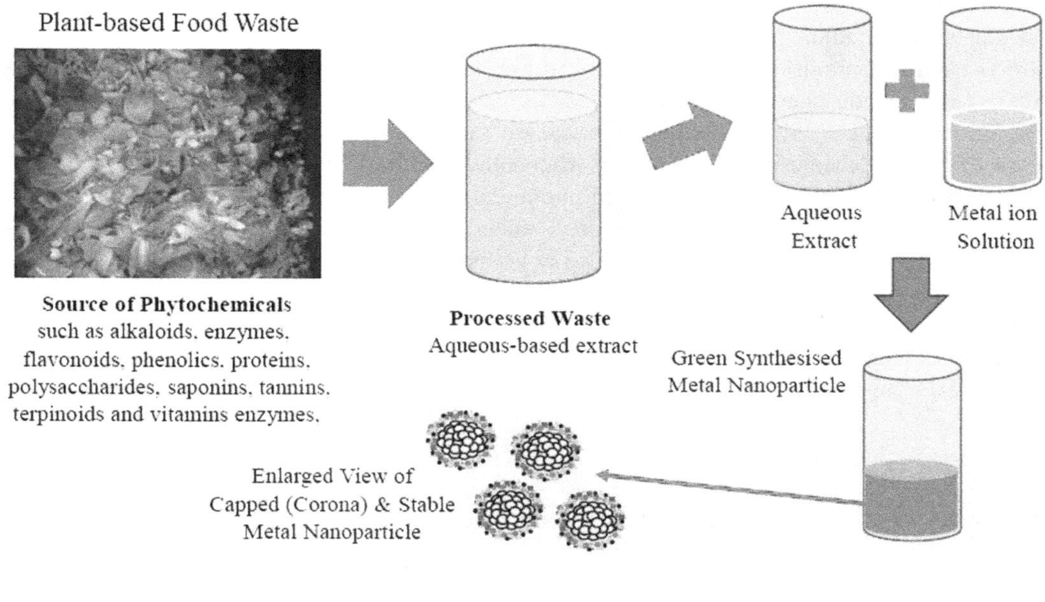

FIGURE 1.5 Schematic representation of the green synthesis of metallic nanoparticles using plant-based food waste [Reproduced with permission from Poinern and Fawcett (2019) [63]].

be significantly influenced by the pH of the reaction mixture. When compared to high pH levels, lower, more acidic pH values typically result in bigger particles. When A. sativa (oat) biomass was used to make Au nanoparticles, for instance, the resultant nanoparticles had the form of rods and changed in size with pH. The sizes of the nanoparticles produced at pH 2 varied from 23 to 93 nm, but those produced at pH 3 and pH 4 were found to be smaller (4–18 nm) due to the greater availability of functional groups. At pH 2, however, there were much fewer functional groups available, which led to more particles clustering together to create bigger nanoparticles [65]. According to studies, the reduction of aqueous chloroaurate solutions can yield Au nanoparticles by using the phytochemicals found in plant-based waste extracts. Similar to this, pseudo-spherical Au nanoparticles with an average size of 300 nm were created using M. paradisiaca (banana) peel extracts. Similarly, Au nanoparticles with a size range of 6 to 18 nm were produced using M. indica (mango) peel extracts. Additionally, a variety of polysaccharides present in diverse plant-based sources, like gum Arabic obtained from acacia trees, may be employed to create Au nanoparticles. Additionally, both red algae (Lemanea fluviatilis) and freshwater green algae (Prasiola crispa) may be utilized to make gold nanoparticles [66, 67]. In recent years, scientists have also looked into producing Ag nanoparticles using green synthesis techniques. Ag^+ ions must first undergo reduction, followed by steps to start nucleation and encourage particle development. Phytochemicals found in plant-based food waste, including vitamins, proteins, phenols, flavonoids and carbohydrates decrease and stabilize precursor Ag^+ ions, which are typically produced from silver nitrate. Many researchers have considered creating environmentally friendly synthesis processes to create Ag nanoparticles with repeatable physical characteristics, such as size and shape. For instance, spherical Ag nanoparticles with diameters ranging from 3 to 12 nm have been made using extracts from the peel of C. sinensis (orange). Citrus unshiu (mandarin) peel extracts have also been utilized to create spherical Ag nanoparticles with sizes between 5 and 20 nm, while A. comosus (pineapple) extracts have produced Ag nanoparticles with sizes between 5 and 35 nm. Similar to this, Ag or Au nanoparticles with a typical size range of 10–25 nm have been produced using extracts of Emblica officinalis (Indian gooseberry). The synthesis of either spherical gold (Au) nanoparticles with a mean size of 11 nm or triangular gold (Ag) nanoparticles with a mean size of 16 nm employed extracts from the tansy fruit T. vulgare [63]. Additionally, Cheviron et al. (2014) created spherical biodegradable starch/silver nanocomposites with sizes ranging from 20–50 nm using Solanum tuberosum (potato) starch [68]. Other than Ag and Au, the green production of metal nanoparticles is also being investigated right now. Recently, palladium (Pd) nanoparticles with a mean particle size of 50 nm were created using extracts from M. paradisiaca (banana) peels. Similar to leftover watermelon rind, lignin has been utilized to create nonspherical Pd nanoparticles with a mean particle size of 96 nm, as well as platinum (Pt) nanoparticles. Similar to this, spherical Ag and Pd nanoparticles with diameters of 5–100 nm have been produced using waste extracts from both tea and coffee. Additionally, magnetic ferric oxide (Fe_3O_4) nanoparticles with cubic and pyramidal forms were made using waste tea extracts and ranged in size from 5 to 25 nm. The water-soluble polysaccharide cell walls of the brown seaweed Sargassum muticum, which are made up of amino, carboxyl and hydroxyl functional groups, have been discovered to decrease and stabilize cubic ferric oxide (Fe3O4) nanoparticles with a mean size of 4-18 nm. Khanehzaei et al. (2015) were also successful in producing Cu cored Cu_2O nanoparticles from red seaweed (Kappaphycus alvarezii). It was reported that the created spherical nanoparticles had a mean particle size of 53 nm [69].

1.4 CONCLUSIONS

The amount of food that is wasted globally—in both developed and developing countries—is between 30 and 40%, although the causes may differ from region to region. Despite the scarcity of data, losses in underdeveloped countries are mostly caused by the lack of a proper food chain infrastructure and understanding or investment in agricultural storage systems. On the other hand, pre-retail losses are substantially lower in the developed countries; however, retail, food service and residential phases of the food system have recently seen a tremendous increase in food waste. A significant amount of methane, a more potent greenhouse gas than CO_2, is produced when food waste is dumped in landfills. Excessive levels of GHG, such as chlorofluorocarbons, CO_2 and methane heat up the earth's atmosphere by absorbing infrared radiation. This process results in global warming and climate change. As a result, government initiatives have largely concentrated on using legislation, taxation and public awareness to divert the food waste away from landfills. Changes must be made at every step of the process, from farmers and food processors to supermarkets and individual customers, in order to reduce the amount of food waste. Priority should be placed on balancing the supply with demand as a first step. Second, more effort has to be put into developing better techniques for harvesting, processing, preserving and distributing food. In this context, valorization of food waste can be a very promising step towards a sustainable environment. The effective utilization of food waste towards extraction of valuable bioactive compounds (phenolics and dietary fibres), production of organic acids (lactic acid and citric acid), bioenergy generation (bioethanol, biodiesel and biogas), nanoparticle synthesis (metal and metal oxide nanoparticles) and composting have played a significant role in reducing the overall environment and economic burden of both the developed and the developing nations. While yard waste management programmes and facilities are well established, food waste management in composting facilities is less advanced and may even be in its infancy stage. Despite this, composting food waste has attracted a lot of interest, and efforts to divert more food waste will probably continue. Taking into consideration the above concluding remarks, this book provides a concise presentation of crucial issues relating to sustainable consumption, food supply chain and the most recent advancements in the field of food sector and post-harvest waste valorization.

ACKNOWLEDGEMENTS

The study is supported by the Indian National Academy of Engineering (INAE/121/AKF/22), Gurgaon, India. The authors are solely responsible for all of the opinions, results and conclusions expressed in this study; INAE's viewpoints are not necessarily reflected in any of these aspects.

REFERENCES

1. A.A. Kader, Handling of horticultural perishables in developing vs. developed countries, *Acta Hortic.* 877 (2010) 121–126. https://doi.org/10.17660/ActaHortic.2010.877.8.
2. S. Otles, S. Despoudi, C. Bucatariu, C. Kartal, *Food waste management, valorization, and sustainability in the food industry*, Elsevier Inc., 2015. https://doi.org/10.1016/B978-0-12-800351-0.00001-8.
3. R. Akkerman, D.P. van Donk, Development and application of a decision support tool for reduction of product losses in the food-processing industry, *J. Clean. Prod.* 16 (2008) 335–342. https://doi.org/10.1016/j.jclepro.2006.07.046.
4. C. Mena, B. Adenso-Diaz, O. Yurt, The causes of food waste in the supplier-retailer interface: Evidences from the UK and Spain, *Resour. Conserv. Recycl.* 55 (2011) 648–658. https://doi.org/10.1016/j.resconrec.2010.09.006.
5. C.M. Galanakis, *Food Waste Recovery: Processing Technologies and Industrial Techniques*, 2015. https://doi.org/10.1016/C2013-0-16046-1.
6. J. Premanandh, Factors affecting food security and contribution of modern technologies in food sustainability, *J. Sci. Food Agric.* 91 (2011) 2707–2714. https://doi.org/10.1002/jsfa.4666.
7. I. Tomlinson, Doubling food production to feed the 9 billion: A critical perspective on a key discourse of food security in the UK, *J. Rural Stud.* 29 (2013) 81–90. https://doi.org/10.1016/j.jrurstud.2011.09.001.
8. R. Campos-Vega and Dave Oomah, *Food wastes and by-products*, 2020. https://doi.org/10.1002/9781119534167.
9. M. Blakeney, *Food loss and food waste: Causes and solutions*, 2019. https://doi.org/10.4337/9781788975391.
10. T. Minowa, T. Kojima, Y. Matsuoka, Study for utilization of municipal residues as bioenergy resource in Japan, *Biomass Bioenergy.* 29 (2005) 360–366. https://doi.org/10.1016/j.biombioe.2004.06.018.

11. P. Morone and F. Papendiek, *Food waste reduction and valorisation: Sustainability assessment and policy analysis*, 2017. https://doi.org/10.1007/978-3-319-50088-1.
12. R.J. Hodges, J.C. Buzby, B. Bennett, Postharvest losses and waste in developed and less developed countries: Opportunities to improve resource use, *J. Agric. Sci.* 149 (2011) 37–45. https://doi.org/10.1017/S0021859610000936.
13. M. Al-Dairi, P.B. Pathare, R. Al-Yahyai, U.L. Opara, Mechanical damage of fresh produce in postharvest transportation: Current status and future prospects, *Trends Food Sci. Technol.* 124 (2022) 195–207. https://doi.org/10.1016/j.tifs.2022.04.018.
14. R. Porat, A. Lichter, L.A. Terry, R. Harker, J. Buzby, Postharvest losses of fruit and vegetables during retail and in consumers' homes: Quantifications, causes, and means of prevention, *Postharvest Biol. Technol.* 139 (2018) 135–149. https://doi.org/10.1016/j.postharvbio.2017.11.019.
15. D. Adams, J. Donovan, C. Topple, Achieving sustainability in food manufacturing operations and their supply chains: Key insights from a systematic literature review, *Sustain. Prod. Consum.* 28 (2021) 1491–1499. https://doi.org/10.1016/j.spc.2021.08.019.
16. E. Desiderio, L. García-Herrero, D. Hall, A. Segrè, M. Vittuari, Social sustainability tools and indicators for the food supply chain: A systematic literature review, *Sustain. Prod. Consum.* 30 (2022) 527–540. https://doi.org/10.1016/j.spc.2021.12.015.
17. G.P. Agnusdei, B. Coluccia, Sustainable agrifood supply chains: Bibliometric, network and content analyses, *Sci. Total Environ.* 824 (2022) 153704. https://doi.org/10.1016/j.scitotenv.2022.153704.
18. N.D. Barnard, Trends in food availability, 1909–2007, *Am. J. Clin. Nutr.* 91 (2010) 1530–1536. https://doi.org/10.3945/ajcn.2010.28701G.
19. G. Rayner, D. Barling, T. Lang, Circular food supply chains – Impact on value addition and safety, *J. Hunger Environ. Nutr.* 3 (2008) 145–168. https://doi.org/10.1080/19320240802243209.
20. P.P. Das; M.K. Purkait, Treatment of steel plant generated biological oxidation treated (BOT) wastewater by hybrid process, *Sep. Purif. Technol.* 258 (2021) 118013. https://doi.org/10.1016/j.seppur.2020.118013.
21. V. Lavelli, Circular food supply chains – Impact on value addition and safety, *Trends Food Sci. Technol.* 114 (2021) 323–332. https://doi.org/10.1016/j.tifs.2021.06.008.
22. M.R. Kosseva, C. Webb, Food industry wastes: Assessment and recuperation of commodities, 2013. https://doi.org/doi.org/10.1016/C2011-0-00035-2.
23. P.C. Slorach, H.K. Jeswani, R. Cuéllar-Franca, A. Azapagic, Assessing the economic and environmental sustainability of household food waste management in the UK: Current situation and future scenarios, *Sci. Total Environ.* 710 (2020) 135580. https://doi.org/10.1016/j.scitotenv.2019.135580.
24. Y. Wang, Z. Yuan, Y. Tang, Enhancing food security and environmental sustainability: A critical review of food loss and waste management, Resour. *Environ. Sustain.* 4 (2021) 100023. https://doi.org/10.1016/j.resenv.2021.100023.
25. M. Al-Obadi, H. Ayad, S. Pokharel, M.A. Ayari, Perspectives on food waste management: Prevention and social innovations, *Sustain. Prod. Consum.* 31 (2022) 190–208. https://doi.org/10.1016/j.spc.2022.02.012.
26. K. Munir, Sustainable food waste management strategies by applying practice theory in hospitality and food services- a systematic literature review, *J. Clean. Prod.* 331 (2022) 129991. https://doi.org/10.1016/j.jclepro.2021.129991.
27. K. Hofvendahl, B. Hahn-Hägerdal, Factors affecting the fermentative lactic acid production from renewable resources, *Enzyme Microb. Technol.* 26 (2000) 87–107. https://doi.org/10.1016/S0141-0229(99)00155-6.
28. F.J. Carr, D. Chill, N. Maida, The lactic acid bacteria: A literature survey, *Crit. Rev. Microbiol.* 28 (2002) 281–370. https://doi.org/10.1080/1040-840291046759.
29. B. Gullón, R. Yáñez, J.L. Alonso, J.C. Parajó, l-Lactic acid production from apple pomace by sequential hydrolysis and fermentation, *Bioresour. Technol.* 99 (2008) 308–319. https://doi.org/10.1016/j.biortech.2006.12.018.
30. C.P. Kubicek, M. Röhr, H.J. Rehm, Citric acid fermentation, *Crit. Rev. Biotechnol.* 3 (1985) 331–373. https://doi.org/10.3109/07388558509150788.
31. S.A. Shojaosadati, V. Babaeipour, Citric acid production from apple pomace in multi-layer packed bed solid-state bioreactor, *Process Biochem.* 37 (2002) 909–914. https://doi.org/10.1016/S0032-9592(01)00294-1.
32. D. Kumar, V.K. Jain, G. Shanker, A. Srivastava, Utilisation of fruits waste for citric acid production by solid state fermentation, *Process Biochem.* 38 (2003) 1725–1729. https:// doi.org/10.1016/S0032-9592(02)00253-4.

33. D. Kumar, V.K. Jain, G. Shanker, A. Srivastava, Citric acid production by solid state fermentation using sugarcane bagasse, *Process Biochem.* 38 (2003) 1731–1738. https://doi.org/10.1016/S0032-9592(02)00252-2.
34. S.B. Imandi, V.V.R. Bandaru, S.R. Somalanka, S.R. Bandaru, H.R. Garapati, Application of statistical experimental designs for the optimization of medium constituents for the production of citric acid from pineapple waste, *Bioresour. Technol.* 99 (2008) 4445–4450. https://doi.org/10.1016/j.biortech.2007.08.071.
35. A. Schieber, F.C. Stintzing, R. Carle, By-products of plant food processing as a source of functional compounds — Recent developments, *Trends Food Sci. Technol.* 12 (2002) 401–413. https://doi.org/10.1016/S0924-2244(02)00012-2.
36. A. Schieber, M. Saldaña, Potato Peels : A source of nutritionally and pharmacologically interesting compounds – A review, *Food.* 3 (2009) 23–29. https://doi.org/10.1016/j.yfood.2015.06.003.
37. J.K. Prasain, S.H. Carlson, J.M. Wyss, Flavonoids and age-related disease: Risk, benefits and critical windows, *Maturitas.* 66 (2010) 163–171. https://doi.org/10.1016/j.maturitas.2010.01.010.
38. G. Laufenberg, B. Kunz, M. Nystroem, Transformation of vegetable waste into value added products: (A) the upgrading concept; (B) practical implementations, *Bioresour. Technol.* 87 (2003) 167–198. https://doi.org/10.1016/S0960-8524(02)00167-0.
39. Y. Shan, X.H. Ye, H. Xin, Effect of the grape seed proanthocyanidin extract on the free radical and energy metabolism indicators during the movement, *Sci. Res. Essays.* 5 (2010) 148–153. https://doi.org/10.1016/C2010-0-00044-3.
40. J.G. Feng, L.Q. Chen, Determination of procyanidin in grape seed extracts, *China Food Addit.* 6 (2003) 103–105. https://doi.org/10.1017/C2003-0-00037-2.
41. C. Santos-Buelga, A. Scalbert, Proanthocyanidins and tannin-like compounds – Nature, occurrence, dietary intake and effects on nutrition and health, *J. Sci. Food Agric.* 80 (2000) 1094–1117. https://doi.org/10.1002/(SICI)1097-0010(20000515)80:7<1094::AID-JSFA569>3.0.CO;2-1.
42. M. Champ, A.-M. Langkilde, F. Brouns, B. Kettlitz, Y.L.B. Collet, Advances in dietary fibre characterisation. 1. Definition of dietary fibre, physiological relevance, health benefits and analytical aspects, *Nutr. Res. Rev.* 16 (2003) 71. https://doi.org/10.1079/nrr200254.
43. M.C. Garau, S. Simal, C. Rosselló, A. Femenia, Effect of air-drying temperature on physico-chemical properties of dietary fibre and antioxidant capacity of orange (Citrus aurantium v. Canoneta) by-products, *Food Chem.* 104 (2007) 1014–1024. https://doi.org/10.1016/j.foodchem.2007.01.009.
44. N. Grigelmo-Miguel, O. Martín-Belloso, Comparison of dietary fibre from by-products of processing fruits and greens and from cereals, *LWT - Food Sci. Technol.* 32 (1999) 503–508. https://doi.org/10.1006/fstl.1999.0587.
45. V. Benítez, E. Mollá, M.A. Martín-Cabrejas, Y. Aguilera, F.J. López-Andréu, R.M. Esteban, Effect of sterilisation on dietary fibre and physicochemical properties of onion by-products, *Food Chem.* 127 (2011) 501–507. https://doi.org/10.1016/j.foodchem.2011.01.031.
46. V. Stojceska, P. Ainsworth, A. Plunkett, E. Ibanoğlu, Ş. Ibanoğlu, Cauliflower by-products as a new source of dietary fibre, antioxidants and proteins in cereal based ready-to-eat expanded snacks, *J. Food Eng.* 87 (2008) 554–563. https://doi.org/10.1016/j.jfoodeng.2008.01.009.
47. S. Woess-Gallasch, Energy-and greenhouse gas-based LCA of biofuel and bioenergy systems: Key issues, ranges and recommendations, *Resour Conserv Recycl.* 53 (2009) 434447De.
48. A. Singh, D. Pant, N.E. Korres, A.S. Nizami, S. Prasad, J.D. Murphy, Key issues in life cycle assessment of ethanol production from lignocellulosic biomass: Challenges and perspectives, *Bioresour. Technol.* 101 (2010) 5003–5012. https://doi.org/10.1016/j.biortech.2009.11.062.
49. H. von Blottnitz, M.A. Curran, A review of assessments conducted on bio-ethanol as a transportation fuel from a net energy, greenhouse gas, and environmental life cycle perspective, *J. Clean. Prod.* 15 (2007) 607–619. https://doi.org/10.1016/j.jclepro.2006.03.002.
50. S. González-García, L. Luo, M.T. Moreira, G. Feijoo, G. Huppes, Life cycle assessment of hemp hurds use in second generation ethanol production, *Biomass and Bioenergy.* 36 (2012) 268–279. https://doi.org/10.1016/j.biombioe.2011.10.041.
51. D. Khatiwada, S. Silveira, Greenhouse gas balances of molasses based ethanol in Nepal, *J. Clean. Prod.* 19 (2011) 1471–1485. https://doi.org/10.1016/j.jclepro.2011.04.012.
52. M.N. Siddiquee, S. Rohani, Lipid extraction and biodiesel production from municipal sewage sludges: A review, *Renew. Sustain. Energy Rev.* 15 (2011) 1067–1072. https://doi.org/10.1016/j.rser.2010.11.029.
53. T. Thamsiriroj, J.D. Murphy, The impact of the life cycle analysis methodology on whether biodiesel produced from residues can meet the EU sustainability criteria for biofuel facilities constructed after 2017, *Renew. Energy.* 36 (2011) 50–63. https://doi.org/10.1016/j.renene.2010.05.018.

54. J. Dufour, D. Iribarren, Life cycle assessment of biodiesel production from free fatty acid-rich wastes, *Renew. Energy.* 38 (2012) 155–162. https://doi.org/10.1016/j.renene.2011.07.016.
55. M. Berglund, P. Börjesson, Assessment of energy performance in the life-cycle of biogas production, *Biomass Bioenergy.* 30 (2006) 254–266. https://doi.org/10.1016/j.biombioe.2005.11.011.
56. P. Börjesson, M. Berglund, Environmental systems analysis of biogas systems-Part II: The environmental impact of replacing various reference systems, *Biomass Bioenergy.* 31 (2007) 326–344. https://doi.org/10.1016/j.biombioe.2007.01.004.
57. M. Guo, A.P. Trzcinski, D.C. Stuckey, R.J. Murphy, Anaerobic digestion of starch-polyvinyl alcohol biopolymer packaging: Biodegradability and environmental impact assessment, *Bioresour. Technol.* 102 (2011) 11137–11146. https://doi.org/10.1016/j.biortech.2011.09.061.
58. G. Finnveden, J. Johansson, P. Lind, Å. Moberg, Life cycle assessment of energy from solid waste - Part 1: General methodology and results, *J. Clean. Prod.* 13 (2005) 213–229. https://doi.org/10.1016/j.jclepro.2004.02.023.
59. S. Lundie, G.M. Peters, Life cycle assessment of food waste management options, *J. Clean. Prod.* 13 (2005) 275–286. https://doi.org/10.1016/j.jclepro.2004.02.020.
60. L. Ruggieri, E. Cadena, J. Martínez-Blanco, C.M. Gasol, J. Rieradevall, X. Gabarrell, T. Gea, X. Sort, A. Sánchez, Recovery of organic wastes in the Spanish wine industry. Technical, economic and environmental analyses of the composting process, *J. Clean. Prod.* 17 (2009) 830–838. https://doi.org/10.1016/j.jclepro.2008.12.005.
61. G. Davis, J.H. Song, Biodegradable packaging based on raw materials from crops and their impact on waste management, *Ind. Crops Prod.* 23 (2006) 147–161. https://doi.org/10.1016/j.indcrop.2005.05.004.
62. F. Razza, M. Fieschi, F.D. Innocenti, C. Bastioli, Compostable cutlery and waste management: An LCA approach, *Waste Manag.* 29 (2009) 1424–1433. https://doi.org/10.1016/j.wasman.2008.08.021.
63. J. Poinern, D. Fawcett, *Sustainable utilization of renewable plant-based food wastes for the green synthesis of metal nanoparticles*, Elsevier Inc., 2019. https://doi.org/10.1016/B978-0-12-813892-2.00001-X.
64. S. Kaviya, J. Santhanalakshmi, B. Viswanathan, J. Muthumary, K. Srinivasan, Biosynthesis of silver nanoparticles using citrus sinensis peel extract and its antibacterial activity, *Spectrochim. Acta - Part A Mol. Biomol. Spectrosc.* 79 (2011) 594–598. https://doi.org/10.1016/j.saa.2011.03.040.
65. V. Armendariz, I. Herrera, J.R. Peralta-Videa, M. Jose-Yacaman, H. Troiani, P. Santiago, J.L. Gardea-Torresdey, Size controlled gold nanoparticle formation by Avena sativa biomass: Use of plants in nanobiotechnology, *J. Nanoparticle Res.* 6 (2004) 377–382. https://doi.org/10.1007/s11051-004-0741-4.
66. A. Bankar, B. Joshi, A.R. Kumar, S. Zinjarde, Banana peel extract mediated novel route for the synthesis of silver nanoparticles, *Colloids Surfaces A Physicochem. Eng. Asp.* 368 (2010) 58–63. https://doi.org/10.1016/j.colsurfa.2010.07.024.
67. N. Yang, L. Weihong, L. Hao, Biosynthesis of Au nanoparticles using agricultural waste mango peel extract and its in vitro cytotoxic effect on two normal cells, *Mater. Lett.* 134 (2014) 67–70. https://doi.org/10.1016/j.matlet.2014.07.025.
68. P. Cheviron, F. Gouanvé, E. Espuche, Green synthesis of colloid silver nanoparticles and resulting biodegradable starch/silver nanocomposites, *Carbohydr. Polym.* 108 (2014) 291–298. https://doi.org/10.1016/j.carbpol.2014.02.059.
69. H. Khanehzaei, M.B. Ahmad, K. Shameli, Z. Ajdari, Synthesis and characterization of Cu@Cu2O core shell nanoparticles prepared in seaweed Kappaphycus alvarezii media, *Int. J. Electrochem. Sci.* 10 (2015) 404–413.

2 Current status and future trends of various food industry waste processing for synthesis of bioactive compounds

2.1 INTRODUCTION

Food and Agriculture Organization (FAO) estimate that one-third of all food produced is lost to wastage. Because of the massive volumes of agro-food waste, landfilling is no longer feasible, which is a problem for food processing industries as well as an environmental and economic concern [1, 2]. Agro-food wastes must be viewed as a renewable supply of high-value bioactive compounds, according to recent studies. The present linear economy model is based on ideas from the industrial revolution that said there will always be a supply of goods with a limited shelf life, necessitating an ever-increasing level of production to meet consumer demand [3]. An important environmental and economic dilemma arises as a result of this linear approach to resource management. A circular economy, on the other hand, encourages waste to be valorized so that new elements may be extracted and returned to the supply chain, so increasing the economy while decreasing environmental effects.

A substantial impact on biodiversity, climate change and human health, is implied by the fact that the majority of the world's food losses and waste originate in the United States (US), which accounts for about 40% of the entire food supply chain. Central and East Asia and North Africa are next with 32%, followed by Europe with 20% and Latin America with 6%. In developing food additives, functional foods, nutritional supplements and nutraceuticals, the high concentration of bioactive components in these food wastes may be valuable. Food waste must be properly managed in order to promote a circular economy model, which is seen as an efficient choice in the long term to transform and add value to the food waste [4, 5].

The food business has concentrated its efforts on creating nutritious options during the past few decades as a result of the sharp rise in ailments linked to the intake of high-calorie, low-nutrient diets. Food producers, therefore, created innovative functional foods by incorporating one or more bioactive substances into a typical food matrix. A substance that has a biological activity essentially means that it has the capacity to influence one or more metabolic processes, hence promoting improved health conditions. Numerous bioactive substances, including enzymes, probiotics, saponins, fibres, prebiotics, phytosterols, peptides, proteins, isoflavones and phytic acid, have been investigated for their beneficial effects on human health [6, 7]. Recently, there has been a lot of interest in the recovery of bioactive compounds from diverse industrial food wastes, including leaves, peels, seeds, hulls, husks, pomaces, algae, wasted grain, the carapace of crabs and shrimp, as well as other fish by-products [7, 8]. Due to the strict rules imposed by the European Union to protect the environment, the necessity to reduce the high costs of waste disposal and the EU's desire to reduce the negative effects of waste, attitudes regarding industrial food waste have altered in recent years. The rising understanding of the advantages resulting from potentially marketable components included in food wastes and co-products is the foundation of the emerging trend in food waste recovery [7].

DOI: 10.1201/9781003315469-2

The possibility of recovering bioactive substances from the waste produced by the food industry is discussed in this chapter. The potential of a wide variety of bioactive chemicals available in the tea and coffee processing industries is highlighted. With the current state of research, the potential of waste generated in the tobacco, dry fruits and winery sectors is underlined. In addition, the availability of bioactive substances in the industries of processing meat and sea food is brought to your attention. Last but not least, the chapter draws a conclusion by addressing a variety of obstacles and future prospects associated with the recuperation of bioactive compounds from the waste produced by the food processing sector.

2.2 BIOACTIVE COMPOUNDS IN VARIOUS FOOD INDUSTRY WASTE

2.2.1 Tea and coffee processing industry

The most popular and widely used non-alcoholic beverage in the world is tea. An estimated 18 to 20 billion cups of tea are consumed every day across the globe. In China and India, two Asian nations, the popularity of tea is skyrocketing. Around 6.3 million metric tonnes of tea were consumed globally in 2020, and by 2025, that number is projected to increase to 7.4 million metric tonnes. In a similar vein, the coffee business is a significant global industry. Huge quantities of waste from the deep-processing of tea and coffee are produced; the majority of these wastes are dumped in landfills or are burned or composted. As a result, given how hard it is to degrade such waste, several environmental issues might develop. Several research have shown that tea and coffee waste may be used to extract antioxidant compounds with therapeutic benefits, such as flavanols, catechin and caffeine [9].

Tea waste is a rich source of polyphenolic compounds, which have remarkable health-promoting properties. Consequently, tea waste may be used as a significant source for the production of such medically significant compounds as polyphenols and polysaccharides. Figure 2.1 illustrates the chemical structures of bioactive compounds that can be recovered from the tea-processing waste. Natural polyphenolic substances, including flavonols, flavanols and other flavonoids, are abundant in tea. As a direct consequence of this, the waste products that are created during the manufacturing of tea also include a sizeable quantity of the aforementioned bioactive chemicals. According to the findings of research conducted by Abdeltaif et al., (2010) the total flavonoid content in the waste produced during the processing of black tea is around 47.40 mg catechin per gram of used black tea [10]. These compounds are called polyphenols because they are made up of an aromatic ring and one or more -OH groups. Activation of tea manufacturing waste by steam explosion can provide a considerable number of bioactive polyphenols, according to Sui et al. (2019) [11]. Due to their excellent health-promoting properties, polyphenols are regarded as the most significant elements of tea waste.

Flavonoids and non-flavonoids are two general subcategories of polyphenols. The bulk of the polyphenols in tea are dihydroflavonols, which are reduced to generate flavan-3-ols or flavanols, sometimes referred to as catechins. The majority of tea's biological activity is connected to catechins [12]. Inhibiting the formation of cancerous cells and lowering blood pressure are both benefits of tea flavonoids. Black tea flavonoids were shown to have the strongest antihypertensive and cancer-fighting properties, which were linked to quercetin, kaempferol and patuletin [9]. According to research by Serdar et al. (2017), black tea waste has the same chemical constituents as black tea, including polyphenols, amino acids, saponin, caffeine and tannins. Tea's antioxidant, antimutagenic, anticarcinogenic and heart disease prevention characteristics are mostly attributed to catechins [13]. Employing a sequential supercritical fluid extraction process, Sokmen et al. (2018) retrieved catechin from green tea before using the best technology to extract catechin from black tea trash. Using this technique, a catechin extract yield of 0.70% was attained [14].

Polysaccharides, which are found in the leaves and buds of tea plants, are bioactive components. Many health advantages have been linked to polysaccharides contained in tea. A wide range of

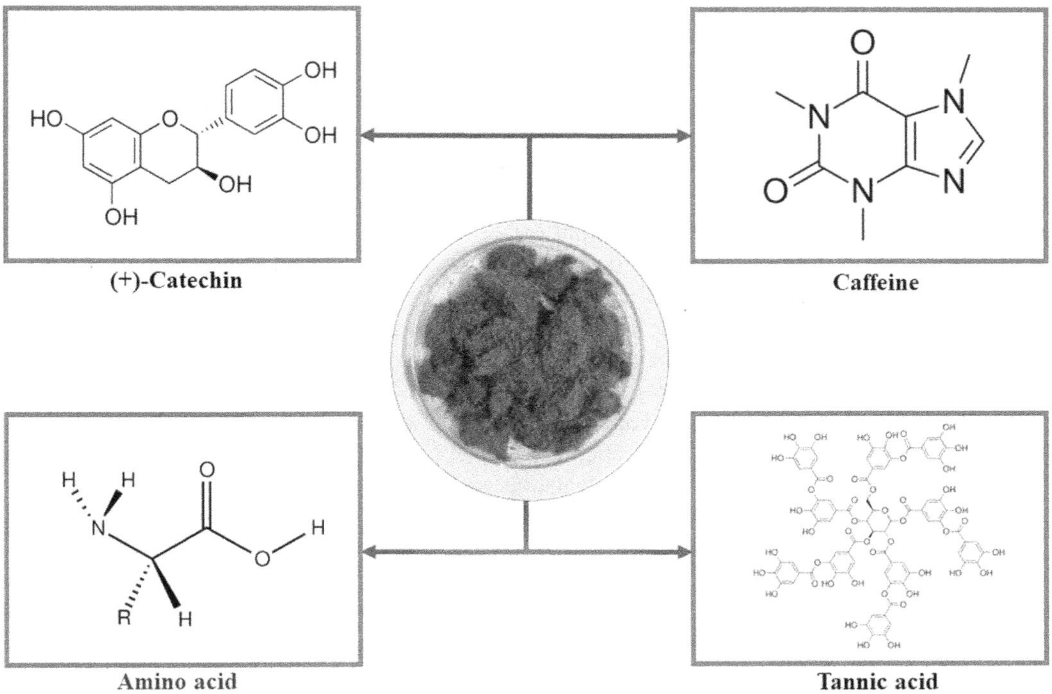

FIGURE 2.1 Examples of bioactive compounds that can be extracted from tea-processing waste.

health advantages may be found in polysaccharide conjugates isolated from green, oolong and black teas, including antioxidative stress, immunostimulatory properties, anti-tumour and anti-obesity actions, as well as the ability to inhibit pathogenic bacteria adhesion [15]. Polysaccharides from green tea leaves were tested for antioxidant activities by Li et al. (2019). Human and animal bodies suffer from oxidative stress due to free radical and free radical scavenging capabilities demonstrated by polysaccharides. Polysaccharides, on the other hand, had a considerable impact on the chicken's weight and antioxidant capacity. Green tea polysaccharides can be employed as natural antioxidants and might be used in the feed sector, according to this study [16]. Because used tea often consists of the leaves, stems and buds that have been thrown away, these components can also be investigated as potential sources of bioactive polysaccharides.

The shell and mucilaginous components of coffee cherries are removed during industrial processing in order to separate coffee powder. Various by-products such as coffee pulp (CP), cherry husk (CH), parchment husk (PH), silver skin (SS) and seed waste (SW) are produced throughout the process of pulping, washing, drying, curing, roasting and brewing coffee. The term "coffee silverskin" refers to the solid residue that is produced during the process of roasting coffee beans, and this residue is collected by cyclone separation while the coffee beans are being roasted. It has been stated that the moisture content is 5–7% [17]. Protein, lipids and minerals account for 16–19%, 1.6–3.3% and 7%, respectively, of the proximate composition. Bioactive compounds with antioxidant potential may be found in this form of residue, according to new research [17].

It has been said that CS possesses high levels of dietary fibre (50–60%), which may be broken down into soluble fibre (15%) and insoluble fibre (85%). This coffee by-product has been called out as a potential alternative source of insoluble dietary fibres due to the fact that its level of fibre is higher than that of regularly used fibre sources, such as wheat and oat brans (29–42%) [17]. Bresciani and colleagues examined the CS's phenolic composition, caffeine concentration and antioxidant potential. Coffeeoylquinic acid (CQA) was the most significant phenolic, with 5- and 3-CQA being the most abundant forms (respectively, 199 and 148 mg/100 g). The three feruloylquinic acids found accounted for 23% of the CGA at 143 mg/100 g. There were just a few phenolics, two coumaroylquinic acids and two caffeoylquinic acid lactones (only 3% of total hydroxycinnamates). The overall antioxidant capacity was 139 mmol Fe^{2+}/kg, while the caffeine concentration was 10 mg/g. Similar to other foods previously recognized as rich sources of antioxidants, such as dark chocolate, herbs and spices, its value is comparable to other dietary items [18].

Studies on chemical compositions and antioxidant potentials have been carried out on various coffee manufacturing leftovers, such as husks (powdered and non-powdered), pulp and silverskin. For the extraction of phenolic chemicals from coffee pulp and husk, silverskin and spent ground coffee (SCG), Murthy et al. (2012) used 60:40 v/v isopropanol and water. DPPH-based antioxidant activity was 65–70%, and hydroxyl radical scavenging activity varied from 59% to 85% for coffee pulp [19]. Coffee husk and wasted coffee ground extracts were used by Andrade et al. (2012) to study the chemical composition and antioxidant activity of supercritical fluid extraction with CO_2 or with CO_2 and co-solvent (high-pressure techniques) [20]. It was also shown that low-pressure extraction methods, such as ultrasound and the Soxhlet method, yielded more total phenolic content (TPC) than the supercritical fluid extraction (SFE) method. TPC readings of 151 (coffee husks, Sohxlet) and 588 mg chlorogenic acid equivalent/g were the highest (spent coffee grounds, ultrasound). Coffee husk extracts with low-pressure extraction had the strongest antioxidant activity.

CS and SCG were compared in terms of their chemical components, functional characteristics and structural features, and Ballesteros et al. (2014) came to the conclusion that both residues are sugar-rich lignocellulosic materials made up of significant amounts of insoluble, soluble and total dietary fibres [21]. In terms of total dietary fibre, SCG had a larger content than CS (60.46% vs. 54.11%), and both substances had equal antioxidant potentials (DPPH values, 20 and 21 mol TE/g). However, when the ferric reducing antioxidant power (FRAP) technique was used, SCG had twice the antioxidant potential of CS: 0.102 mmol Fe^{2+}/g compared to 0.045 mmol Fe^{2+}/g. A study by Jiménez-Zamora et al. (2015) evaluated the prebiotic, antibacterial and antioxidant effects of both the spent coffee grounds and the coffee silverskin, both pure and combined with SCG melanoidins [22]. Regular and sugar-added roasted coffees yielded the same amount of residue. For both forms of residue, prebiotic action was shown to be important, and coffee melanoidins were found to interfere with this beneficial property. Melanoidins, on the other hand, greatly boosted antibacterial action. The antioxidant and antibacterial activity of the residues was enhanced by the addition of melanoidins and sugar during the roasting process.

2.2.2 Tobacco industry

Tobacco is one of the largest crops grown worldwide, with tremendous importance from economic, agricultural and social perspectives. Its most common use is smoking, chewing and sniffing, although it is also occasionally used for inhalation. Only two of the more than 600 kinds of tobacco exist in today's commercially available products for human use. Although it originated in South America, tobacco is grown worldwide, including in the Republic of Croatia, with China being the world's largest producer [23]. Because the tobacco plant's life cycle from seed to seeding only lasts three months, it is frequently utilized in biotechnology as a model plant for creating cell culture and genetic engineering. Additionally, tobacco is among the earliest transgenic plants [24].

One thousand of the approximately 4000 particles and gases included in tobacco are released while smoking [25]. Throughout maturing, drying, fermenting, processing and storing, the elemental composition of tobacco leaves changes and is influenced by several factors, including the type of production, the climate and many others [23]. The starch content drops throughout the drying process, but the reducing sugar level increases. In addition, polyphenol and carbohydrate concentrations were reduced during the fermentation process [26]. In addition to nicotine, tobacco contains alkaloids (such as polyphenols, terpenoids and important fragrance components, such as limonene and indole), fatty alcohols and phytosterols that have been identified by several studies. Because tobacco includes so many distinct chemical compounds that might be removed together with the target component, it is challenging to develop procedures that are specifically designed to extract desired compounds from tobacco and materials connected to tobacco. Therefore, there is a need for innovative bioactive chemical extraction techniques that are quick, easy, repeatable, affordable and environmentally friendly. Conventional extraction methods are known for their lengthy extraction times, heavy solvent and heating loads and inferior extraction yields [27].

According to Wang et al. (2010) [28], the tobacco business generated 1.25 million metric tons of agro-industrial waste in 2005, but landfills can't accept it due to its high nicotine and total organic carbon content. There appears to be no standard definition of tobacco waste in the literature. Low-grade tobacco and tobacco powder are commonly classified as solid waste [29]. Stem, leaf vein and roots; faulty tobacco leaves; discarded picadura; and a combination of powder and leaf fragments are commonly referred to as waste in scientific articles. Generally, tobacco waste is solid waste accumulated during the processing of tobacco and mostly contains tobacco leaf particles that have consequently similar content to bioactive compounds [30]. Scrap is defined as little fragments of tobacco leaf broken during processing in the Glossary of Tobacco Terms. It's critical to identify scrap from leaf breaks that happen prior to manufacturing. The midrib of a tobacco leaf is removed during production. It is the shortest stem that has been stretched from the stalk. The tiniest fragments of tobacco waste, known as dust or offal, are thrown away as non-useful waste since they are unsuitable for any use. Examples of bioactive compounds that can be extracted from tobacco industry waste are depicted in Figure 2.2.

Tobacco processing has a significant impact on the chemical composition of the waste. As a heterogeneous substance, tobacco waste can have a substantial impact on the final product's quality. Chlorogenic acid (CA) and rutin (RU) concentration were the topics of several studies, which concluded that these two compounds could differentiate between various tobacco wastes [28]. This substance has a high moisture content (50–80%), a pH range of 5.1–6.7, a carbon to nitrogen ratio (C:N) of 17:1–21:1 and a considerable level of nicotine [31]. Tobacco waste is an excellent source of alkaloids, aromatic compounds and protein fractions, but it is also a significant source of nicotine.

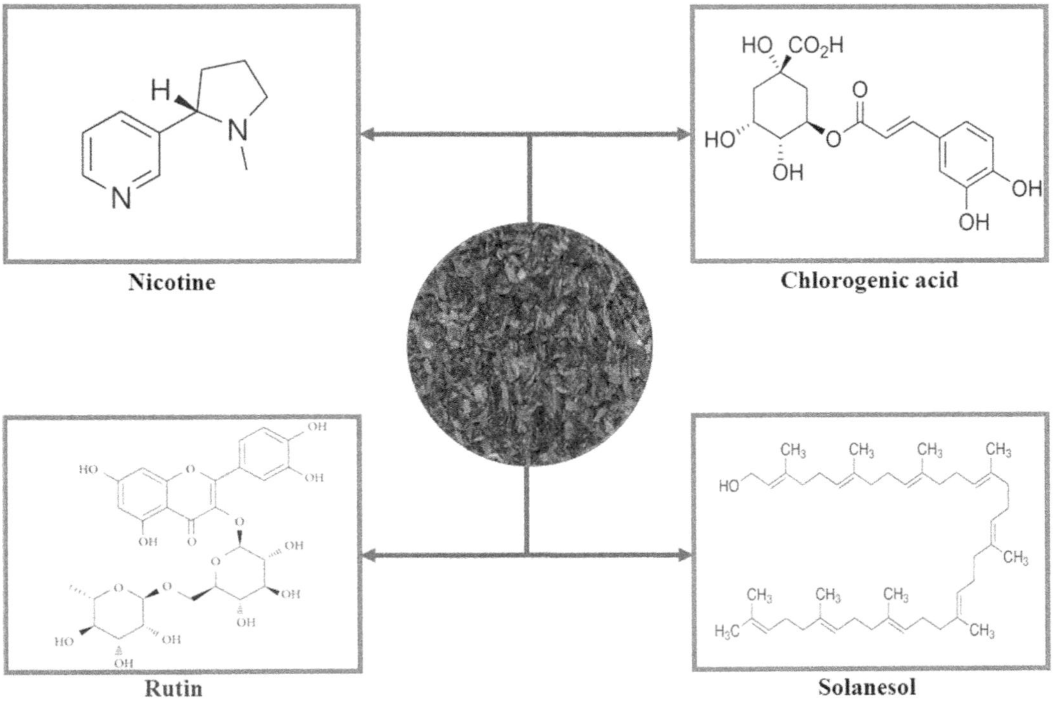

FIGURE 2.2 Examples of bioactive compounds that can be extracted from tobacco industry waste.

Tobacco waste may be used in a variety of ways. A number of studies have indicated that tobacco waste may be used for aerobic composting, organic fertilizer and energy briquettes due to its high calorific content [30]. Ineffective and harmful to the environment, these methods of reusing tobacco waste have been demonstrated time and time again. It's possible that reconstituted tobacco made from tobacco waste will end up being utilized in lower-quality goods. According to Zeng et al. (2012) [32], extraction of water-soluble components is necessary to obtain the desired physical and chemical qualities of a paper-like product known as reconstituted tobacco. Some researchers also looked at employing microbes to remove nicotine from tobacco waste. This sort of waste is better suited for use as fertilizer, fibreboards and pulp after the extraction of several bioactive substances, including nicotine and solanesol [33].

Genetic and environmental variables influence the number of alkaloids a plant produces. *Nicotiana spp.* is known for their pyridine alkaloids, which are weak organic bases. Approximately 95% of the alkaloids in tobacco are made up of nicotine, while the other 5% are made up of minor alkaloids. Minor alkaloids include nornicotine, anabasine, nicotein, myosmine, anatabine, nicotyrine, quinine, 2,3-dipyridyl, muscarine, atropine, morphine, strychnine and nicotimine, to name just a few [34]. Troje et al. (1997) [35] presented a study showing the relevance of tobacco alkaloids, although their principal role in tobacco remains a mystery. There are 3.12% alkaloids in tobacco leaves, and 2.53% of them are nicotine. Other alkaloids, such as nicotinic acid, have also been shown to have important physiological effects on the human body [36]. As a result, pure minor tobacco alkaloids must be isolated for use in pharmacokinetics, pharmacology and toxicology studies. To synthesize minor tobacco alkaloids, Crooks [37] outlined possible routes. Except for extracting these substances from human fluids for toxicity assessments, extraction of minor alkaloids hasn't gotten as much attention as it should.

One alkaloid in tobacco is 3(1-Methyl-2-pyrrolidinyl) Pyridine. Ornithine decarboxylase/metabolism initiates nicotine production in the root, which accumulates in the leaves as the plant grows. The fact that tobacco seeds do not contain nicotine is intriguing [38]. Tobacco leaves, tomato leaves, potato leaves and other Solanaceae crops can be used to extract it for commercial reasons. In both plant and human samples, the amount of nicotine present has been extensively studied. The concentration in leaves is typically 0.3–3%, but additional extremes have been documented [25].

The chemical properties of nicotine are well understood and have been well researched. It is a hygroscopic oily liquid that is water soluble, colourless or less than light yellow, and its boiling point is between 246°C and 247°C. The empirical formula of this substance is $C_{10}H_{14}N_2$, and its molecular weight is 162.23. There are three different forms of nicotine that may occur. Depending on the acidity of the solution, 3(1-methyl-2-pyrrolidinyl) pyridine, l-methyl-2-(3-pyridyl) pyrrolidine or -pyridyl—N-methylpyrrolidine can be produced. The most frequent form is 3(1-methyl-2-pyrrolidinyl) pyridine [35].

Although there are additional applications, cigarettes and cigars are the most popular ways to ingest nicotine. Insecticides like nicotine are frequently used when growing fruits and vegetables. New nicotine-containing products, like nicotine patches, chewing gums, nazal sprays, inhalers and even nicotine water, are being developed by the pharmaceutical sector. These goods substitute for tobacco through inventive methods for supplying nicotine to the body. In addition to helping patients break the cycle, these nicotine substitutes prevent them from ingesting the other dangerous elements of tobacco smoke [39]. Patients with postencephalitic parkinsonism, Alzheimer's disease and Tourette's syndrome have been shown to benefit therapeutically from nicotine, according to other research. However, this type of study should be done with great caution and consideration for the amount employed, as nicotine may be harmful to human health and even fatal in high doses [40].

All plants in the Solanaceae family, including tobacco, contain the polyisoprenoid alcohol known as solanesol. One of the most valuable resources for the chemical production of quinones and vitamin K, as well as a major precursor to polycyclic aromatic hydrocarbons (PAHs), is this plant's abundance of isoprene units. Solanesol is present in tobacco leaves in two states: as a free compound and bound to esters of palmitic, linolenic, myristic, oleic and linoleic acids. Food additive and medicinal component solanesol are of high value to the consumer. Solanesol is found in a variety of plants, including tobacco, tomatoes, potatoes, eggplants and chilli peppers, all of which belong to the Solanaceae family [41]. Tobacco leaves and waste contain solanesol in concentrations ranging from 0.044% to 3.6%. The pharmacological properties of solanesol have been established. Researchers have discovered that solanesol has antiproliferative and antibacterial properties, and they've recommended that it be used to make medications to cure a variety of ailments. It has also been shown to have cardio-stimulant characteristics. In addition, solanesol was discovered to have strong antioxidant action by Hang et al. (2008) [42].

Numerous investigations have discovered that tobacco and tobacco waste are abundant sources of chlorogenic acids and flavonoids (mostly rutin). They aid in the unique flavour of tobacco as well as the colour changes that occur during curing. In plants, phenylalanine, tyrosine and malonate combine to generate flavonoids, which are phenolic chemicals that typically appear as glycosylated derivatives. Flavonoids are capable of scavenging free radicals and help plants fight oxidative stress. The flavonoids in tobacco are abundant. According to Ru et al. (2012) [43], tobacco leaves contain 49.82 mg/g of flavonoids on a dry weight basis, and according to Fathiazad et al. (2010) [44], three flavonoids—rutin, apigenin and quercetin—were extracted from tobacco waste. The primary flavonoid in tobacco is rutin, also known as quercetin-3-rhamnosyl glucoside. Tobaccoin pharmaceuticals has been proven to offer anti-inflammatory, vasodilatory and vasodilatory qualities as well as anti-tumour antibacterial, antiviral and antiprotozoal capabilities, according to Docheva et al. (2014) [45].

Chlorogenic acids are phenolic acids that make up more than 400 phenolic acids. Chlorogenic acids were found in tobacco and tobacco waste. Food and cosmetics manufacturers can use chlorogenic acid as an ingredient. Pharmaceutical manufacturers can use it as a solvent. Herbalists often use Eucommia ulmoides Oliver and Lonicera japonica Thunb as their primary sources of the substance for extraction. Chlorogenic acid must be produced from new sources since the existing ones are not cost-effective or big enough. One of the principal phenolic chemicals found in tobacco waste is chlorogenic acid, suggesting that tobacco waste may be a valuable source of this substance [23].

The quality and flavour of tobacco as well as tobacco products, are influenced by the presence of carbohydrates, which are significant tobacco leaf ingredients. Together with amino acids, they participate in Maillard reactions and create glycosides with phenols, alcohols and sterols. There are 30.25% of sugars in tobacco leaves, and 22.58% of them are reducing sugars, according to Troje et al. (1997) [35]. Research on the sugars found in tobacco has been ongoing for the last few decades. Sugars such as fructose and glucose were found in tobacco leaves as well as raffinose, maltose and sucrose. Tobacco quality benefits from sugars and amino acids, whereas starch and protein have the opposite effect. It has been determined that reducing sugars and amino acids can be produced by biocatalysis of the aforementioned substances. There hasn't been much research done on tobacco polysaccharides' bioactivities. For the first time, tobacco polysaccharides were found to exhibit antioxidant action by Xu et al. (2013) [46]. They established that tobacco's polysaccharides contribute to the antioxidant activity of tobacco extracts. The radical is thought to be reduced by the hydroxyl, COOH, C=O and O groups due to their ability to donate electrons, produce more stable forms or interact with free radicals. Complex chemical structures of carbohydrate-based bioactive chemicals have hampered their growth, but the diversity of their bioactivities and their incorporation in several physiological and pathological functions of organisms have been acknowledged [27].

Cholesterol, campesterol, stigmasterol and β-sistosterol are the primary sterols found in tobacco leaves and make up about 0.1–0.3% of the dry weight of the plant. Tobacco sterols were discovered by Severson et al. (1978) [47] in several parts of the tobacco plant. Sterols were identified in all regions of the tobacco plant except for the stem and flowers, where the concentration was much lower. Tobacco sterols, precursors to nitrosamines, are considered undesirable tobacco constituents. To get tobacco sterols, Shen et al. (2005) [48] used rapid solvent extraction, ultrasonic aided extraction and Soxhlet extraction. Soxhlet extraction was shown to be the most effective method for extracting sterols.

2.2.3 Winemaking industry

Grape pomace, stalks and lees are the primary biowaste products of the winemaking industry. Figure 2.3 illustrates the production of different kinds of waste during the winemaking process from white and red grapes. Grape pomace and lees are the principal waste formed after de-stemming, respectively, whereas grape stalks are the primary residues generated after crushing and harvesting. According to reports, grape pomace makes up 15–20% of the by-products produced during the winemaking process. During the winemaking procedure, yeast lees make up between 3.5% and 8.5%, while grape stems and seeds make up between 2.5% and7.5% and 3% and 6%, respectively [49]. Due to a lack of substitute markets and financial support, grape pomace has been undervalued among these by-products for a considerable time. Traditional wine distillation uses wine pomace to create a variety of wine alcohols as well as distilled spirits, liqueurs and liquors. These winemaking by-products are rich in dietary fibre and bioactive substances with high antioxidant properties, according to published research [50]. However, it is essential to investigate appropriate suitable environmentally friendly methods and ideal circumstances for maximizing the extraction of bioactive substances from winery by-products. After extensive study into extraction methods to recover the bioactive components from winery waste that include significant levels of antioxidant phytochemicals, winery waste is still discarded throughout the world on a regular basis. In contrast, winery waste can also be fed to pets and used as fertilizer. Additionally, the possible physiological health advantages of these winemaking by-products, such as anticancer, cardioprotective, anti-inflammatory and anti-ageing, have been well researched. Additionally, it has been observed that dietary and pharmaceutical supplements include antioxidants as well as technologically useful features, including colouring and texturizing abilities.

Nutrients found in grape pomace include carbohydrates, fibre, minerals and vitamins. Dietary fibre ranks high on the list of nutrients because of its role in digestion and absorption. Many studies have found that grape pomace contains up to 70% of total dietary fibre, with cellulose and

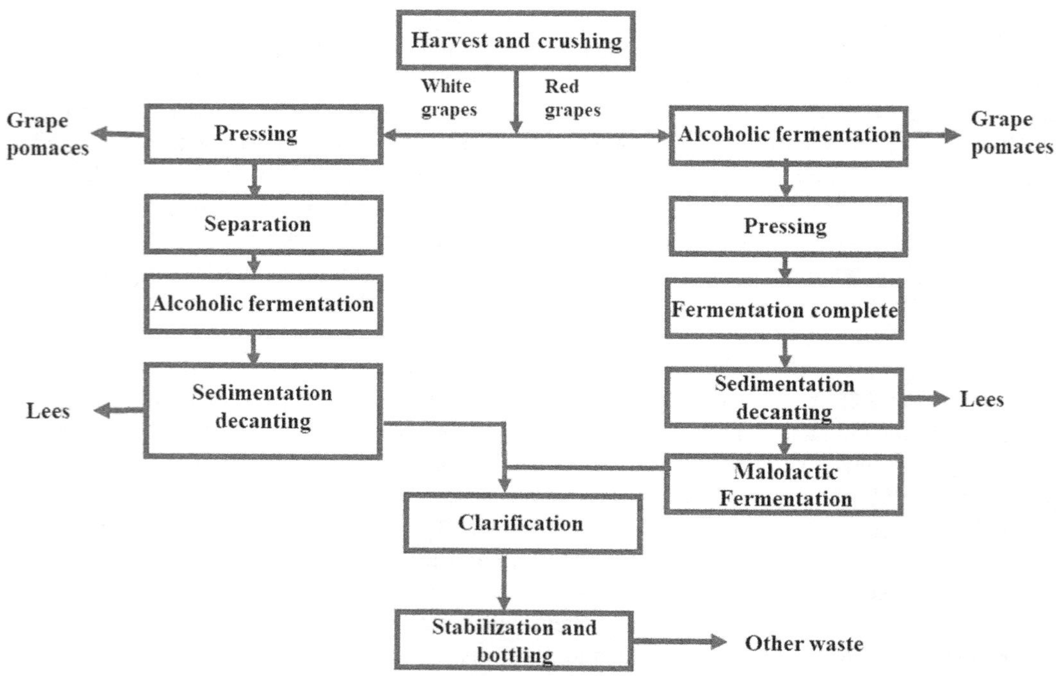

FIGURE 2.3 Schematic representation of the various wastes produced during the production of wine from red and white grapes.

hemicellulose accounting for between 26% and 78% of this total. In contrast, 9–11% of pomace is made up of water-soluble dietary fibre (DF), such as β-glucans, gums, pectins and the like [51, 52]. The health benefits of pomace fibre are directly linked to its ability to ferment in the colon and form short-chain fatty acids that serve as prebiotics. Recent research established the use of grape pomace as a source of DF in food products and found an increase in DF content. Vitamin C is a naturally occurring antioxidant and heat-labile substance. As such, its presence indicates that the remaining nutrients may possibly be kept in the meal, and therefore, it is regarded as a nutritive value indicator of foods. According to Pinheiro et al. (2009) [53], grapes' edible component contains roughly 11 mg of vitamin C per 100 grams, but the amount for grape pomace appears to be 4.90–26.25 mg per 100 grams, as noted in the research.

Gonzalez et al. (2013) [54] assert that even after processing, the phenolic chemicals found in grapes will still be present in the pomace. Compared to other fruit and vegetable waste, they are present in significant numbers. There are a broad variety of phenolic compounds in grapes, including monomeric phenolic acids, oligomeric proanthocyanidins and glycosylated anthocyanins, which have antioxidant, antibacterial and other properties [55]. Catechins, epicatechins, dimers and trimers of procyanidins, resveratrol and other phytonutrients may be found in these plants. Pigments called anthocyanins have been found in grape pomace, ranging in concentration from 131 mg/100 g to 179 mg/100 g, according to published research [56]. The anthocyanin concentration of guava (3.2 µg/100g) and acerola (8.4 µg/100g) residues is significantly lower. According to Rockenbach et al. (2011), anthocyanin levels in red grape bagasse range from 385 mg/100g to 934 mg/100g. Physical variables, such as pH and temperature, environmental conditions, cultivation methods and cultivars can affect the concentration of anthocyanin pigment in grapes [55].

Similarly, a significant amount of total phenolic content has been reported in different varieties of grapes. The phenolic compounds found in winery by-products include hydroxycinnamic acids (267 mg/100g), procyanidins (1046 mg/100g), tannins (2860.00–31,549.72 mg CTE/100g (Catechin equivalents), proanthocyanidins (250–1580 mg CTE/100g), catechin (0.384–151 mg/100g), flavonols (189–6120 mg/100g) and epicatechin (44.36–122 mg/100g) [55]. It has been shown that the natural stilbenoid phenol, resveratrol, as well as the non-flavonoid phenolic acid resveratrol gallic acid (4.59–396 mg/100g) are abundant in grape pomace, according to Iora et al. (2015) [57]. In prior investigations, the aforementioned values were reported by various authors. The presence of these phenolic compounds is likely responsible for the antioxidant activity that is exhibited by the pomace extracts. Grape pomace and the derivatives of grape pomace can be utilized in a broad variety of food products as a source of bioactive elements due to the physiological advantages that they provide.

Cancer, cardiovascular disease and diabetes can all be prevented by eating a diet rich in fruits and vegetables. In addition, grape waste, pomace in particular, has been observed to have potential health-promoting actions. As a consequence of this, they are utilized for the purpose of producing selected therapeutic effects in the maintenance of human health. A person who suffers from hyperglycaemia has sugar levels in their blood that are significantly elevated. This condition has been tied to the development of type 2 diabetes [55]. A 10 g/mL concentration of red and white grape pomace extract reduced the activity of the -glucosidase enzyme in yeast cells by 63% and 43%, respectively. Similar outcomes against the rat intestinal -glucosidase enzyme were noted, with inhibitory activities of 47% (red grape pomace) and 39% (white grape pomace). However, postprandial hyperglycemia is reduced in STZ-induced rats when red grape pomace extract at a dose of 40 mg/100 g is added to the diet [58]. When given grape pomace extract (0.5%) and omija fruit extract (0.05%), male c57BL/KsJ-db/db mice had less hyperglycemia and had lower levels of glycosylated haemoglobin and plasma insulin. Glucose-6-phosphatase and phosphoenolpyruvate carboxykinase enzymes in the liver have reduced activity. Peroxisome proliferator-activated receptor G (PPARg) and glucose transporter protein 4 (GLUT4) were shown to be upregulated in adipose tissue by Costabile et al. (2019) [59], resulting in improved insulin sensitivity, modulated adipogenesis and increased adipose glucose uptake. This was a human clinical trial in which the participants were given a 250 mL red grape pomace drink, and their insulin sensitivity was found to improve.

According to World Health Organization (WHO) figures, cardiovascular diseases (CVDs) are the number one killer in the world, causing 17.9 million deaths annually. Oxidative stress is a crucial component of the pathogenesis of CVDs and is regarded as the primary disease marker for therapeutic interventions [60]. The potential lipid-lowering capability of grape pomace had been demonstrated by in vivo experiments on Wistar rats. Consequently, the use of grape pomace as a cheap option to treat coronary heart disease is recommended [61]. Condensed tannins and anthocyanins, as well as polyphenols from the red grape variety Fetească neagră, displayed a comparable impact on the isoprenaline-induced lesion, decreasing enzyme indicators of the cardiovascular system. Malondialdehyde and other oxidative stress indicators were also shown to be decreased, while serum antioxidant levels were increased. Polyphenols, including rutin, quercetin, trans-cinnamic acid, kaempferol, malvidin, catechin and delphinidin are found in the pomace of Cabernet, Sauvignon, Marselan and Syrah grapes, have been shown to have a cardioprotective effect when utilized in concentrations of 0.25–2 mg/100g [62].

In vitro and in vivo research has shown that bioactive substances isolated from grape pomace have anticancer effects—the bioactive chemicals found in grape pomace display cancer protection through a number of different methods. Angiogenesis, invasion and metastasis are a few metabolic pathways inhibited by the polyphenols in grape pomace as well as proteases and phase I and II drug-metabolizing enzymes. Additionally, they influence apoptosis, cell-cycle checkpoints and receptor-mediated activities [55]. There is a significant body of research indicating that pomace extracts have anticancer properties. Grape seed proanthocyanidins were shown to inhibit the invasion of cancer cells in a dose-dependent manner when tested on cell lines at doses of 0, 10, 20 and 40 g/mL. It was determined that inhibiting epidermal growth factor receptor (EGFR) expression and reversing the process of epithelial-to-mesenchymal transition were responsible for this action taken against human cutaneous HNSCC cells. Alguacil et al. (2014) [63] revealed that grape pomace extract has antiproliferative effects on fibroblasts and colon cancer cell lines (Caco-2, HT-29) at various doses of 5–250 g/mL. Through the upregulation of Ptg2 in Caco-2 cells and the downregulation of Myc gene expression in HT-29, the study had shown that grape seed extract exhibited anti-tumour potential. Grape seed extract's non-anthocyanin component has demonstrated potential activity against colorectal cancer cells.

The composition of the microflora in the gut is still considered a sign of intestinal health. They not only improve the health of the brain, but they also improve the immune system, tolerance to allergens and food metabolism. They also lower the number of harmful bacteria, amongst other benefits. Foods that influence the microbiota's ability to regulate the host's metabolic homeostasis are becoming more widely accepted. Testing on the effects of phenolic component-rich extracts of grape pomace from Cabernet Sauvignon, Marselan and Syrah on the intestinal microbiota of rats was done. Selective development of the gut microbial community was improved by feeding at 2.5 mg/kg/d and 5 mg/kg/d [64]. Due to the presence of prebiotic compounds, such as fructans and grape skin polysaccharides, beneficial microorganisms like Lactic bacteria and Bifidobacteria have also been shown to proliferate more rapidly in the stomach of pigs. Another research, however, found that giving rats grape seed proanthocyanidins resulted in a tendency towards increased development of Bacteroidetes and a trend towards decreased growth of Firmicutes.

An increased amount of triglyceride or cholesterol in the blood is referred to as hyperlipidaemia. Long-term hyperlipidaemia results in micro- and macrovascular problems such as cerebrovascular, microangiopathy, cardiovascular and other metabolic illnesses [65]. Numerous research has been done to show how grape seed extract helps to manage postprandial hyperlipidaemia. It has been demonstrated that male Wistar rats receiving 5 mL of grape seed extract per kg of body weight have lower triglyceride and cholesterol levels. Numerous processes are responsible for the grape seed extract's anti-hyperlipidaemic effects. By reducing the pancreatic lipase, cholesterol esterase, cholesterol micellization and bile acid-binding ability, for example, grape seed extract hinders the digestion and absorption of lipids. Ishimoto et al. (2020) [66] showed similar outcomes in male golden Syrian hamsters on a 20g/100g grape pomace flour diet. The authors proposed that the

3-HMG-CoA reductase enzyme may be inhibited, and that decreasing the plasma concentration of chylomicrons might also reduce the absorption of dietary fat in the gut.

2.2.4 Dry fruit processing

The European or sweet chestnut plant, *Castanea sativa*, yields excellent fruits that are highly and commercially significant in southern Europe. Different *C. sativa* plant components have long been used in traditional medicine to treat conditions like cough, diarrhoea and infertility. In hilly areas, the *C. sativa* tree is a priceless ethnic and historic legacy that yields nuts that are typical of Mediterranean cuisine and have special sensory, nutritional and health-promoting qualities [67]. Chestnuts are a great source of essential nutrients fatty acids, fibre, starch, minerals (primarily K, P, Ca and Mg) and vitamins (C, B9 and E). In producing nations, chestnuts are recognized as a resource with added value. In 2018, the chestnut crop produced the most (2.4 million tonnes), covering 612,877 acres. In addition to China, which produces 83% of the world's chestnuts, Europe contributes 154,612 tonnes, a significant amount. The chestnut industry has shown consistent development in Europe over the past several years, underscoring the substantial expenditures made in this sector by governments and corporations. Exports of chestnuts, which account for 77.6% of total production, mostly to Spain, Brazil, France, Italy, United States and Switzerland, increase these countries' economic output [68]. There are three steps in the preparation of chestnuts: calibration, high-temperature shell removal, peeling of the inner shell using water vapour and mechanical peeling are the first four steps [68]. Despite the fact that a sizeable amount of the crop is intended for fresh consumption, a sizeable portion of the chestnut harvest is transformed into chestnut-derived products such as frozen fruit and chestnut-based goods including marron glacé, jam, purée and flour. These nuts and their derivatives are excellent for celiac disease sufferers since they are gluten-free, which reduces the body's immunological reaction. The production of chestnut by-products, mostly shells, is rising as a result of the ongoing expansion of the farming and processing of chestnuts [67].

Around 20% of the weight of the fruit is made up of *C. sativa* shells (CSS), a plentiful and underutilized residue produced during peeling of the chestnuts, of which 6.3–10.1% are interior shells and 9.0–13.5% are outer shells [69]. Chestnut shells are an unmarketable type of lignocellulosic biomass; however, they are most commonly used as a fuel for the generation of energy in order to counteract environmental difficulties brought on by the toxic compounds that are formed. Tannins, which are essential to the maturation process of wine, may be extracted from the shells. Antioxidant polyphenols in CSS range from 2.7% to 5.2% (w/w) according to prior research, whereas sugars that may be utilized for biofuel production make up roughly 36% (w/w) [70]. Flavonoids (catechin, quercetin, epicatechin, rutin and apigenin) and phenolic acids (gallic and ellagic acids) are abundant in this agricultural residue and provide an extensive range of beneficial biologic effects, including anti-inflammatory, antioxidant, anticancer and antibacterial qualities [69]. The use of shells as bio-adsorbents, phenol replacements for wood glue formulations, and chromium substitutes for leather tanning have all been documented by various authors [71]. However, these applications do not entirely exhaust the supply of shells left over from the processing of chestnuts, and there are still many of these underappreciated agricultural by-products that have no marketable use. Therefore, continuing studies have concentrated on coming up with new, intriguing uses for CSS [67]. The abundance of bioactive substances, such as lignin, polyphenols and vitamin E, as well as nondigestible oligosaccharides, such as fibre, hemicellulose and cellulose, supports the potential for value-added products with appealing uses in the food and nutraceutical industries to be created from chestnut shells.

Chestnuts are made up of the fruit, the bur that envelops the fruit and the inner (integument) and outside (pericarp) shells. Various publications claim that CSS are exceptional sources of bioactive chemicals, especially polyphenols such as ellagitannins, condensed tannins, phenolic acids and flavonoids. Tannins have been cited as the most prevalent polyphenolic class by a number of publications. CSS includes 60% of tanning active compounds such as vescalagin, castalin, castalagin and vescalin, which are readily hydrolyzable, according to Comandini et al. [72]. Additionally, the researchers discovered 1-O-galloyl castalagin, gallic acid and ellagic acid as additional polyphenols to consider. In four commercial CSS samples, the total polyphenol content ranged from 4.8 mg/g to 16.7 mg/g [72]. Gallocatechin, galloyl glucosides and catechin in CSS aqueous extract were correctly identified by Fernández et al. (2014) [73] using RP-HPLC-ESI-TOF analysis. Likewise, catechin, epigallocatechin, epicatechin, gallocatechin, galloyl glucosides, dicatechin structures and ellagic acid were reported as the major monomeric compounds in aqueous extracts by Vázquez et al. (2013) [74]. Procyanidins, prorobinetidins and prodelphinidins were also found in trace amounts. Numerous other research has also been published that support the existence of different phenolic chemicals and caffeine and their extraction processes.

Additionally, various literature discussed the process of removing tannins from CSS using alkali solvents. For instance, Aires et al. (2016) [75] demonstrated that moderate concentrations of alkali solvents used over longer times produced the greatest outcomes. One percent Na_2SO_3 was mostly used to extract condensed tannins, whereas one percent NaOH effectively extracted hydrolyzable tannins [75]. The existence of high molecular weight polyphenols in CSS, as determined by Fourier-transform infrared (FTIR) spectroscopy, has been previously validated in earlier research. This might be because CSS has a high concentration of hydrolyzable tannins [76].

Chestnut shells were used to extract phenolic chemicals using environmentally friendly methods. There was a significant amount of ellagic acid in the sample (40.4 g/mg DW), followed by catechin monomers and polymers, caffeic acid derivatives and epigallocatechin (all 15.3 g/mg DW), according to Lameirão et al. (2020) [77]. Fatty acid derivatives, polyphenols (including procyanidin and flavonol) and sugar moieties were found in the ^{1}H NMR spectrum. The response surface method (RSM), was applied in order to analyze the effects of factors such as time (4–46 minutes) and temperature (34–76°C) on the levels of antioxidants found in the extracts. For the ideal ultrasound-assisted extraction (UAE) extract, Pinto et al. (2020) reported a comparable profile to that of the CSS-SFE extract. Furthermore, polyphenol-rich extracts derived from CSS employing CO_2 as the supercritical fluid and ethanol as the co-solvent were also produced by SFE utilizing this method [78]. Phytochemical content and antioxidant properties of extracts were studied using an RSM at temperatures ranging from 40°C to 60°C, at pressures ranging from 150 bar to 350 bar and at co-solvent percentages of 7–15%. SFE extracts, which had a phenolic content equal to that of UAE extracts, included echinacea (1.2 mg/g), a caffeic acid derivative (0.3 mg/g), epigallocatechin (0.4–0.4 mg/g), catechin/epicatechin (0.3–0.3 mg/g) and epigallocatechin (0.4–0.4 mg/g). Proanthocyanidins, 7-O-rutinoside, apigenin- and luteolin-7-O-rutinoside were among the other polyphenols found. The presence of fatty acids and their esters, ellagic acid derivatives and a sugar or an ester connected OCH of sugars were demonstrated using ^{1}H nuclear magnetic resonance (NMR) analysis [78]. Ellagic acid, gallic acid, pyrogallol and protocatechuic acid were all found in the best CSS extract made by subcritical water extraction (SWE) (220°C/30 min) in another investigation, with ellagic acid being the most abundant [67]. The 80°C/10 min SWE extract was rich in tannins (12.5 mg/g DW), phenolic acids (4.6 mg/g DW) and flavonoids (2.1 mg/g DW). RSM also utilized time (between 6 and 34 minutes) and temperature (between 51 and 249°C) to validate the SWE conditions [67]. Microwave-assisted extraction (MAE) was also used to change CSS by Pinto, Silva, Freitas, Vallverdú-Queralt, Delerue-Matos and Rodrigues [79]. The aqueous extract had a lot of tannins, phenolic acids and flavonoids [79]. Pyrogallol and protocatechuic acid have been found in CSS extracts by certain authors, perhaps because gallic acid and catechin have been degraded at high temperatures. The phytochemical content of CSS extracts obtained by conventional extraction, UAE and MAE, was also compared by Cacciola et al. (2019). Gallic and protocatechuic acids were found to be the most prominent polyphenols in the three extracts, with the UAE having the highest concentrations. Chlorogenic, sinapic, ellagic, p-coumaric, ferulic and syringic acids, as well as epicatechin and scopoletin were among the other polyphenols found [80].

In addition, the phenolic compositions of the inner and outer shells were distinct from one another. In their study, Squillaci et al. (2018) evaluated the levels of bioactivity in extracts obtained from a mixture of inner and outer shells (IOS) and isolated inner shells (IS), both of which were made by boiling water for one hour [81]. IOS extract had the greatest total phenolic content (205.9 mg GAE/g DW), hydrolyzable tannins (12.9 mg GAE/g DW), condensed tannins (78.9%), total flavonoid content (40.9 mg CE/g DW) and ortho-diphenols (98.1 mg CAE/g DW). Gallic acid was the main phenolic component, while protocatechuic acid was the second-most common. Additionally, p-coumaric acid, catechin, ellagic acid, epicatechin and scopoletin were shown to be polyphenols [81].

Similarly, Vella et al. (2019) studied the phenolic profile of aqueous extracts from several Italian cultivars of inner and outer shells. The polyphenols, flavonoids, ortho-diphenols and tannins in the inner shells were found to be higher than in the outer shells, according to the study's authors. One kind of polyphenol was found in both shells, with larger concentrations in the innermost part of the shell, namely gallic and ellagic acids [82]. As an added bonus, Vasconselos et al. (2010) [83] examined the inner and outer shells from four varieties of Portuguese chestnuts and found polyphenols with low molecular weight (gallic and ellagic acids), condensed tannins and ellagitannins (acutissimin A and B, castalagin and vescalagin). Inner shells contained higher quantities of total polyphenols and gallic acid than outer shells. Condensed tannins and procyanidins were more concentrated in inner shells than in outer shells. Silva et al. (2020) [84] made traditional extracts of the inner and outer shells of chestnuts at room temperature for two hours while using ethanol as the solvent. While tannin concentrations were 35 g EE/mg DW and 9 g EE/mg DW, the TPC for the inner and outer shells, respectively, was 321 g EE/mg DW and 240 g EE/mg DW. It is important to emphasize that this by-phenolic product's profile directly depends on the extraction techniques utilized, as well as the extraction conditions and solvents used.

However, the phenolic content of CSS may be affected by meteorological circumstances (such as light and temperature), cultivars, geographical location, soil nutrients and the availability of water. As an example, Rodrigues et al. (2015) evaluated hydroalcoholic extracts made at 50°C for 30 minutes from three distinct districts of Portugal, namely Minho, Trás-os-Montes and Beira-Alta, which are all located in the north of the country. The Trás-os-Montes area had the greatest total flavonoid content (TFC) (43.3 mg CE/g DW) and TPC (796.8 mg GAE/g DW) concentrations, respectively. Trás-os-Montes (12.6%) and Minho samples had similar extraction yields, though (13.7%) [85]. Barreira et al. (2010) [86] emphasized the various phytochemical content of CSS extracts from several cultivars of chestnut, including Aveleira, Boa Ventura, Judia and Longal. The Boa Ventura variety showed the greatest TFC (146.1 mg CE/g DW) and TPC (805.7 mg GAE/g DW).

Electrochemical biosensors based on laccase are one way already used to detect polyphenols. This method, which used the laccase substrate 2,2′-azino-bis(3-ethylbenzothiazoline-6-sulfonic acid) (ABTS) to measure the gallic acid levels, may be an alternative to spectrophotometric assays like Folin–Ciocalteu's. Gomes et al. (2019) [87] also produced molecularly imprinted polymers (MIPs) that improved the adsorption capability of polyphenols from natural matrices, specifically the extraction of ellagic acid from CSS.

A lack in reporting of micronutrients, including vitamins and minerals, in the shells of chestnuts has been noted. Nutritional value is influenced by climate changes in the amount of vitamin E found in CSS (e.g., temperature and humidity). The antioxidant defence mechanism of cells relies heavily on this vitamin, which is a strong lipid-soluble antioxidant that is exclusively obtained by food. Bioactive vitamer α-tocopherol is present in most plant tissues, boosting the immune system, enhancing cardiovascular health, aiding skin restoration and protecting cells from oxidative stress, among other benefits [67].

On the other hand, substantial concentrations of calcium (1.3 g/kg), magnesium (1.0 g/kg), sodium (47.4 mg/kg), manganese (122.4 mg/kg) and potassium (48.9 mg/kg), make up the majority of the mineral components in CSS. But there are also considerable concentrations of other minerals. Despite the potential of these minerals being acknowledged, no literature reports on their CSS benefits could be located. Additionally, another group examined the mineral makeup of the Longal and Judia kinds of Portuguese chestnut shells. The most abundant mineral was found to be calcium.

While exterior shells from the Judia variety had greater levels of K, inner shells from the Longal type were richer in Ca and P. Low concentrations of minor and trace elements such as arsenic, barium, bromine, chromium, nickel, lead, rubidium, strontium and titanium were also discovered [67, 88].

2.2.5 Meat processing industries

Slaughtering is the primary source of waste in the meat industry. Waste from slaughterhouses is the animal's carcass that cannot be sold as meat or turned into meat products. Bones, tendons, skin, digestive system contents, blood and internal organ waste are examples of this type of waste. Depending on the type, they can be rather diverse. For the meat business to be profitable, meat by-products must be used effectively. Derivatives are thought to account for 11.4% of the gross revenue from beef and 7.5% of the gross revenue from pork. By-products were previously a popular cuisine in Asia, but because of health concerns, there is now a greater emphasis on non-food purposes, such as pet foods, medications, cosmetics and animal feed [89]. Slaughterhouses, wholesalers, rendering plants and meat processors all create meat by-products. Due to low-cost and safety concerns, traditional markets for edible beef by-products have been steadily vanishing. Meat processors have focused their marketing and development efforts on non-food purposes in response to these issues.

Numerous important nutrients may be found in edible animal by-products. Because they include unique nutrients such vitamins, amino acids, fatty acids, hormones and minerals, some are utilized as medications. Blood is not the only meat by-product with a higher moisture content than meat. Lung, tripe, kidney, spleen and brain are a few examples. Some organ meats, such as liver and kidney, have a greater carbohydrate content than other meat products.

The fat content and moisture level of pork tails are the lowest of any meat by-products. Cattle's liver, feet, tail and ears have protein levels comparable to those of lean meat, but the ears and feet contain a high amount of collagen, making them less desirable for human consumption [90]. The brain, chitterlings and fatty tissue have the lowest levels of protein. Mechanically deboned beef and pig must include at least 14% protein and a maximum of 30% fat, according to the US Department of Agriculture. Because of the substantial presence of connective tissue, the amino acid content of meat by-products differs from that of lean tissue. This results in a higher concentration of proline, hydroxyproline and glycine in by-products, including ears, lungs, stomach and tripe, whereas tryptophan and tyrosine are depleted. Organ meats often have a higher nutritional compared to lean meat. For instance, the amount of riboflavin in the kidney and liver is 5–10 times higher (1.697–3.630 mg/100 g). Niacin, vitamin B12, folacin, ascorbic acid and vitamin A are all abundant in the liver, making it an optimal food source for these nutrients. Vitamin B6, B12 and folacin are also found in kidneys. Thirty-seven percent (37%) of the recommended daily allowance (RDA) for ascorbic acid is available in a 100-gram serving from pork or beef liver, and vitamin A is available at 450%–1,100% of the RDA. Iron and vitamins abound in lamb kidneys and hog liver, lungs and spleen. The livers of beef, lamb and veal contain the most copper. In terms of the RDA for copper (2 mg/day), they provide 90–350%. Manganese (0.128–0.344 mg/100 g) is found at the highest concentration in the liver. Phosphorus and potassium concentrations are higher in meat by-products, such as the thymus and sweetbreads, which have concentrations of 393–558 mg/100 g [91]. Sodium levels in most organ by-products are lower than in lean tissue, except for the brain, kidney, lung, spleen and ears. In terms of calcium content, mechanically deboned beef is the best option (315–485 mg/100 g).

Polyunsaturated fatty acids are more prevalent in organ meats than in lean tissue. All of these organs contain the lowest concentration of monounsaturated fat and the highest concentration of polyunsaturated fat. Organ meats contain three to five times as much cholesterol (260–410 mg/100 g) as lean meat, as well as large amounts of phospholipids (fat-soluble vitamins). The brain has the most cholesterol (1352–2195 mg/100 g) and the most phospholipids compared to other beef by-products, making it an excellent source of cholesterol. As a result, the US Department of Health advises people to consume these by-products in moderation due to potential health risks. Another reason to limit intake is the high cholesterol content of many other organ meats, as well as the potential build-up of pesticides, drug residues and harmful heavy metals.

Protein and heme iron is found in animal blood, making it a significant food source. Blood sausages, biscuits and blood pudding have long been traditional European foods made with animal blood. Blood curd, blood cake and blood pudding are all popular desserts in Asia [92]. It is also employed for non-food products such as binders, feeds and fertilizers. According to the US Meat Inspection Act, blood that has been drawn from an animal that has undergone inspection and been given the "all-clear" for use in meat food items is permitted.

Typically, an animal in good health has sterile blood. It contains a high protein content of 17.0% and a decently balanced amino acid profile (2.4–8.0% of the animal's live weight) and makes up a sizeable portion of the animal's body mass. The blood recovery rates of pigs, cattle and lambs are typically 3.0–4.0, 3.0–4.0 and 3.5–4.0 percent, respectively. However, using blood during the preparation of meat might result in a product that is unappealingly black in hue. The part of blood that is of most interest is the plasma because of its useful characteristics and lack of colour.

Several therapeutically useful fractions of blood may be extracted from whole blood. The biggest proportion is liquid plasma (63.0%). Albumin (3.5%), globulin and fibrinogen make up its composition (4.0%). Numerous blood products are utilized in the laboratory as nutrients for tissue culture medium, an essential component of blood agar, and as peptones for microbiological usage [93]. For biological assays, glycerophosphates, albumins, globulins, sphingomyelins and catalase are also employed. Numerous blood components, including plasminogen, fibrinogen, fibrinolysin, serotonin, kalikreninsa and immunoglobulins, are separated for use in drugs or chemicals. Animals whose blood or fluids have been lost can benefit from the use of purified bovine albumin. Vaccines can be stabilized by using it as a stabilizer for the Rh factor. Antibiotic sensitivity tests also utilize it.

Humans have been using animal skins for shelter, clothes and containers since the beginning of human history. From 4% to as much as 11% of a live animal's weight can be attributed to its skins. Animal hides and skins are among the most sought-after by-products. Leather shoes and purses, rawhide, sporting equipment, reformed sausage casing and cosmetics, sausage skins, edible gelatine and glue are examples of completed items made from cattle, pigs and sheep pelts. The hydrolysis of water-insoluble collagen generated from protein produces gelatine. Fresh, edible raw materials (hides or bone) are used to make this food product. Bones and skins both contain a significant amount of collagen. Three primary processes are involved in the production of gelatine from the hide. Non-collagenous material must be removed from the raw material in the first phase. Hydrolysis of collagen to gelatine is the next step. Finally, the finished product is recovered and dried.

Animal skins and hides may be used to make gelatine, which can be utilized in culinary products. Lard may be made from basic material. Pork skin is submerged, boiled, dried and then fried to form a snack dish (pork rinds) in the United States, Latin America, Europe and certain Asian nations. Collagen from the skin and hides of animals is an emulsifier in meat products, as it can bind huge amounts of fat. This makes it useful as a meat substitute or filler. Collagen sausage may also be made from the skins of cattle.

In addition to being a key component in jellies and aspic, gelatine is used in a broad variety of other foods. In addition to jellied sweets, it is used in a variety of meat items, including meat pies, because of its "melt-in-the-mouth" characteristic. Ice cream and other frozen sweets can also benefit from the use of gelatine as a stabilizer. Ice cream, yogurt and cream pies benefit from the use of high-bloom gelatine, which serves as a protective colloid. During storage, gelatine is supposed to prevent ice crystals from forming and lactose from recrystallizing.

The pharmaceutical business consumes 6.5% of the world's total gelatine production. The majority of it is employed in the production of gelatine capsules. Medicinal tablets and pastilles can also be made using gelatine as a binding and compounding ingredient. For ulcerated varicose veins, zinc gelatine is a vital element in a protective ointment. It creates a sterile gelatine sponge: just beat the gelatine into a foam, add formaldehyde and allow it to air dry [94]. These sponges are employed during surgery as well as for direct implantation of medications, such as antibiotics. Gelatine, a protein, is used to expand blood plasma in situations of extremely severe shock and damage. For many emulsions and foams, gelatine works great as an emulsifier and stabilizer. It is utilized in cosmetics as well as printing processes, including silk screen printing and photogravure.

Although collagen casing goods were created in Germany in the 1920s, American consumers didn't start using them until the 1960s. As with gelatine, the procedure does not convert the collagen into a soluble substance. Instead, it creates a product that is robust enough to be utilized as a casing for sausages and other items while yet retaining a significant amount of the natural collagen fibre. Water is added to the extracted collagen to make a dough, which is then extruded either wet or dry. To precipitate the collagen, the extruded tube is passed through a strong salt solution and an ammonia chamber. An elastic film is formed from the swelling gel. Glycerin can be added to make it more pliable. It is then dried to a water content of 10.0–15.0%.

Pork carcasses have 11% bone; beef carcasses include 15% bone; lamb carcasses contain 16% bone. Meat sticking to the bone raises these numbers even further. Some of the bones' marrow can also be eaten. The marrow's weight in the percentage of the carcass typically 4–6%. Bones have been used for soup and gelatine for ages. For the meat industry, new procedures have emerged in recent years in an effort to extract more flesh from the bones. Tissue that has been "mechanically separated," "mechanically deboned," or "mechanically removed" comes from the deboning of cattle, hogs or lambs. Many nations have now permitted the use of this meat in meat products (either in combination with other meats or alone).

Products with a high percentage of mechanically separated red meat tend to not perform well and don't taste as good. With more water in the meat, the colour darkens and the flesh gets mushy. Therefore, mechanically separated meat is often kept to a minimum. Some nations have a negative view of mechanically recovered meat because of health worries regarding the presence of bovine spongiform encephalopathy (BSE) [95]. The meat industry has recommended a threshold of 5.0–20.0% in hamburgers and ground beef and 10.0–40.0% in sausages.

Products that contain mechanically separated red meat are already subject to laws in several nations. Meat that has been mechanically separated cannot be used to make hamburgers, baby food, ground beef or meat pies in the US. The maximum concentration for sausage emulsion is 20%. In Denmark, a product's label must make a note of the presence of mechanically separated red meat if it accounts for more than 2% weight of the product. When items are exported from Australia, if they contain beef or mutton that has been mechanically deboned, the quantity must be indicated on the label. Additionally, the maximum level of calcium, the moisture content and the minimum protein level must also be included.

As a result of the high levels of readily accessible essential amino acids, minerals and vitamin B12 that it contains, meat and bone meal (also known as MBM) was widely advocated for and utilized in animal nutrition as a protein source in place of protein-rich diets. MBM and similarly rendered protein commodities have the potential to be used for purposes other than animal feed, such as the production of fuel or phosphorus fertilizer.

Many countries, including Japan, China and India, have long employed animal glands and organs as traditional medicines. Hormones (enzymes that control the body's metabolism) are produced by the endocrine glands. Other organs affected include the pancreas and intestines, the pituitary and thyroid glands and the pancreato-adrenal axis. Only healthy animals are used to harvest the glands for study. Some practice is required to find the glands, as their size and the surrounding tissue make them difficult to detect [96].

Various animals have glands that are crucial, depending on the animal's species, sex and age. Rapid freezing is the best strategy for preserving most glands in order to prevent tissue degradation due to bacterial development. A thorough cleaning and trimming of fat and connective tissue is required before freezing. At a temperature of −18°C or below, the glands are put on waxed paper. For extraction, glands are first examined before being diced and combined with various solutions or dried in a vacuum drier. Solutions including ethylene, light petroleum, gasoline or acetone can be used to remove excess fat from desiccated glands. The glands or extracts are dried and defatted before being crushed into a powder and put into capsules or used as a drink. Before they can be sold, they must pass stringent safety and potency tests.

Cholesterol, the essential component for the production of vitamin D3, may be recovered from a variety of sources, such as the brain, the neurological system and the spinal cord. In the cosmetics industry, cholesterol is frequently employed in the role of an emulsifier [97]. To accomplish the same goal, other components from the hypothalamus can be separated. It is being tested for schizophrenia, sleeplessness and other conditions, including mental retardation, by extracting melatonin from the pineal gland. The gall bladder is where we get our bile, which is made up of various acids, pigments, proteins, cholesterol and so on. Indigestion, constipation and bile tract diseases can be treated with it. Also, the liver's secretory function can be boosted by taking this supplement. Both dry and liquid extracts are available for purchase. Prednisone and cortisone, two hormones found in bile, can be isolated and utilized as medications. Anecdotally, gallstones have been linked to erectile dysfunction, and they can be sold for high prices. Necklaces and pendants often have them as decorative accents [96].

The liver is the biggest gland in an animal. A mature cow's liver weighs roughly 5 kilograms, but a pig's liver weighs about 1.4 kilograms. Raw powdered liver and heated, slightly acidified water are combined to make liver extract. Concentrated stock is utilized as a raw material in the pharmacological business after being concentrated in a vacuum at low temperatures. To acquire vitamin B12, pig and cattle liver extracts have long been used as a dietary supplement to treat various forms of anaemia. The lungs and the lining of the small intestines, in addition to the liver, are both potential sources of heparin for extraction. It acts as an anticoagulant, which means that it slows down the process of blood clotting. Additionally, it is used to avoid the clotting of blood during surgical procedures and organ transplants by thinning the blood.

Pig ovaries are a potential source of the hormones progesterone and oestrogen. It is possible that it might be utilized to address reproductive issues that some women experience. Relaxin is a hormone that is commonly utilized during labour and delivery. The hormone is extracted from the ovaries of pregnant sows [96].

Insulin is produced by the pancreas. Insulin controls the body's metabolism of sugar and is an important component in the treatment of diabetes. Glucagon, which is derived from the cells of the pancreas, is used to treat insulin overdoses or low blood sugar caused by alcoholism. It is also used to raise the level of sugar in the blood. Both chymotrypsin and trypsin are used to expedite the healing process following surgical procedures or injuries.

The production of catgut, which is utilized for the creation of internal surgical sutures, requires the utilization of the intestines of sheep and calves. Pig and cattle small intestinal lining can be removed and conserved for use in other products during the process of converting intestines into casings. It is kept either in its natural condition, or it is treated into a dry powder before being sent to companies that make heparin [96].

2.2.6 Other industries

Many industries use fruits and vegetables to produce beverages and other supplements. Biodegradable solids, such as stems, skins and seeds, are among the waste products from the beverage industry. These materials contain a significant quantity of bioactive chemicals that may be recovered for use in other industries. For instance, phenolic compounds, dietary fibre, vitamin C and carotenoids are among the bioactive substances found in mango peels. This group of compounds has long been known to help reduce the risk of several diseases, like cancer, ageing-related cognitive decline, Alzheimer's and Parkinson's. Adding these substances to functional foods, nutraceuticals and cosmetics is an excellent method for preventing degenerative processes. These are extensively employed for the production of medications and as food additives to boost the functioning of foods. In terms of fruit production, citrus is the most plentiful crop in the world. One-third of the harvest is used for processing. A total of 98% of the total industrialized crop is made up of oranges, lemons, grapefruits and mandarins. Citrus fruits are not only juiced but are also processed into jams and mandarin segments in the fruit canning sector. A great number of phytochemicals can also be recovered from the waste of citrus fruits [7].

Waste from the dairy and seafood industries is included in the category of animal-derived waste. Wastes like shells, heads, guts, scales, bones, fins and others are produced in enormous quantities as a result of the commercial processing of seafood for various products. Additionally, fishing operations that target popular species also result in the capture of substantial quantities of fish that are considered by-catch because they are commercially unviable. The majority of these materials are now dumped in landfills or only partially processed into goods like animal feed and leather, which poses considerable environmental risks. Given the quick development of biotechnology, the prospect for utilizing wastes as sources of beneficial nutraceuticals and other substances is significant. These ingredients include proteins like collagen, protein hydrolyzates and gelatine bioactive peptides, lipids rich in polyunsaturated squalene, fatty acids and carotenoids; polysaccharides, like chitin glycosaminoglycans and chitosan as well as their derivatives; and nutraceuticals, among others [98].

2.3 COMMERCIAL ASPECTS

The rising demand for nutraceuticals and nutritional supplements, as well as increased public awareness of the health advantages of natural antioxidants, are driving the growth of the bioactive compound industry, such as the natural antioxidant market. Since there has been a significant increase in the consumption of nutritional supplements in North America and Europe, there is projected to be an increased need for natural antioxidants. These regions are likely to see an increase in demand for natural antioxidants as a result of the growing popularity of natural goods. The market for nutraceuticals is expected to grow as a consequence of the increasing use of natural antioxidants. Over half of the market's revenue is expected to come from the sale of natural antioxidants used in nutraceuticals, foods and drinks by 2029. In addition, it is projected that the utilization of natural antioxidants will continue to be utilized in the meat processing business throughout the course of the upcoming years. The various bioactive substances that can be retrieved from the waste produced by the food industry hold a significant amount of promise as well. When used on a commercial scale, the synthesis of bioactive compounds from waste products generated by the food industry has the potential to significantly reduce the costs of producing bioactive compounds.

2.4 CHALLENGES AND FUTURE PERSPECTIVES

Sustainability, which refers to employing practices, materials and systems that won't disrupt natural cycles or deplete natural resources, is the major direction the food business is moving towards. This idea encompasses "green engineering" and "green chemistry" ideas. Researchers have been made aware, by this backdrop, of the necessity to explore innovative environmentally friendly procedures and protocols to extract bioactive materials from food industry waste. For the reuse of food industry waste to gain momentum with consumers and businesses alike, researchers must dig deep into innovative extraction methods that use less solvents and, as a result, contribute to a healthier food supply chain. One of the primary aims of green extraction techniques is to replace ecologically harmful solvents like acetonitrile, methanol, dichloromethane and toluene with more environmentally friendly solvents like ethanol and water. It is important for the food processing industry to be aware of the importance of food safety, yet the laws in each nation are different. Natural supplement side effects are rare, but authorities must be careful to protect public safety due to the rising frequency of goods containing bioactive chemicals.

While there are no regulations or rules for food waste use, this analysis compares European and American regulatory frameworks as well as the risk assessment criteria that food-related bioactive chemicals must follow in order to guarantee product safety. Food waste is altering consumer attitudes and behaviours, and their focus is turning to "natural" components for a better lifestyle. As a result, corporations and national authorities have an increased need to provide a safer product.

2.5 CONCLUSIONS

Natural bioactive compounds are a unique class of pharmaceutical molecules that have potential applications in a wide range of food classes, including functional foods, nutraceuticals, dietary supplements and food additives, which offer consumers healthy and sustainable alternatives to synthetic ones. Food loss and waste may provide these high-value chemicals, but incorporating them into food additives, nutraceuticals and nutritional supplements would present several hurdles, chief among them the need to ensure their safety and toxicity. Encapsulation technology is now being used to improve component stability, bioactive chemical release control and product shelf life.

To the best of our knowledge, there is very little production of bioactive compounds from food processing waste on a commercial scale, despite the fact that numerous studies on the utilization of bioactive compounds derived from food processing waste are currently being reported on a laboratory scale. Because of this, it is of the highest importance to discover a suitable extraction process for the biotransformation of these wastes into useful products that are low-cost and have high nutritional value. The use of waste products unquestionably eliminates not only the problem of illegal dumping but also the issue of pollution-related issues. Therefore, additional endorsements from regulatory bodies, in addition to primary financial backing, are required in order to introduce these value-added items to the commercial market. The transformation of waste from the food processing industry into useful compounds has the potential not only to open up new avenues of inquiry for academics but also to reduce the environmental risks that are already present.

ACKNOWLEDGEMENT

The study is supported by the Indian National Academy of Engineering (INAE/121/AKF/22), Gurgaon, India. The authors are solely responsible for all of the opinions, results and conclusions expressed in this study; INAE's viewpoints are not necessarily reflected in any of these aspects.

REFERENCES

1. A.A. Vilas-Boas, M. Pintado, A.L.S. Oliveira, Natural bioactive compounds from food waste: Toxicity and safety concerns, *Foods* 10(7) (2021) 1564.
2. S. Emani, R. Uppaluri, M.K. Purkait, Preparation and characterization of low cost ceramic membranes for mosambi juice clarification, *Desalination* 317 (2013) 32–40.
3. P. Duarah, D. Haldar, M.K. Purkait, Technological advancement in the synthesis and applications of lignin-based nanoparticles derived from agro-industrial waste residues: A review, *Int. J. Biol. Macromol.* 163 (2020) 1828–1843.
4. B. Debnath, D. Haldar, M.K. Purkait, Environmental remediation by tea waste and its derivative products: A review on present status and technological advancements, *Chemosphere* 300 (2022) 134480.
5. D. Haldar, P. Duarah, M.K. Purkait, Chapter 16 - Progress in the synthesis and applications of polymeric nanomaterials derived from waste lignocellulosic biomass, in: D. Giannakoudakis, L. Meili, I. Anastopoulos (Eds.), *Advanced materials for sustainable environmental remediation*, Elsevier, 2022, pp. 419–433.
6. T.N. Baite, B. Mandal, M.K. Purkait, Ultrasound assisted extraction of gallic acid from Ficus auriculata leaves using green solvent, *Food Bioprod. Process.* 128 (2021) 1–11.
7. K. Kumar, A.N. Yadav, V. Kumar, P. Vyas, H.S. Dhaliwal, Food waste: A potential bioresource for extraction of nutraceuticals and bioactive compounds, *Bioresour. Bioprocess.* 4(1) (2017) 18.
8. B. Debnath, D. Haldar, M.K. Purkait, A critical review on the techniques used for the synthesis and applications of crystalline cellulose derived from agricultural wastes and forest residues, *Carbohydrate Polym.* 273 (2021) 118537.

9. B. Debnath, D. Haldar, M.K. Purkait, Potential and sustainable utilization of tea waste: A review on present status and future trends, *J. Environ. Chem. Eng.* 9(5) (2021) 106179.
10. S.A. Abdeltaif, K.A. SirElkhatim, A.B. Hassan, Estimation of phenolic and flavonoid compounds and antioxidant activity of spent coffee and black tea (processing) waste for potential recovery and reuse in Sudan, *Recycling* 3(2) (2018) 27.
11. W. Sui, Y. Xiao, R. Liu, T. Wu, M. Zhang, Steam explosion modification on tea waste to enhance bioactive compounds' extractability and antioxidant capacity of extracts, *J. Food Eng.* 261 (2019) 51–59.
12. R.K. Singla, A.K. Dubey, A. Garg, R.K. Sharma, M. Fiorino, S.M. Ameen, M.A. Haddad, M. Al-Hiary, *Natural polyphenols: Chemical classification, definition of classes, subcategories, and structures*, Oxford University Press, 2019, pp. 1397–1400.
13. G. Serdar, E. Demir, M. Sökmen, Recycling of tea waste: Simple and effective separation of caffeine and catechins by microwave assisted extraction (MAE), *Int. J. Second. Metabol.* 4(2) (2017) 78–89.
14. M. Sökmen, E. Demir, S.Y. Alomar, Optimization of sequential supercritical fluid extraction (SFE) of caffeine and catechins from green tea, *J. Supercrit. Fluids.* 133 (2018) 171–176.
15. L.-L. Du, Q.-Y. Fu, L.-P. Xiang, X.-Q. Zheng, J.-L. Lu, J.-H. Ye, Q.-S. Li, C.A. Polito, Y.-R. Liang, Tea polysaccharides and their bioactivities, *Molecules.* 21(11) (2016) 1449.
16. X. Li, S. Chen, J.-E. Li, N. Wang, X. Liu, Q. An, X.-M. Ye, Z.-T. Zhao, M. Zhao, Y. Han, Chemical composition and antioxidant activities of polysaccharides from Yingshan cloud mist tea, *Oxid. Med. Cell.Longev.* 2019 (2019). https://doi.org/10.1155/2019/1915967
17. Y. Narita, K. Inouye, Review on utilization and composition of coffee silverskin, *Food Res. Int.* 61 (2014) 16–22.
18. D. Lerda, *Coffee contaminated with otaand genotoxicity, Production, consumption and health benefits*, Nova Science, 2016, 157.
19. P.S. Murthy, M.M. Naidu, Recovery of phenolic antioxidants and functional compounds from coffee industry by-products, *Food Bioprocess Technol.* 5(3) (2012) 897–903.
20. K.S. Andrade, R.T. Gonçalvez, M. Maraschin, R.M. Ribeiro-do-Valle, J. Martínez, S.R. Ferreira, Supercritical fluid extraction from spent coffee grounds and coffee husks: Antioxidant activity and effect of operational variables on extract composition, *Talanta* 88 (2012) 544–552.
21. L.F. Ballesteros, J.A. Teixeira, S.I. Mussatto, Chemical, functional, and structural properties of spent coffee grounds and coffee silverskin, *Food Bioprocess Technol.* 7(12) (2014) 3493–3503.
22. A. Jiménez-Zamora, S. Pastoriza, J.A. Rufián-Henares, Revalorization of coffee by-products. Prebiotic, antimicrobial and antioxidant properties, *LWT-Food Sci. Technol.* 61(1) (2015) 12–18.
23. M. Banožić, D. Šubarić, S. Jokić, Tobacco waste in Bosnia and Herzegovina–problem or high-value material?, *Glasnik Zaštite Bilja* 41(4) (2018) 64–72.
24. T.R. Ganapathi, P. Suprasanna, P. Rao, V. Bapat, Tobacco (Nicotiana tabacum L.)-A model system for tissue culture interventions and genetic engineering, *Indian J. Biotechnol* 3 (2004) 171–184.
25. G. Tayoub, H. Sulaiman, M. Alorfi, Determination of nicotine levels in the leaves of some Nicotiana tabacum varieties cultivated in Syria, *Herba Polonica* 61(4) (2015) 23–30.
26. R.A. Bhisey, Chemistry and toxicology of smokeless tobacco, *Indian J. Cancer* 49(4) (2012) 364.
27. M. Banožić, J. Babić, S. Jokić, Recent advances in extraction of bioactive compounds from tobacco industrial waste-a review, *Ind. Crop. Prod.* 144 (2020) 112009.
28. J. Wang, D. Lu, H. Zhao, B. Jiang, J. Wang, X. Ling, H. Chai, P. Ouyang, Discrimination and classification of tobacco wastes by identification and quantification of polyphenols with LC-MS/MS, *J. Serbian Chem. Soc.* 75(7) (2010) 875–891.
29. J. Chen, X. LIU, X. Xu, F.S.-C. Lee, X. Wang, Rapid determination of total solanesol in tobacco leaf by ultrasound-assisted extraction with RP-HPLC and ESI-TOF/MS, *J. Pharm. Biomed. Anal.* 43(3) (2007) 879–885.
30. F. Briški, N. Horgas, M. Vuković, Z. Gomzi, Aerobic composting of tobacco industry solid waste—Simulation of the process, *Clean Technol. Environ. Policy* 5(3) (2003) 295–301.
31. M. Vuković, I. Ćosić, K. Kolačko, F. Briškia, Kinetika biorazgradnje organskih tvari u procjednoj vodi iz duhanskog otpada, *Kemija u Industriji* 61 (2012) 417–425.
32. J. Zeng, K.F. Chen, J.P. Xie, G. Xu, J. Li, G.H. Rao, F. Yang, W.H. Gao, Study on tobacco stem and tobacco dust making reconstituted tobacco paper-base, Advanced Materials Research, *Trans Tech Publ*, 2012, pp. 3316–3322.
33. R.-S. Hu, J. Wang, H. Li, H. Ni, Y.-F. Chen, Y.-W. Zhang, S.-P. Xiang, H.-H. Li, Simultaneous extraction of nicotine and solanesol from waste tobacco materials by the column chromatographic extraction method and their separation and purification, *Sep. Purif. Technol.* 146 (2015) 1–7.
34. A. Rawat, R.R. Mal, Phytochemical properties and pharmcological activities of Nicotiana tabacum, *Indian J. Pharm. Biol. Res.* 1(02) (2013) 74.

35. Z.Š. Troje, Z. Fröbe, Đ. Perović, Analysis of selected alkaloids and sugars in tobacco extract, *J. Chromatograph. A* 775(1–2) (1997) 101–107.
36. C.A. Vieira, S.A. de Paiva, M.N. Funai, M.M. Bergamaschi, R.H. Queiroz, J.R. Giglio, Quantification of nicotine in commercial brand cigarettes: How much is inhaled by the smoker?, *Biochem. Mol. Biol. Educ.* 38(5) (2010) 330–334.
37. P.A. Crooks, *Chemical properties of nicotine and other tobacco-related compounds, Analytical determination of nicotine and related compounds and their metabolites*, Elsevier, 1999, pp. 69–147.
38. I.T. Stanisavljević, M. Lazić, V. Veljković, Ultrasonic extraction of oil from tobacco (Nicotiana tabacum L.) seeds, *Ultrason. Sonochem.* 14(5) (2007) 646–652.
39. M. Grigg, H. Glasgow, Subsidised nicotine replacement therapy, *Tob. Control* 12(2) (2003) 238–239.
40. A. Charlton, Medicinal uses of tobacco in history, *J. Roy. Soc. Med.* 97(6) (2004) 292–296.
41. Y. Wang, W. Gu, Study on supercritical fluid extraction of solanesol from industrial tobacco waste, *J. Supercrit. Fluids.* 138 (2018) 228–237.
42. W. Huang, Z. Li, H. Niu, J. Wang, Y. Qin, Bioactivity of solanesol extracted from tobacco leaves with carbon dioxide–ethanol fluids, *Biochem. Eng. J.* 42(1) (2008) 92–96.
43. Q.-M. Ru, L.-J. Wang, W.-M. Li, J.-L. Wang, Y.-T. Ding, In vitro antioxidant properties of flavonoids and polysaccharides extract from tobacco (Nicotiana tabacum L.) leaves, *Molecules* 17(9) (2012) 11281–11291.
44. F. Fathiazad, A. Delazar, R. Amiri, S.D. Sarker, Extraction of flavonoids and quantification of rutin from waste tobacco leaves, *Iran. J. Pharm. Res.* 3 (2010) 222–227.
45. M. Docheva, S. Dagnon, S. Statkova-Abeghe, Flavonoid content and radical scavenging potential of extracts prepared from tobacco cultivars and waste, *Nat. Prod. Res.* 28(17) (2014) 1328–1334.
46. C.-P. Xu, Y. Xiao, D. Mao, Antioxidant activities of polysaccharide fractions isolated from burley tobacco flowers, *Croat. J. Food Sci. Technol.* 5(2) (2013) 46–52.
47. R. Severson, J. Ellington, R. Arrendale, M. Snook, Quantitative gas chromatographic method for the analysis of aliphatic hydrocarbons, terpenes, fatty alcohols, fatty acids and sterols in tobacco, *J. Chromatograph. A* 160(1) (1978) 155–168.
48. J. Shen, X. Shao, A comparison of accelerated solvent extraction, Soxhlet extraction, and ultrasonic-assisted extraction for analysis of terpenoids and sterols in tobacco, *Anal. Bioanal. Chem.* 383(6) (2005) 1003–1008.
49. E.T. Nerantzis, P. Tataridis, Integrated enology-utilization of winery by-products into high added value products, *J. Sci. Tech.* 1 (2006) 79–89.
50. V. Silva, G. Igrejas, V. Falco, T.P. Santos, C. Torres, A.M. Oliveira, J.E. Pereira, J.S. Amaral, P. Poeta, Chemical composition, antioxidant and antimicrobial activity of phenolic compounds extracted from wine industry by-products, *Food Control* 92 (2018) 516–522.
51. R.H. Dashwood, E. Ho, Dietary histone deacetylase inhibitors: from cells to mice to man, *Semin. Cancer Biol.*, Elsevier, 2007, pp. 363–369.
52. A. Bassani, N. Alberici, C. Fiorentini, G. Giuberti, R. Dordoni, G. Spigno, Hydrothermal treatment of grape skins for sugars, antioxidants and soluble fibers production, *Chem. Eng. Trans.* 79 (2020).
53. E.S. Pinheiro, J.M. da Costa, E. Clemente, P.H. Machado, G.A. Maia, Physical chemical and mineral stability of grape juice obtained by steam extraction, *Rev. Ciênc. Agron.* 40(3) (2009) 373.
54. M.R. Gonzalez-Centeno, M. Jourdes, A. Femenia, S. Simal, C. Rosselló, P.-L. Teissedre, Characterization of polyphenols and antioxidant potential of white grape pomace byproducts (Vitis vinifera L.), *J. Agric. Food Chem.* 61(47) (2013) 11579–11587.
55. A.K. Chakka, A.S. Babu, Bioactive compounds of winery by-products: Extraction techniques and their potential health benefits, *Appl. Food Res.* 2(1) (2022) 100058.
56. C. Negro, L. Tommasi, A. Miceli, Phenolic compounds and antioxidant activity from red grape marc extracts, *Bioresour. Technol.* 87(1) (2003) 41–44.
57. S.R. Iora, G.M. Maciel, A.A. Zielinski, M.V. da Silva, P.V.d.A. Pontes, C.W. Haminiuk, D. Granato, Evaluation of the bioactive compounds and the antioxidant capacity of grape pomace, *Int. J. Food Sci. Technol.* 50(1) (2015) 62–69.
58. S. Hogan, L. Zhang, J. Li, S. Sun, C. Canning, K. Zhou, Antioxidant rich grape pomace extract suppresses postprandial hyperglycemia in diabetic mice by specifically inhibiting alpha-glucosidase, *Nutr. Metab.* 7(1) (2010) 1–9.
59. G. Costabile, M. Vitale, D. Luongo, D. Naviglio, C. Vetrani, P. Ciciola, A. Tura, F. Castello, P. Mena, D. Del Rio, Grape pomace polyphenols improve insulin response to a standard meal in healthy individuals: A pilot study, *Clin. Nutr.* 38(6) (2019) 2727–2734.
60. N. Vaisman, E. Niv, Daily consumption of red grape cell powder in a dietary dose improves cardiovascular parameters: A double blind, placebo-controlled, randomized study, *Int. J. Food Sci. Nutr.* 66(3) (2015) 342–349.

61. W.P. de Oliveira, A.C.T. Biasoto, V.F. Marques, I.M. Dos Santos, K. Magalhães, L.C. Correa, M. Negro-Dellacqua, M.S. Miranda, A.C. de Camargo, F. Shahidi, Phenolics from winemaking by-products better decrease VLDL-cholesterol and triacylglycerol levels than those of red wine in wistar rats, *J. Food Sci.* 82(10) (2017) 2432–2437.
62. S. Chacar, J. Hajal, Y. Saliba, P. Bois, N. Louka, R.G. Maroun, J.F. Faivre, N. Fares, Long-term intake of phenolic compounds attenuates age-related cardiac remodeling, *Aging Cell* 18(2) (2019) e12894.
63. L. Alguacil, E. Salas, J. Perez-Ortiz, N. Castillo-Munoz, I. Hermosin-Gutierrez, E. Garcia-Romero, S. Gomez-Alonso, C. Gonzalez-Martin, *Anti-proliferative and cytotoxic effects of grape pomace and grape seed extracts on colorectal cancer cell lines, Basic & clinical pharmacology & toxicology*, Wiley-Blackwell, 2014, pp. 45–46.
64. S. Chacar, T. Itani, J. Hajal, Y. Saliba, N. Louka, J.F. Faivre, R. Maroun, N. Fares, The impact of long-term intake of phenolic compounds-rich grape pomace on rat gut microbiota, *J. Food Sci.* 83(1) (2018) 246–251.
65. T.A. Jacobson, M. Miller, E.J. Schaefer, Hypertriglyceridemia and cardiovascular risk reduction, *Clin. Ther.* 29(5) (2007) 763–777.
66. E.Y. Ishimoto, S.J.V. Vicente, R.J. Cruz, E.A.F.d.S. Torres, Hypolipidemic and antioxidant effects of grape processing by-products in high-fat/cholesterol diet-induced hyperlipidemic hamsters, *Food Sci. Technol.* 40 (2020) 558–567.
67. D. Pinto, M.d.l.L. Cádiz-Gurrea, A. Vallverdú-Queralt, C. Delerue-Matos, F. Rodrigues, Castanea sativa shells: A review on phytochemical composition, bioactivity and waste management approaches for industrial valorization, *Food Res. Int.* 144 (2021) 110364.
68. M.D.C.B.M. De Vasconcelos, R.N. Bennett, E.A. Rosa, J.V.F. Cardoso, Primary and secondary metabolite composition of kernels from three cultivars of Portuguese chestnut (Castanea sativa Mill.) at different stages of industrial transformation, *J. Agric. Food Chem.* 55(9) (2007) 3508–3516.
69. M.C. De Vasconcelos, R.N. Bennett, E.A. Rosa, J.V. Ferreira-Cardoso, Composition of European chestnut (Castanea sativa Mill.) and association with health effects: Fresh and processed products, *J. Sci. Food Agric.* 90(10) (2010) 1578–1589.
70. F.M. Vella, B. Laratta, F. La Cara, A. Morana, Recovery of bioactive molecules from chestnut (Castanea sativa Mill.) by-products through extraction by different solvents, *Nat. Prod. Res.* 32(9) (2018) 1022–1032.
71. G. Vázquez, O. Mosquera, M.S. Freire, G. Antorrena, J. González-Álvarez, Alkaline pre-treatment of waste chestnut shell from a food industry to enhance cadmium, copper, lead and zinc ions removal, *Chem. Eng. J.* 184 (2012) 147–155.
72. P. Comandini, M.J. Lerma-García, E.F. Simó-Alfonso, T.G. Toschi, Tannin analysis of chestnut bark samples (Castanea sativa Mill.) by HPLC-DAD–MS, *Food Chem.* 157 (2014) 290–295.
73. A. Fernández-Agulló, M.S. Freire, G. Antorrena, J.A. Pereira, J. González-Álvarez, Effect of the extraction technique and operational conditions on the recovery of bioactive compounds from chestnut (Castanea sativa) bur and shell, *Sep. Sci. Technol.* 49(2) (2014) 267–277.
74. G. Vázquez, A. Pizzi, M.S. Freire, J. Santos, G. Antorrena, J. González-Álvarez, MALDI-TOF, HPLC-ESI-TOF and 13C-NMR characterization of chestnut (Castanea sativa) shell tannins for wood adhesives, *Wood Sci. Technol.* 47(3) (2013) 523–535.
75. A. Aires, R. Carvalho, M.J. Saavedra, Valorization of solid wastes from chestnut industry processing: Extraction and optimization of polyphenols, tannins and ellagitannins and its potential for adhesives, cosmetic and pharmaceutical industry, *Waste Manag.* 48 (2016) 457–464.
76. G. Vázquez, J. González-Alvarez, J. Santos, M. Freire, G. Antorrena, Evaluation of potential applications for chestnut (Castanea sativa) shell and eucalyptus (Eucalyptus globulus) bark extracts, *Ind. Crops Prod.* 29(2–3) (2009) 364–370.
77. F. Lameirão, D. Pinto, E. F. Vieira, A. F. Peixoto, C. Freire, S. Sut, S. Dall'Acqua, P. Costa, C. Delerue-Matos, F. Rodrigues, Green-sustainable recovery of phenolic and antioxidant compounds from industrial chestnut shells using ultrasound-assisted extraction: Optimization and evaluation of biological activities in vitro, *Antioxidants.* 9(3) (2020) 267.
78. D. Pinto, M. de la Luz Cadiz-Gurrea, S. Sut, A.S. Ferreira, F.J. Leyva-Jimenez, S. Dall'Acqua, A. Segura-Carretero, C. Delerue-Matos, F. Rodrigues, Valorisation of underexploited Castanea sativa shells bioactive compounds recovered by supercritical fluid extraction with CO2: A response surface methodology approach, *J. CO2 Util.* 40 (2020) 101194.
79. D. Pinto, A.M. Silva, V. Freitas, A. Vallverdú-Queralt, C. Delerue-Matos, F. Rodrigues, Microwave-assisted extraction as a green technology approach to recover polyphenols from Castanea sativa shells, *ACS Food Sci. Technol.* 1(2) (2021) 229–241.

80. N.A. Cacciola, G. Squillaci, M. D'Apolito, O. Petillo, F. Veraldi, F. La Cara, G. Peluso, S. Margarucci, A. Morana, Castanea sativa Mill. shells aqueous extract exhibits anticancer properties inducing cytotoxic and pro-apoptotic effects, *Molecules* 24(18) (2019) 3401.
81. G. Squillaci, F. Apone, L.M. Sena, A. Carola, A. Tito, M. Bimonte, A. De Lucia, G. Colucci, F. La Cara, A. Morana, Chestnut (Castanea sativa Mill.) industrial wastes as a valued bioresource for the production of active ingredients, *Process Biochem.* 64 (2018) 228–236.
82. F.M. Vella, L. De Masi, R. Calandrelli, A. Morana, B. Laratta, Valorization of the agro-forestry wastes from Italian chestnut cultivars for the recovery of bioactive compounds, *Eur. Food Res. Technol.* 245(12) (2019) 2679–2686.
83. M. Vasconcelos, R.N. Bennett, S. Quideau, R. Jacquet, E.A. Rosa, J.V. Ferreira-Cardoso, Evaluating the potential of chestnut (Castanea sativa Mill.) fruit pericarp and integument as a source of tocopherols, pigments and polyphenols, *Ind. Crops Prod.* 31(2) (2010) 301–311.
84. V. Silva, V. Falco, M.I. Dias, L. Barros, A. Silva, R. Capita, C. Alonso-Calleja, J.S. Amaral, G. Igrejas, I.C.F.R. Ferreira, Evaluation of the phenolic profile of Castanea sativa Mill. by-products and their antioxidant and antimicrobial activity against multiresistant bacteria, *Antioxidants* 9(1) (2020) 87.
85. F. Rodrigues, J. Santos, F.B. Pimentel, N. Braga, A. Palmeira-de-Oliveira, M.B.P. Oliveira, Promising new applications of Castanea sativa shell: Nutritional composition, antioxidant activity, amino acids and vitamin E profile, *Food Funct.* 6(8) (2015) 2854–2860.
86. J. Barreira, I.C. Ferreira, M.B.P. Oliveira, J. Pereira, Antioxidant potential of chestnut (Castanea sativa L.) and almond (Prunus dulcis L.) by-products, *Food Sci. Technol. Int.* 16(3) (2010) 209–216.
87. C.P. Gomes, R. Dias, M.R.P. Costa, Preparation of molecularly imprinted adsorbents with improved retention capability of polyphenols and their application in continuous separation processes, *Chromatographia* 82(6) (2019) 893–916.
88. V. Corregidor, A.L. Antonio, L.C. Alves, S.C. Verde, Castanea sativa shells and fruits: Compositional analysis by proton induced X-ray emission, *Nucl. Instrum Methods Phys. Res. Section B: Beam Interact. Mater. Atoms* 477 (2020) 98–103.
89. J.A. Rivera, J.G. Sebranek, R.E. Rust, L.B. Tabatabai, Composition and protein fractions of different meat by-products used for petfood compared with mechanically separated chicken (MSC), *Meat Sci.* 55(1) (2000) 53–59.
90. M. Ünsal, N. Aktaş, Fractionation and characterization of edible sheep tail fat, *Meat Sci.* 63(2) (2003) 235–239.
91. S. Devatkal, S. Mendiratta, N. Kondaiah, M. Sharma, A. Anjaneyulu, Physicochemical, functional and microbiological quality of buffalo liver, *Meat Sci.* 68(1) (2004) 79–86.
92. R. Ghosh, Fractionation of biological macromolecules using carrier phase ultrafiltration, *Biotechnol. Bioeng.* 74(1) (2001) 1–11.
93. E.B. Kurbanoglu, N.I. Kurbanoglu, Utilization as peptone for glycerol production of ram horn waste with a new process, *Energy Convers. Manag.* 45(2) (2004) 225–234.
94. J. Gómez-Estaca, P. Montero, F. Fernández-Martín, M. Gómez-Guillén, Physico-chemical and film-forming properties of bovine-hide and tuna-skin gelatin: A comparative study, *J. Food Eng.* 90(4) (2009) 480–486.
95. I.S. Arvanitoyannis, D. Ladas, Meat waste treatment methods and potential uses, *Int. J. Food Sci. Technol.* 43(3) (2008) 543–559.
96. K. Jayathilakan, K. Sultana, K. Radhakrishna, A.S. Bawa, Utilization of byproducts and waste materials from meat, poultry and fish processing industries: A review, *J. Food Sci. Technol.* 49(3) (2012) 278–93.
97. E.E. Chukwunonso, N.E. Tufon, Cholesterol concentration in different parts of bovine meat sold in Nsukka, Nigeria: Implications for cardiovascular disease risk, *Afr. J. Biochem. Res.* 3(4) (2009) 095–097.
98. V.V. Menon, S.S. Lele, *Nutraceuticals and bioactive compounds from seafood processing waste, Springer handbook of marine biotechnology*, Springer, 2015, pp. 1405–1425.

3 Technological advancement in the extraction of bioactive compounds from food industry waste

3.1 INTRODUCTION

The biological system of plants consists of both primary and secondary metabolites. Proteins, amino acids and carbohydrates are the major metabolites that are utilized extensively by plant tissues during their development and maturation. During the developmental phase, secondary metabolites are generated to help the plants survive and overcome natural challenges. The biological effects of different active ingredients vary significantly from one another and have been broadly utilized in the production of functional foods and treatment of human ailments [1]. The extraction of biologically active compounds from plant materials has practical implications for the well-being of human society. As such, the use of appropriate extraction techniques for obtaining bioactive compounds from natural sources has become a frontier topic in the pharmaceutical and food industries. Biologically active compounds are currently extracted from a wide range of natural sources, including microbes, fungus, plants and animals, among others. Bioactive compounds are divided into various categories, such as nitrogen-containing compounds, terpenoids, phenolics, alkaloids and organosulfur compounds [2]. Terpenoids are antioxidants that include carotenoids, tocotrienol, limonoids, tocopherols and phytosterols. It also includes lutein, β-cryptoxanthin and α and β-carotene. Alkaloids may be categorized into different types depending on the synergy of different compounds, viz. lycopodium, morphine, pyrrolizidine, atropine, pyrrolidine, quinine, quinolizidine, strychnine, indole, ephedrine, quinoline, nicotine and steroids. Such phytochemicals have been found to possess several health benefit properties including anti-diabetic, anticancerous, anti-inflammatory, improved digestion and enhanced blood circulation rate, among others [3]. There are several parameters which affect the extraction process, such as pressure, matrix properties, treatment time, extraction solvent, temperature and solvent to matrix ratio.

As such, it is crucial to select an appropriate and effective extraction technique to maximize the amount of extract from the plant tissues along with proper quantification and characterization of the active compounds. Some of the most commonly used conventional extraction techniques for obtaining bioactive compounds in recent years includes hot water extraction, impregnation method and Soxhlet extraction [4]. These techniques, however, consist of several drawbacks. For instance, Soxhlet extraction technique displays low process efficiency apart from generating a lot of organic reagent waste. The applicability of Soxhlet extraction in industrial amplification is limited due to such shortcomings. And the impregnation method has the drawbacks of long extraction time and low process efficiency. The extracts produced using impregnation method are susceptible to mould, necessitating the use of a preservative. Also, a considerable amount of extract is obtained in this process, which requires an additional concentration step. Furthermore, low extraction rate is still one of the major drawbacks in the hot water extraction process. The temperature during the process is very high, and heat-sensitive compounds may be subjected to the risk of denaturation [5, 6]. Due to the aforementioned limitations of the conventional processes, a wide range of advanced extraction techniques for the separation of bioactive compounds have gained significant interest over the

DOI: 10.1201/9781003315469-3

last few decades. Advanced extraction techniques have been developed that are considered eco-friendly and sustainable due to the reduced utilization of organic and synthetic chemicals, shorter treatment time and increased yield and extract quality. These techniques have been extensively used, as they are effective in increasing the overall selectivity and yield of bioactive compounds obtained from plant sources. A few advanced recovery techniques are also characterized as "green recovery methods" since they consume less energy, require minimal amounts of organic solvents and causes no adverse effect on the environment [7]. The food industry in particular requires the use of advanced extraction techniques for the effective separation of bioactive compounds from natural sources. Besides, several studies have shown the effectiveness of advanced extraction techniques for faster and more efficient extraction [8].

The current chapter focuses on various advanced extraction techniques, viz. microwave-assisted extraction, ultrasound-assisted extraction, enzyme-assisted extraction, pulsed-assisted extraction, pressurized liquid extraction and supercritical liquid extraction for the separation of food-based bioactive compounds in addition to the existing conventional techniques. Detailed descriptions and various experimental conditions associated with each of these processes that affect the extraction efficiency are described. The benefits and limitations of both conventional and advanced extraction techniques are covered in detail. In addition, the commercial applications of the extraction techniques are also discussed and summarized in this chapter.

3.2 CONVENTIONAL METHODS FOR THE EXTRACTION OF BIOACTIVE COMPOUNDS

Extraction methods encompass a variety of ways, each of which is used to extract biologically active compounds for a distinct purpose. The extraction capacity of different solvents, as well as the method of heating and/or stirring, are used to develop these procedures. On application of a long extraction period, polyphenol loss has been seen due to hydrolysis, ionization and oxidation throughout the extraction technique [9]. Table 3.1 represents the different organic solvents used for the extraction of bioactive compounds.

Soxhlet extractor: In 1879, a German scientist named Franz Ritter von Soxhlet created the Soxhlet extractor (SE). SE was first employed for lipid extraction, but it was later expanded to include a wide range of products. It entails continuous extraction of active components from a solid mixture [11]. During the solvent extraction process, the boiling solvent fumes rise up through the larger side-arm. The required component of the solid mixture is dissolved via condensed droplets of solvent falling into a porous cup. A syphoning action takes place when the smaller side-arm becomes overloaded. The solvent containing the dissolved component is then pumped down into the boiler. As new solvent droplets fall into the porous cup, the residual solvent drains away and the cycle begins again. It may be used to compare the effectiveness of novel extraction techniques. Some of the benefits of SE include low start-up costs, applicability at high temperatures, improved process kinetics, simplicity, non-requirement of filtration step and the continuous presence of sample and solvent during the extraction process. By constantly bringing new solvent into contact with the solid matrix, the transfer equilibrium is displaced. Using heat from the distillation flask, a reasonably high extraction temperature can be maintained. Besides, the process does not necessitate the use of filtration [12]. The most significant drawbacks of the SE technique, however, are the utilization of numerous solvents, lower extraction efficiency and extended operating procedure. Solvents allow the target components to evaporate and decompose, apart from increasing the extraction kinetics, thus heating the compounds at higher temperatures and losing the thermolabile molecules. Also, the Soxhlet apparatus does not permit agitation. Furthermore, as the extraction process typically takes place for a longer period of time, the risk of the target compounds being thermally degraded cannot be overlooked [13].

Hydrodistillation: Hydrodistillation is one of the conventional extraction techniques, which extracts bioactive compounds and essential oils from various foods or food waste. It does not necessitate the use

TABLE 3.1
Bioactive compounds extracted by different organic solvents Reproduced with permission from Jha et al. (2022) [10]

Sl No.	Bioactive compounds	Solvents
1.	Phenolic acids, flavanols, anthocyanins	Ethyl acetate
2.	Flavanones, anthocyanins, catechins and phenolic acids	Different aqueous forms and methanol (60–90%, v/v)
3.	Flavones, flavanols, chlorogenic acids, rutin, ellagic acids, procyanidins	Different aqueous forms and ethanol (10–80%, v/v)
4.	Flavanols, free phenolic acids	Chloroform
5.	Flavanols, phenolic acids	Diethyl ether
6.	Proanthocyanins, phenolic acids	Hot water (75–100°C)
7.	Tannins, bound phenolic acids	Sodium hydroxide (3–10 N)
8.	Phenolic compounds, phenolic acids	Petroleum ether
9.	Flavanols, phenolic acids, hydroxycinnamic acids, coumarins, Flavanols Xanthones	Acetone/water 10–90% (v/v)
10.	Polyphenols, terpenoids, flavones, anthocyanin, tannins, saponins	Methanol
11.	Flavanols, phenolic acids, simple phenolics, anthocyanins	*n*-Hexane, isooctane, ethyl acetate
12.	Polyphenols from olive leaves, oleuropein and rutin	Acetone, ethanol and their aqueous forms (10–90%, v/v)
13.	Flavanols, quercetin 3,40- diglucoside and quercetin 40-monoglucoside	Methanol/water 70% v/v
14.	Anthocyanins, Tannins, Saponins,	Water
15.	Tannins, Polyphenols, Flavonol, Terpenoids, Alkaloids	Ethanol

of organic solvents and the process can be conducted prior to the drying of target material. The materials are packed into a still chamber, followed by the addition of an ample amount of water. The mixture is then heated to a boil. Alternately, direct steam can be injected into the sample. The vapour mixture of oil and water is condensed by cooling it with water. The condensed mixture is transported from the condenser to a separator, which automatically separates the bioactive compounds and oil from the water [14]. Some of the benefits of hydrodistillation include a rapid extraction period, ease in separation and non-requirement of organic solvents in the process. However, a major drawback of this technique is the restriction in the use of heat-sensitive phenolic compounds owing to the limitations on high-temperature usage. Also, there is a possibility of the samples undergoing combustion periodically [10].

Maceration: Maceration is an extraction method that is carried out at ambient temperature. To ensure adequate mixing of samples with the solvent, such as oil, water and alcohol, the whole or coarse food product is ground or sliced to enhance the surface area. The process is conducted in a closed vessel with the addition of a suitable solvent (menstruum). The solvent is then strained out, and the solid residue (termed as "marc") of the extraction process is pressed to retrieve an ideal amount of occluded solution. The strained solvent and the pressed-out liquid are mixed together and filtered to remove any undesirable substances [15]. During maceration, extraction of bioactive compounds is aided by frequent agitation via two processes: (i) improvement in diffusion and (ii) addition of new solvent to the menstruum to separate the concentrated solution from the surface of the sample. This technique has the potential to extract thermolabile substances. It can also suppress the growth of spoilage microorganisms prior to the onset of active fermentation, apart from affecting the subsequent production of yeast flavorants throughout fermentation. However, the drawbacks of this technique include low extraction efficiency and prolonged operating time [16].

Digestion: Digestion is the improved version of maceration that includes moderate heating throughout the recovery process to enhance the menstruum efficacy. Ethanol is used as the extraction solvent, and the time taken for extraction is around 24 h. The solvent weakens the cell wall, allowing the extract to pass through the membrane more easily. Mild heating can be used to further increase the extraction yield. However, if the phytochemicals are heat stable, they may not be easily extracted from the target material [17].

Conventional reflux extraction: Conventional reflux is an extraction technique that is more efficient than maceration, as it requires less solvent and shorter extraction time. However, it lacks the potential to extract thermolabile natural compounds. This technique has been used to extract bioactive compounds from various food products. During the extraction of total phenolic content from dried *Terminalia chebula Retz.* fruit, ethanol (76.5%) was used as the extraction solvent. The other operating parameters considered were treatment time (82 min), temperature (76°C) and solid–liquid ratio (150 ml/g). The complete depletion of the total pore volume of the target material was achieved by the use of an additional extraction cycle. The extract was achieved under vacuum condition followed by filtration. When 50% ethanol was utilized as the extraction solvent, the conventional reflux technique was shown to achieve the highest yields for puerarin and baicalin [18].

3.3 PROGRESS IN THE EXTRACTION OF BIOACTIVE COMPOUNDS

One of the significant drawbacks of the conventional extraction process is that it requires prolonged operating time to complete the process, resulting in the loss of thermosensitive compounds and the need for expensive and pure solvents that evaporate rapidly. Such limitations have instigated the development of various novel and advanced recovery processes. The benefits and drawbacks of various conventional, advanced and combined extraction techniques are shown in Table 3.2. There is a trend to limit the use of typical organic solvents, which are often toxic, difficult to separate from the extract and environmentally hazardous. The use of new green extraction techniques provides high extraction yield, shorter extraction time, low extraction temperature, improved extraction efficiency and ease in separation and purification of the extract. In general, the environmental, health and safety characteristics have been significantly improved by the application of such techniques. Nevertheless, "zero solvent use" is still unattainable, as it has irreversible effects on mass transfer during extraction [19]. To conclude, advanced extraction techniques have a lot of potential to replace the conventional approaches, however, the optimization and application of such techniques should be discussed in detail. Figure 3.1 shows the schematic diagram of some of the conventional, advanced and combined extraction techniques.

3.3.1 Microwave-assisted extraction

Microwave-assisted extraction (MAE) is a cost-effective, eco-friendly and sustainable technique for extracting bioactive compounds from various plant sources. Ganzler reported the use of MAE for the first time in 1986. MAE technique fits the criteria of green analytical chemistry and has the benefits of reduced energy usage, shorter operating time, less utilization of solvent and low CO_2 emissions. The technique was developed for the isolation of high value-added constituents from solid samples [20]. According to Alupului et al. (2012), MAE offers several economic benefits, including low operating cost, high extraction efficiency and ease in separation and purification [21]. Microwaves consist of magnetic and electric fields that are perpendicular to one another. The electric field generates heat via two simultaneous mechanisms: ionic conduction and dipolar rotation. Dipolar rotation is caused by adjusting the electric field of molecules having a dipolar moment in the solid sample and the solvent. Microwave heating results in the breakdown of weak hydrogen bonds induced by the dipolar rotation of the molecules. The mechanism for MAE occurs in two steps: The first step involves the desorption of solutes at high temperature and pressure from the active sites in the sample matrix. In the second step, the extraction fluid is diffused into the matrix. The solutes

TABLE 3.2
Conventional, advanced and combined techniques for the extraction of bioactive compounds

Extraction method		Advantage	Limitations
Conventional	Maceration	Selectivity modification via solvent choice, lower cost of investment	Few compounds may get thermally destroyed
	Hydrodistillation	Automatic separation of phytochemicals from water	Loss of volatile constituents if the operating temperature is high
	Digestion	Heating can be applied to increase the extraction yield	Requires prolonged extraction time
	Soxhlet	The kinetics of the process can be improved via simplicity and applicability at elevated temperature	Extraction efficacy is poor. Time consuming process
	Conventional reflux extraction	Requires less solvent and extraction time	Few compounds may undergo thermal destruction
Advanced	Supercritical fluid extraction	Operates at lower temperature, higher yields; suitable for compounds with low polarity	Higher cost
	Microwave-assisted extraction	The processing time and utilization of solvents are reduced	Polar solvent, elevated temperature
	Ultrasound-assisted extraction	Reduces the treatment time and can operate at lower temperature	The plant materials may get swelled
	Pressurized liquid extraction	An increase in the mass transfer rate and solubility reduces the surface tension and viscosity of the solvent, leading to higher extraction rate	Elevated temperature, higher cost of investment and lower throughput
	Pulsed electric fields	Reduces the treatment time, lower use of solvent	Necessitates the use of enzymes and high conductivity
	Enzyme-assisted extraction	Facilitated extraction from a plant tissue	Additional long operation in wet conditions
Combined	Ultrasound–Microwave-assisted extraction	Produces higher yield, lower solvent use and faster extraction time	With the increase in alkyl chain length, the rate of extraction decreases
	Enzyme–Ultrasound–Microwave-based extraction	Increase in recovery yield and faster extraction time	Elevated maintenance costs
	Supercritical fluid-assisted pressurized fluid extraction	Provides 2–3 times faster extraction compared to conventional processes, leading to improved extraction yield	Higher cost
	Supercritical carbon dioxide—pressure swing technique	Improved process efficacy in terms of quality and amount of the extract. Most effective method in recovering phenolic compounds	High capital cost

Modified with permission from Jha et al. (2022) [10].

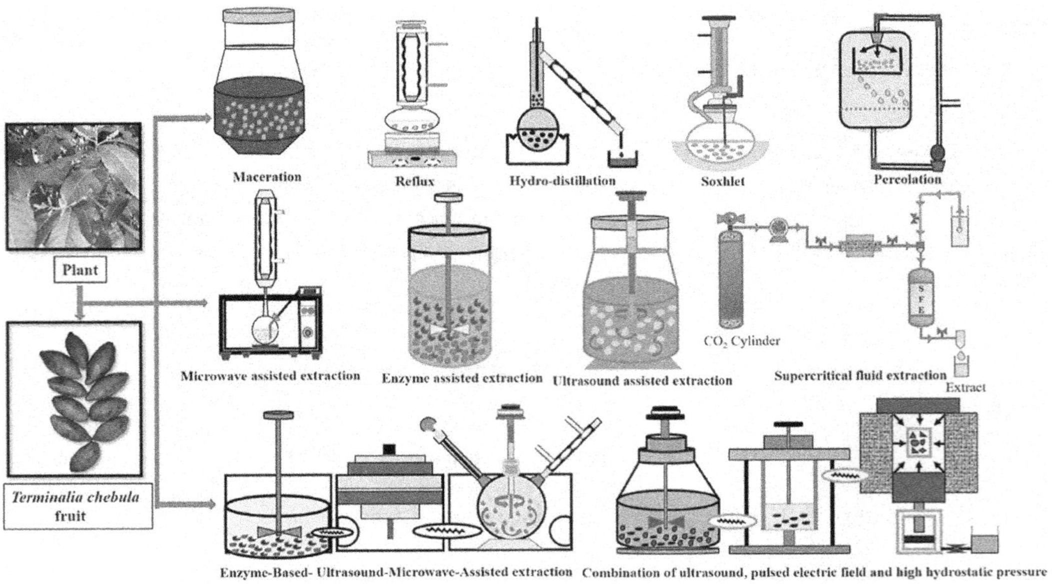

FIGURE 3.1 Schematic representation of the conventional, advanced and combined extraction techniques (Modified with permission from Jha et al. (2022) [10]).

can then be separated from the matrix in the extraction fluid, depending upon the sample matrix. The high performance of MAE is due to the higher surface balance disturbance and increase in the mass transfer effect and solubility [22]. Mandal et al. (2007) reported that factors viz. extraction time, nature and volume of the solvent, matrix properties, temperature and microwave power significantly influence the efficiency of the MAE process [23]. According to Belwal et al. (2018) [24], regulation of the solvent mixture ratio and the temperature/potency ratio assist in improving the solubility of the target bioactive compounds. MAE process can easily facilitate the extraction of volatile (solvent-based extraction) and non-volatile (dry extraction) compounds. In some studies, microwave energy has also been used to extract essential oils, providing improved oil quality and higher extraction yield compared to the typical distillation process. Alupului et al. (2012) studied the effects of extraction yield, microwave power, temperature and operating time on the extraction of polyphenols, glycosides and flavonoids from Cynara scolymus leaves [21]. It was reported that the extraction efficiency in a closed microwave system is significantly affected by pressure, temperature, sample size, extraction time, microwave power and the type of solvent utilized. The extraction technique was performed in a closed environment with a variation in extraction time and temperature range of 70–100°C and 1–5 min, respectively. The study concluded that the highest recovery yield of the target constituents was obtained at a temperature of 70°C and an extraction time of 3 min. Boukroufa et al. (2015) designed a sustainable approach by utilizing microwave and ultrasound techniques with high extraction yields of isolate polyphenols, essential oil and pectin from orange peels [25]. It was observed that there were no significant changes between distillation and microwave extraction techniques during the extraction of essential oil. A microwave power of 500 W was sufficient to separate the essential oil within 15 min of operating time. The water utilized in the extraction of essential oil was recycled as a solvent to isolate pectin and polyphenols. The pectin and polyphenols present in the matrix were extracted via ultrasound and microwave processes, respectively. MAE was shown to be the most promising technique for extracting bioactive compounds and thermolabile substances from tobacco and tobacco waste. This technique can also be utilized for isolating a wide range of high-value constituents, viz. volatile compounds, organic acids, solanesol and phenolic compounds, with a promising application for producing cosmetics, drugs and food additives.

3.3.2 Ultrasound-assisted extraction

Another advanced technique for the extraction of bioactive compounds is the one established by the application of ultrasound. Ultrasound-assisted extraction (UAE) was first introduced in 1950 to carry out brewing from hops. This technique has been extensively used since then, especially in dealing with the extraction of secondary metabolites from a wide range of plant sources. Ultrasound can be considered a crucial technique for achieving the long-term goals of green chemistry. UAE assists in the complete extraction of bioactive compounds from different types of matrices, viz. foods, microalgae, plants and yeasts, within a short period of time. Also, it results in low extraction temperature, high reproducibility, low solvent usage and enhanced purity of the final product. Due to the high extraction efficiency of the UAE process, it has also been used in the separation of phenolic acids [26]. During the extraction of bioactive compounds, UAE employs ultrasonic energy (>20 kHz) with the help of either an ultrasonic probe or an ultrasonic bath. The effects induced by ultrasound may be attributed to the production, growth and collapse of gas bubbles in the liquid caused by the compression and rarefaction of molecules that form the medium. In particular, when the UAE process was applied to plant materials, the suspended powder sample caused the bubbles to collapse asymmetrically, resulting in more effective extraction. The use of different pre-treatment methods may also contribute to higher extraction efficiency. As per Armenta et al. (2015), "green recovery methods" are crucial alternates to the conventional processes that had more detrimental effects in the past, and their research aims to increase both the selectivity and sensitivity of the analytical methods [27]. The study indicated that the sample preparation stage involving the utilization of ultrasound and microwaves assists in the development of sustainable and efficient techniques, particularly for digesting and dissolving samples. Moreover, the extraction of anthocyanins from rosemary herb (Hibiscus sabdariffa L.) via UAE (40 kHz and 40 min) and maceration (24 h) processes at ambient temperature was effectively investigated by Aryanti et al. (2019) [28]. In this study, the effects of different solvents, viz. ethanol and water, and the solute/solvent ratios (1:10 and 1:5 m/v) were analyzed. A comparison of both methods showed that UAE with a 1:10 solute-solvent ratio is more effective than the maceration process. Also, the use of water as a solvent provided better extraction efficiency than ethanol. These findings can be attributed to the fact that anthocyanins, the pigments that give fruits and plants their colour, are water-soluble and pH-dependent. In another study, Sousa et al. (2020) used an optimized ultrasonic bath to extract phenolic compounds from Croton heliotropiifolius leaves [29]. Parameters such as extraction time (39.5 min), solvent volume (11.5 ml), solvent composition (37.5% v/v of ethanol) and extraction temperature (54.8°C) were optimized by using Doehlert Matrix. The bioactive compounds extracted from the sample were quercetin, syringic, catechin, *p*-coumaric, gallic, ferulic, *trans*-cinnamic, ellagic, vanillic and chlorogenic acids. The use of ultrasonic bath was found to be simple, fast and effective, and it met the criteria of the green extraction technique. It was concluded that the extract obtained via the UAE process is rich in phenolic content, such as rutin and chlorogenic acids, which are considered to be valuable compounds for pharmaceutical industries. Given the scarcity of resources, there is an increasing demand for bioactive compound extraction from new sources, such as waste generated from the tobacco industry. Also, it is feasible to obtain solanesol-rich extracts via UAE by employing saponification as a pre-treatment step. However, selectivity of the UAE process and quality of the obtained extract are areas that require further research.

3.3.3 Enzyme-assisted extraction

Enzyme-assisted extraction (EAE) works on the principle of enzymatic pre-treatment of raw samples to release components bound to their cell walls, thereby increasing the overall extract yield and recovering the target compounds. The key steps of EAE are as follows: (i) sample selection and preliminary preparation; (ii) size reduction, homogenization, drying and powdering; (iii) pH and temperature adjustments; (iv) addition and incubation of enzymes; (v) inactivation of enzymes; (vi) filtration and/or centrifugation; (vii) collection of the enzymatic extract's aqueous phase; (viii) other processes based on the product requirement [30]. The most commonly used enzymes are ligninases, cellulases, pectinases and hemicellulases, however, proteinases may also be used in some specific cases. The use of enzymes in combination with different extraction methods has proven to be effective in recovering several bioactive compounds from natural sources. Macedo et al. (2021) investigated the possibility of combining EAE with microwaves to separate phenolic constituents from olive pomace [31]. The study reported higher extraction yield by the combined microwave–enzyme-assisted extraction (MEAE) compared to the conventional techniques. Domínguez-Rodríguez et al. (2021) evaluated the efficiency of the EAE technique via three distinct enzymes to isolate polyphenols from the extraction residue of sweet cherry pomace [32]. The concentration of proanthocyanidins extracted by EAE at optimum conditions was higher than that obtained by acid and alkaline hydrolysis. In another study, a combination of EAE with ultrasound was used by Rezende et al. (2021) for the extraction of bioactive compounds from monguba seed (*Pachira Aquatica Aubl*) [33]. The enzymatic treatment significantly aided the phenolic extraction process, making it a viable alternative to solvent-assisted extraction methods while also being safer due to the reduced use of ethanol. Although the enzymes are expensive, its wide application has considerably enhanced the extraction process. Thus, the higher cost involved upholds its utilization, thereby making it a beneficial investment. The study concluded that the enzyme extracts, particularly those derived from monguba seed can be effectively utilized to create novel functional constituents. Several scientific papers have already been published on the use of enzymes to extract different types of bioactive compounds from plant materials. Choudhari and Ananthanarayan (2007) exhibited that the utilization of pectinase and cellulase enzymes on the extraction of lycopene from tomatoes produces a higher yield [34]. Also, Boulila et al. (2015) found that the application of xylanase, cellulase, hemicellulase and a ternary combination of these enzymes significantly improves the extraction yield of essential oil from bay leaves [35]. In another study, it was reported that a combination of amyloglucosidase and α-amylase enzymes considerably improved the yield of curcumin extraction from turmeric [36]. Even though EAE offers a number of benefits, there are still few technical and commercial limitations associated with it. For instance, complete hydrolysis of the cell wall matrix is still not possible with the currently available enzyme preparation methods, thus, limiting the extraction yield of phyto-bioactive compounds. Also, if the enzymes are prepared with a high enzyme-to-substrate ratio, the overall manufacturing cost may substantially increase during industrial or large-scale applications.

3.3.4 Pulsed-assisted extraction

Pulse-assisted extraction (PAE) is an advanced non-thermal process that applies high voltage pulses (kV) to the target material for a short period of time. The most crucial properties that influence the PAE process include pulse frequency, electric field strength, pulse wave shape, pH and treatment time. An electroporation phenomenon occurs due to the high electric field, which increases the cell membrane permeability, thus, resulting in high extraction yields. The major benefits of the PAE process are as follows: (i) potential to develop online operations in automated systems; (ii) selective separation of intracellular constituents; and (iii) minimal degradation of thermolabile bioactive compounds via low energy usage and short treatment time [37]. PAE has evolved as a novel technique for pressing, drying, extraction and metabolite diffusion in the field of food preservation and processing. In the last two decades, a fair share of beneficial applications related to PAE has been reported [38]. PAE treatment typically involves the use of simple circuits with exponential decay pulses. It improves the extraction efficiency by deteriorating the cell membrane of the target material. The electric potential travels across the cell membrane, which greatly assists in the extraction process based on the membrane charge and the dipole characteristics of the membrane molecules. When the transmembrane potential between the two membranes surpasses 1 V, the charge-bearing molecules repel each other, resulting in higher permeability [39]. However, some of the primary drawbacks of PAE include high instrumentation cost, extraction constraints for lipophilic compounds and the inability to use organic solvent. PAE has been broadly utilized to recover polyphenols from grape seed, purple-fleshed potato, orange peel, grape by-products and blueberry by-products. Pashazadeh et al. (2020) investigated the effect of PAE process on the extraction of phenolic compounds from cinnamon [40]. The study showed that the operating parameters, viz. pulse number and voltage, significantly improved the phenolic and antioxidant activity of the cinnamon extract due to the increased cell permeability of the plant. It also contributed to the separation of intracellular compounds from the damaged cells. In another research, Lon˘cari´c et al. (2020) analyzed the isolation of phenolic constituents from blueberry pomace by green extraction techniques, viz. pulsed-assisted extraction, high voltage electric discharge (HVED) and ultrasound-assisted extraction [37]. It was found that PAE significantly increased the polyphenol extraction efficiency from blueberry pomace compared to other techniques. PAE consumes less energy, as it does not necessitate the use of heat, thereby maintaining the product quality. Also, various food items, such as soup, dairy products and juices, have been effectively pasteurized via the PAE technique. Fincan (2015) reported a decomposition index of 0.86 when 99 pulses of 3 kV/cm with a specific energy input of 4100 ± 240 J/kg were utilized to concentrate phenolic compounds from borage, sesame seed cake and spearmint [41]. The findings showed improved recovery yields of antioxidant property and total phenolic content, and the results can be compared to those of other treatment techniques, including microwave-assisted extraction. However, there are a few criteria for the PAE technique to work effectively: the product should be free of air bubbles with low electrical conductivity, and the product should possess smaller particle size than the gap in the treatment region to ensure proper extraction.

3.3.5 Pressurized liquid extraction

Pressurized liquid extraction (PLE) is known by different names, viz. high-pressure solvent extraction (HSPE), pressurized fluid extraction (PFE), enhanced solvent extraction (ESE) and accelerated solvent extraction (ASE) [42]. During PLE, high pressure is applied to maintain a liquid at a constant temperature even after it has reached its boiling point. The extraction procedure becomes easier when high pressure is used. Also, the extraction temperature can improve higher analyte solubility, which enhances the rate of extraction, by increasing both solubility and mass transfer rate. Thus, faster extraction can be achieved via PLE method due to high pressure and temperature. This trend, along with a lower solvent requirement and shorter extraction time, is the driving force behind the greater proliferation of the PLE method. By increasing the rate of mass transfer and solubility, the surface tension and viscosity of solvents can be decreased, which results in a rapid extraction rate [43]. PLE entails the following steps: (i) use of extraction solvent to moisten the sample; (ii) compound desorption from the matrix; (iii) compound dissolution in the extraction solvent; (iv) compound dispersion outside of the matrix; and (v) diffusion around the matrix to reach the bulk solvent via the closest solvent layer. Accelerated solvent extraction/pressurized liquid extraction is more beneficial than the conventional extraction methods such as solid–liquid extraction. It enhances the mass transfer rate, consumes less solvent, improves solubility and necessitates less time and easy sample handling during extraction [44]. However, this method is not suitable for thermolabile phenolic compounds as the use of high temperatures may adversely affect their functional and structural activity. Besides, this process can be also referred to as pressured hot water extraction or subcritical water extraction due to the utilization of water at its critical point [45]. The primary factors affecting the performance of the extraction process are solvent/sample ratio, flow rate, matrix type, pressure, temperature, treatment time and dispersant. The solvents are selected depending upon the solubility properties of the required solute. The physico-chemical properties of pressurized solvents, viz. viscosity, diffusivity and density, can be modified by changing the pressure and temperature of the extraction process. The need for a faster, less hazardous, sustainable and eco-friendly extraction technique has made PLE a crucial method, particularly in the food and pharmaceutical industries [46].

3.3.6 Supercritical fluid extraction

In 1880, Hannay and Hogarth developed the supercritical fluid extraction (SFE) method for the extraction of bioactive compounds from natural sources. Later, the application of the SFE method was further modified by Zosel, who patented the use of SFE for decaffeination of coffee [10, 47]. The physical properties of a gas, such as CO_2, changes when it is heated and compressed, resulting in a supercritical fluid. Under such conditions, it shows the diffusivity and solvating properties of a gas and a liquid. Thus, it possesses the characteristics of both gases and liquids. As such, supercritical fluids can be applied as a processing medium for a wide variety of polymer, biological and chemical extraction techniques. In SFE, a supercritical fluid is used as an extraction medium in substances containing the target compounds, and extraction takes place depending upon the solubility differences. SFE method offers several benefits over the conventional extraction processes, viz. (i) allows for selective extraction; (ii) operates at a low temperature; (iii) uses safe CO_2; (iv) employs an automated process; (v) has a faster extraction period; (vi) does not require organic solvent; and (vii) prevents sample oxidation [48]. However, there are also some drawbacks associated with this method, such as high power consumption, non-extraction of polar compounds, operation at elevated pressure, greater cost and complex equipment. In SFE, convection in the supercritical solvent phase serves as the primary transport mechanism. Also, the adaptability of SFE is enhanced by its favourable transport properties. Supercritical fluids possess low viscosity compared to the liquid solvent used in typical extraction methods, allowing them to effortlessly spread within the solid matrix. It also has low surface tension, which permits a faster solvent penetration into the solid sample, resulting in higher extraction efficiency [49]. Since density is proportional to solubility, the force of the fluid may be altered by adjusting the extraction pressure. Under supercritical conditions (pressure >7.28 MPa and temperature >31°C), fluids with desirable properties, viz. low surface tension, high diffusivity and low viscosity are utilized in the recovery of constituents from solid matrices [50]. Several plant sources are used to extract nutraceutical and bioactive compounds using supercritical fluid. As such, the use of supercritical fluids, viz. nitrous oxide, carbon dioxide, methanol, ethane, water and propane, have been extensively studied in various investigations. According to Da Silva et al. (2016), the density of the solvent/fluid significantly affects the solubility of the extracts [51]. Due to its low cost and easy availability, CO_2 is the most frequently used solvent during supercritical fluid extraction. Moreover, a few more benefits of CO_2 as a solvent have been reported, such as easy solvent regeneration, low surface tension, improved efficiency, high diffusivity and prevention of oxidation during the extraction process, allowing a higher mass transfer rate from the plant cells. Therefore, SFE method has gained much attention in the last two decades and is applied to a wide range of separation processes for extracting phytochemicals from natural sources [52].

3.4 COMMERCIAL UTILIZATION OF THE EXTRACTION PROCESS

Numerous studies to date have demonstrated the ability of plant matrices, such as fruit, seed, flower, roots, leaves and other parts, to synthesize phytochemicals of industrial interest. Over the course of 2016–2020, the global market for phytochemicals grew at a pace of 7%. Demand for plant-based chemicals in the food and beverage sector is expected to rise as will the use of antioxidants in nutraceuticals and the cosmetic advantages of plant-based phytochemicals. This is due to increasing worldwide awareness of the benefits of natural and chemical-free goods. North America is the second-largest market after Europe. As a result, the area has a strong foothold in the food and beverage and cosmetic sectors. The bulk of the procedures described above is still being tested in the laboratory. For large-scale use, these extraction methods require a lot of fine-tuning. Research and development along with innovation activities by leading phytochemical providers are increasingly being used to identify new uses for phytochemicals. In order to attract a larger client base, players are also focused on assuring quality, safety and customer satisfaction through the use of technology. By focusing on research and innovation, BASF SE (the largest chemical-producing company) ensures that its products are the highest quality. The firm has developed three specific divisions for research, including advanced materials & system research, process research & chemical engineering and bioscience research, which are constantly working on new product development. Natural colours manufacturer Chr Hansen Holding A/S extended its manufacturing facilities in North America in 2018 so that it could better serve its clients. Separate research and development, application, quality assurance, manufacturing, warehousing and marketing/sales units have been constructed to ensure that the company's clients receive high-quality goods and services. Absorption, Distribution, Metabolism and Excretion (ADME) is continuously emphasized in Phyto Life Sciences Pvt. Ltd.'s research and development programme. Particles that are denser, more soluble, less bitter and dryer are all goals of the company's several R&D activities.

3.5 CONCLUSIONS

The ever-increasing demand for bioactive phytochemicals fuels the continual search for better extraction procedures. There has been a lot of advancement in terms of extraction procedures during the last several decades. The non-traditional extraction procedures discussed above were primarily employed to increase the yield of the targeted compounds. However, each strategy has its own set of challenges. Non-uniform sound wave dispersion, the formation of free radicals and metal contamination are some of the disadvantages of using an ultrasonic probe in ultrasound-assisted extraction. Moderate processing parameters in aqueous solutions are one of the most difficult challenges in enzyme-assisted extraction. The high cost of enzymes necessary for large-scale extraction is another disadvantage. The capacity to effectively extract bioactive compounds may be limited if enzyme behaviour varies as a result of various environmental factors, such as temperature or nutrition availability. Microwave-assisted extraction also has disadvantages, such as the recovery of nonpolar molecules and changes in the chemical structure of target compounds, which might impact bioactivity and limit future uses. As such, focus should be on scaling up of such techniques to overcome the existing limitations. Scale-up approach is required for other techniques as well, such as pressurized liquid extraction, supercritical fluid extraction and pulse-assisted extraction, in order to examine the process yield, operational parameters and cost–benefit analyses in industrial settings. These activities will need standardized research and development facilities. Knowledge of the plant matrix properties as well as the chemistry of the target bioactive molecule is also essential for the optimization of these extraction strategies.

ACKNOWLEDGEMENTS

The study is supported by the Indian National Academy of Engineering (INAE/121/AKF/22), Gurgaon, India. The authors are solely responsible for all of the opinions, results and conclusions expressed in this study; INAE's viewpoints are not necessarily reflected in any of these aspects.

REFERENCES

1. J. Azmir, I.S.M. Zaidul, M.M. Rahman, K.M. Sharif, A. Mohamed, F. Sahena, M.H.A. Jahurul, K. Ghafoor, N.A.N. Norulaini, A.K.M. Omar, Techniques for extraction of bioactive compounds from plant materials: A review, *J. Food Eng.* 117 (2013) 426–436. https://doi.org/10.1016/j.jfoodeng.2013.01.014.
2. A. Altemimi, N. Lakhssassi, A. Baharlouei, D.G. Watson, D.A. Lightfoot, Phytochemicals: Extraction, isolation, and identification of bioactive compounds from plant extracts, *Plants.* 6 (2017). https://doi.org/10.3390/plants6040042.
3. J.J. Zhang, Y. Li, T. Zhou, D.P. Xu, P. Zhang, S. Li, H. B. Li, Bioactivities and health benefits of mushrooms mainly from China, *Molecules.* 21 (2016) 1–16. https://doi.org/10.3390/molecules21070938.
4. J. Zhang, C. Wen, H. Zhang, Y. Duan, H. Ma, Recent advances in the extraction of bioactive compounds with subcritical water : A review, *Trends Food Sci. Technol.* 95 (2020) 183–195. https://doi.org/10.1016/j.tifs.2019.11.018.

5. S. Armenta, F.A. Esteve-Turrillas, S. Garrigues, M. de la Guardia, Green analytical chemistry: The role of green extraction techniques, *Compr. Anal. Chem.* 76 (2017) 1–25. https://doi.org/10.1016/bs.coac.2017.01.003.
6. Y. Chi, Y. Li, G. Zhang, Y. Gao, H. Ye, J. Gao, P. Wang, Effect of extraction techniques on properties of polysaccharides from Enteromorpha prolifera and their applicability in iron chelation, *Carbohydr. Polym.* 181 (2018) 616–623. https://doi.org/10.1016/j.carbpol.2017.11.104.
7. K. Ghafoor, F.Y. AL-Juhaimi, Y.H. Choi, Supercritical fluid extraction of phenolic compounds and antioxidants from grape (Vitis labrusca B.) seeds, *Plant Foods Hum. Nutr.* 67 (2012) 407–414. https://doi.org/10.1007/s11130-012-0313-1.
8. F. Chemat, N. Rombaut, A.G. Sicaire, A. Meullemiestre, A.S. Fabiano-Tixier, M. Abert-Vian, Ultrasound assisted extraction of food and natural products. Mechanisms, techniques, combinations, protocols and applications. A review, *Ultrason. Sonochem.* 34 (2017) 540–560. https://doi.org/10.1016/j.ultsonch.2016.06.035.
9. H. Li, B. Chen, S. Yao, Application of ultrasonic technique for extracting chlorogenic acid from Eucommia ulmodies Oliv. (E. ulmodies), *Ultrason. Sonochem.* 12 (2005) 295–300. https://doi.org/10.1016/j.ultsonch.2004.01.033.
10. A.K. Jha, N. Sit, Extraction of bioactive compounds from plant materials using combination of various novel methods : A review, *Trends Food Sci. Technol.* 119 (2022) 579–591. https://doi.org/10.1016/j.tifs.2021.11.019.
11. G.L. Teixeira, S. Ávila, P.S. Hornung, R.C.T. Barbi, R.H. Ribani, Sapucaia nut (Lecythis pisonis Cambess.) flour as a new industrial ingredient: Physicochemical, thermal, and functional properties, *Food Res. Int.* 109 (2018) 572–582. https://doi.org/10.1016/j.foodres.2018.04.071.
12. D. Grigonis, P.R. Venskutonis, B. Sivik, M. Sandahl, C.S. Eskilsson, Comparison of different extraction techniques for isolation of antioxidants from sweet grass (Hierochloë odorata), *J. Supercrit. Fluids.* 33 (2005) 223–233. https://doi.org/10.1016/j.supflu.2004.08.006.
13. W. Xiao, L. Han, B. Shi, Microwave-assisted extraction of flavonoids from Radix Astragali, *Sep. Purif. Technol.* 62 (2008) 614–618. https://doi.org/10.1016/j.seppur.2008.03.025.
14. L. V. Silva, D.L. Nelson, M.F.B. Drummond, L. Dufossé, M.B.A. Glória, Comparison of hydrodistillation methods for the deodorization of turmeric, *Food Res. Int.* 38 (2005) 1087–1096. https://doi.org/10.1016/j.foodres.2005.02.025.
15. N. Srivastava, A. Singh, P. Kumari, J.H. Nishad, V.S. Gautam, M. Yadav, R. Bharti, D. Kumar, R.N. Kharwar, *Advances in extraction technologies: Isolation and purification of bioactive compounds from biological materials*, Elsevier Inc., 2021. https://doi.org/10.1016/b978-0-12-820655-3.00021-5.
16. M.B. Soquetta, L. de M. Terra, C.P. Bastos, Green technologies for the extraction of bioactive compounds in fruits and vegetables, *CYTA - J. Food.* 16 (2018) 400–412. https://doi.org/10.1080/19476337.2017.1411978.
17. M. K. Hussain, M. Saquib, M. Faheem Khan, *Techniques for extraction, isolation, and standardization of bio-active compounds from medicinal plants BT - natural bio-active compounds: Volume 2: Chemistry, pharmacology and health care practices*, in: M.K. Swamy, M.S. Akhtar (Eds.), Springer, 2019: pp. 179–200. https://doi.org/10.1007/978-981-13-7205-6_8.
18. Q.W. Zhang, L.G. Lin, W.C. Ye, Techniques for extraction and isolation of natural products: A comprehensive review, *Chinese Med.* 13 (2018) 1–26. https://doi.org/10.1186/s13020-018-0177-x.
19. M. Banožić, J. Babić, S. Jokić, Recent advances in extraction of bioactive compounds from tobacco industrial waste-a review, *Ind. Crops Prod.* 144 (2020). https://doi.org/10.1016/j.indcrop.2019.112009.
20. R.F. da Silva, C.N. Carneiro, C.B. Cheila, F. J. V. Gomez, M. Espino, J. Boiteux, M. de los Á. Fernández, M.F. Silva, F. de S. Dias, Sustainable extraction bioactive compounds procedures in medicinal plants based on the principles of green analytical chemistry: A review, *Microchem. J.* 175 (2022). https://doi.org/10.1016/j.microc.2022.107184.
21. A. Alupului, I. Călinescu, V. Lavric, Microwave extraction of active principles, *U.P.B. Sci. Bull., Ser. B.* 74 (2012) 129–142.
22. L. Wang, C.L. Weller, Recent advances in extraction of nutraceuticals from plants, *Trends Food Sci. Technol.* 17 (2006) 300–312. https://doi.org/10.1016/j.tifs.2005.12.004.
23. V. Mandal, Y. Mohan, S. Hemalatha, Microwave Assisted Extraction – An Innovative and Promising Extraction Tool for Medicinal Plant Research, *Pharmacogn. Rev.* 1 (2007) 7–18.
24. T. Belwal, S.M. Ezzat, L. Rastrelli, I.D. Bhatt, M. Daglia, A. Baldi, H.P. Devkota, I.E. Orhan, J.K. Patra, G. Das, C. Anandharamakrishnan, L. Gomez-Gomez, S.F. Nabavi, S.M. Nabavi, A.G. Atanasov, A critical analysis of extraction techniques used for botanicals: Trends, priorities, industrial uses and optimization strategies, *TrAC - Trends Anal. Chem.* 100 (2018) 82–102. https://doi.org/10.1016/j.trac.2017.12.018.

25. M. Boukroufa, C. Boutekedjiret, L. Petigny, N. Rakotomanomana, F. Chemat, Bio-refinery of orange peels waste: A new concept based on integrated green and solvent free extraction processes using ultrasound and microwave techniques to obtain essential oil, polyphenols and pectin, *Ultrason. Sonochem.* 24 (2015) 72–79. https://doi.org/10.1016/j.ultsonch.2014.11.015.
26. M. Vinatoru, T.J. Mason, I. Calinescu, Ultrasonically assisted extraction (UAE) and microwave assisted extraction (MAE) of functional compounds from plant materials, *TrAC - Trends Anal. Chem.* 97 (2017) 159–178. https://doi.org/10.1016/j.trac.2017.09.002.
27. S. Armenta, S. Garrigues, M. de la Guardia, The role of green extraction techniques in Green Analytical Chemistry, *TrAC - Trends Anal. Chem.* 71 (2015) 2–8. https://doi.org/10.1016/j.trac.2014.12.011.
28. N. Aryanti, A. Nafiunisa, D.H. Wardhani, Conventional and ultrasound-assisted extraction of anthocyanin from red and purple roselle (Hibiscus sabdariffa L.) calyces and characterisation of its anthocyanin powder, *Int. Food Res. J.* 26 (2019) 529–535.
29. do C. de Sousa, G.L. dos Anjos, R.S.A. Nóbrega, A. da S. Magaton, F.M. de Miranda, F. de S. Dias, Greener ultrasound-assisted extraction of bioactive phenolic compounds in Croton heliotropiifolius Kunth leaves, *Microchem. J.* 159 (2020). https://doi.org/10.1016/j.microc.2020.105525.
30. T. Belwal, F. Chemat, P.R. Venskutonis, G. Cravotto, D.K. Jaiswal, I.D. Bhatt, H.P. Devkota, Z. Luo, Recent advances in scaling-up of non-conventional extraction techniques: Learning from successes and failures, *TrAC - Trends Anal. Chem.* 127 (2020) 115895. https://doi.org/10.1016/j.trac.2020.115895.
31. G.A. Macedo, Á.L. Santana, L.M. Crawford, S.C. Wang, F.F.G. Dias, J.M.L.N. de Mour Bell, Integrated microwave- and enzyme-assisted extraction of phenolic compounds from olive pomace, *Lwt.* 138 (2021). https://doi.org/10.1016/j.lwt.2020.110621.
32. G. Domínguez-Rodríguez, M.L. Marina, M. Plaza, Enzyme-assisted extraction of bioactive non-extractable polyphenols from sweet cherry (Prunus avium L.) pomace, *Food Chem.* 339 (2021) 128086. https://doi.org/10.1016/j.foodchem.2020.128086.
33. Y.R.R.S. Rezende, J.P. Nogueira, T.O.M. Silva, R.G.C. Barros, C.S. de Oliveira, G.C. Cunha, N.C. Gualberto, M. Rajan, N. Narain, Enzymatic and ultrasonic-assisted pretreatment in the extraction of bioactive compounds from Monguba (Pachira aquatic Aubl) leaf, bark and seed, *Food Res. Int.* 140 (2021) 109869. https://doi.org/10.1016/j.foodres.2020.109869.
34. S.M. Choudhari, L. Ananthanarayan, Enzyme aided extraction of lycopene from tomato tissues, *Food Chem.* 102 (2007) 77–81. https://doi.org/10.1016/j.foodchem.2006.04.031.
35. A. Boulila, I. Hassen, L. Haouari, F. Mejri, I. Ben Amor, H. Casabianca, K. Hosni, Enzyme-assisted extraction of bioactive compounds from bay leaves (Laurus nobilis L.), *Ind. Crops Prod.* 74 (2015) 485–493. https://doi.org/10.1016/j.indcrop.2015.05.050.
36. F. Sahne, M. Mohammadi, G.D. Najafpour, A.A. Moghadamnia, Enzyme-assisted ionic liquid extraction of bioactive compound from turmeric (Curcuma longa L.): Isolation, purification and analysis of curcumin, *Ind. Crops Prod.* 95 (2017) 686–694. https://doi.org/10.1016/j.indcrop.2016.11.037.
37. A. Lončarić, M. Celeiro, A. Jozinović, J. Jelinić, T. Kovač, S. Jokić, J. Babić, T. Moslavac, S. Zavadlav, M. Lores, Green extraction methods for extraction of polyphenolic compounds from blueberry pomace, *Foods.* 9 (2020) 1521. https://doi.org/10.3390/foods9111521.
38. E. Vorobiev, N.I. Lebovka, Extraction of Intercellular Components by Pulsed Electric Fields BT - Pulsed Electric Fields Technology for the Food Industry: Fundamentals and Applications, in: J. Raso, V. Heinz (Eds.), Springer US, Boston, MA, 2006: pp. 153–193. https://doi.org/10.1007/978-0-387-31122-7_6.
39. G. Bryant, J. Wolfe, Electromechanical stresses produced in the plasma membranes of suspended cells by applied electric fields, *J. Membr. Biol.* 96 (1987) 129–139. https://doi.org/10.1007/BF01869239.
40. B. Pashazadeh, A.H. Elhamirad, H. Hajnajari, P. Sharayei, M. Armin, Optimization of the pulsed electric field -assisted extraction of functional compounds from cinnamon, *Biocatal. Agric. Biotechnol.* 23 (2020) 101461. https://doi.org/10.1016/j.bcab.2019.101461.
41. M. Fincan, Extractability of phenolics from spearmint treated with pulsed electric field, *J. Food Eng.* 162 (2015) 31–37. https://doi.org/10.1016/j.jfoodeng.2015.04.004.
42. A. Nieto, F. Borrull, E. Pocurull, R.M. Marcé, Pressurized liquid extraction: A useful technique to extract pharmaceuticals and personal-care products from sewage sludge, *TrAC - Trends Anal. Chem.* 29 (2010) 752–764. https://doi.org/10.1016/j.trac.2010.03.014.
43. E. Ibañez, M. Herrero, J.A. Mendiola, M. Castro-Puyana, Extraction and Characterization of Bioactive Compounds with Health Benefits from Marine Resources: Macro and Micro Algae, Cyanobacteria, and Invertebrates BT - Marine Bioactive Compounds: Sources, Characterization and Applications, in: M. Hayes (Ed.), Springer US, Boston, MA, 2012: pp. 55–98. https://doi.org/10.1007/978-1-4614-1247-2_2.
44. C.M. Ajila, S.K. Brar, M. Verma, R.D. Tyagi, S. Godbout, J.R. Valéro, Extraction and analysis of polyphenols: Recent trends, *Crit. Rev. Biotechnol.* 31 (2011) 227–249. https://doi.org/10.3109/07388551.2010.513677.

45. A. Mena-García, A.I. Ruiz-Matute, A.C. Soria, M.L. Sanz, Green techniques for extraction of bioactive carbohydrates, *TrAC - Trends Anal. Chem.* 119 (2019) 115612. https://doi.org/10.1016/j.trac.2019.07.023.
46. G. Alvarez-Rivera, M. Bueno, D. Ballesteros-Vivas, J.A. Mendiola, E. Ibañez, Pressurized liquid extraction, *Liq. Extr. Handbooks Sep. Sci.* (2019) 375–398. https://doi.org/10.1016/B978-0-12-816911-7.00013-X.
47. A. Zosel, Der Schubmodul von Hochpolymeren als Funktion von Druck und Temperatur, *Kolloid-Zeitschrift Zeitschrift Für Polym.* 199 (1964) 113–125. https://doi.org/10.1007/BF01499216.
48. F. Al Khawli, M. Pateiro, R. Domínguez, J.M. Lorenzo, P. Gullón, K. Kousoulaki, E. Ferrer, H. Berrada, F.J. Barba, Innovative green technologies of intensification for valorization of seafood and their by-products, *Mar. Drugs.* 17 (2019) 1–20. https://doi.org/10.3390/md17120689.
49. Y. Pouliot, V. Conway, P.-L. Leclerc, Separation and concentration technologies in food processing, *Food Process.* (2014) 33–60. https://doi.org/10.1002/9781118846315.ch3.
50. S. Jokić, B. Nagy, K. Aladi, B. Simándi, Supercritical fluid extraction of soybean oil from the surface of spiked quartz sand - Modelling study, *Croat. J. Food Sci. Technol.* 5 (2013) 70–77.
51. B. Vieira da Silva, J.C.M. Barreira, M.B.P.P. Oliveira, Natural phytochemicals and probiotics as bioactive ingredients for functional foods: Extraction, biochemistry and protected-delivery technologies, *Trends Food Sci. Technol.* 50 (2016) 144–158. https://doi.org/10.1016/j.tifs.2015.12.007.
52. M. Cvjetko Bubalo, S. Vidović, I. Radojčić Redovniković, S. Jokić, Green solvents for green technologies, *J. Chem. Technol. Biotechnol.* 90 (2015) 1631–1639. https://doi.org/10.1002/jctb.4668.

4 Recovery of bioactive compounds from fruit and vegetable peel

4.1 INTRODUCTION

The world's population has rapidly expanded over the past ten years, which has also prompted the food processing industry to develop in order to meet the daily rising demand of customers for a variety of healthy food products. As customer desire for cleaner-label, healthier foods has grown, natural food additives stand out as one of the most intriguing and inventive segments of the food industry. Food additives are frequently utilized to improve the nutritional content of foods, their organoleptic (colour, flavour) qualities and to extend their shelf life by adding ingredients that shield foods from deterioration brought on either by microbes or oxidation. Oxidation in food systems produces a number of products that reduce the nutritional content of foods, change their sensory properties (such as flavour, colour, texture and appearance) and shorten their shelf life [1]. Until recently, synthetic antioxidants, like tert-butylhydroquinone (TBHQ), butylated hydroxyanisole (BHA) and butylated hydroxytoluene (BHT), were frequently utilized as food additives to prevent or decelerate oxidation. Nevertheless, because of their potential for toxin and carcinogenic effects, the researchers are now more interested in finding valuable bioactive chemicals from natural sources as an alternative to synthetic chemicals. A number of fruits and vegetables contain phytochemicals, such as dietary fibre, carotenoids and phenolic compounds, which are natural antioxidants. The primary raw materials containing bioactive compounds for food processing are fresh plant materials, which include fruits, vegetables, grains and legumes [2]. But the food processing business invariably produces a number of by-products, including peels, shells, seeds, kernels, pomace, brans and seed coatings, which are frequently thrown away as food waste and have a negative impact on the environment and the economy. These waste products nevertheless include a lot of high-value materials that may be put to good use. As a result, there has been a lot of interest in using agricultural by-products as a source of useful components, especially those from crop plants [3]. Many efforts have been made to return these by-products to the food supply chain or allied businesses in the context of sustainability and the circular economy in order to increase the overall added value. For instance, the use of plant materials such as seeds, peels, bran and pomace has shown significant promise in enhancing the nutritional and functional qualities of foods without compromising consumer acceptance. The peel of various fruits and vegetables, in particular, was shown to have an even higher concentration of important bioactive elements than other plant parts [4].

High levels of phenolic acids (such as gallic, ellagic and caffeic acids), flavonoids (such as catechin, gallocatechin and epicatechin) and hydrolyzable tannins (ellagitannins, gallotannins) are found in pomegranate peels. Peels also contain significant levels of crude protein, including isoleucine, lysine, valine, threonine and aromatic amino acids [5]. Pitaya peels are rich in essential nutrients and phytochemical compounds. According to one study, dried pitaya peels contain a whopping 70% of their weight in dietary fibre and pectin. Additionally, pitaya peels may contain phytochemicals, particularly betacyanins, which have strong bioactivities and may be extracted to create functional foods or increase the sustainability of food items by acting as antioxidants [6]. Mango peels have received significant focus from the research community because of their high concentration of beneficial substances viz. enzymes, polyphenols, vitamin E, phytochemicals,

DOI: 10.1201/9781003315469-4

vitamin C and carotenoids that have predominately functional and antioxidant qualities. A substantial quantity of protein, dietary fibre, pectin, lipids, hemicellulose and cellulose can also be found in mango peels [7]. Additionally, the catecholamines, flavonols, flavan-3-ols and hydroxycinnamic acids found in banana peels may also be generally divided into four groupings. Rutin and its conjugates were the most predominate flavonols among those identified. The flavan-3-ols, which include monomers, dimers and polymers, were the largest category of phenolics discovered in banana peel [8]. Total phenolics, flavonoids and flavonols can all be found in good amounts in onion peels. Additionally, onion peel contains a healthy quantity of insoluble fibre (54.7%), soluble fibre (7.3%) and total dietary fibre (62.1%). Quercetin, kaempferol, vanillic acid, ferulic acid, protocatechuic acid and epicatechin are few of the significant bioactive chemicals found in the peels of the red and yellow varieties of onion [9]. Potato peels are rich in protein, dietary fibre and starch, which give them significant nutritional value; 30–52% dw of the total carbs in it are made up of starch. The phenolic chemicals caffeic acid, catechin and quercetin as well as the glycoalkaloids, viz. α-chaconine and α-solanine, are abundant in potato peel. Organic and non-organic potato peels have dietary fibre content totalling 21.4% and 22.39% dw, respectively [10]. Furthermore, carbohydrates and amino acids are abundant in pumpkin skin. Minerals and vitamin E are among the many nutrients found in pumpkin peels. The two main fatty acids found in pumpkin peels are linoleic acid and oleic acid. Among other things, phenolic acids, carotenoids, dietary fibre and flavonoids are among the significant phytochemicals found in pumpkin peel [11]. A considerable number of bioactive substances, including total phenolics, total flavonoids, total carotenoids, vitamins A and C and hydrolyzable tannins, are also present in papaya peel. Papaya peels also include a sizeable quantity of protein, minerals and carbs. Alkaloids and saponin, which are helpful for the body's general health, are also abundant in papaya peel [12]. Therefore, it is important to examine these fruits and vegetable peels properly in order to extract valuable bioactive chemicals and, subsequently, lessen their negative environmental effects.

This chapter elaborately evaluates the presence and extraction of phytochemicals and bioactive compounds from different fruit and vegetable (F&V) peels, viz. from banana, mango, pitaya, pomegranate, onion, potato, papaya and pumpkin. The different extraction methods for the recovery of bioactive compounds from such F&V waste are also covered in detail. Moreover, the health-promoting benefits and utilization of the bioactive compounds in functional foods and other food-based applications recovered from F&V peels are also extensively discussed and summarized.

4.2 RECOVERY OF BIOACTIVE COMPOUNDS FROM FRUIT PEEL

4.2.1 Banana peel

Banana (Musaceae spp.) is quite nutrient-dense compared to other fruits. Typically, fruit is eaten fresh or processed into a variety of products on a small or large scale, including ice cream, chips, wine, dried fruit, wheat, smoothies, bread and ingredients for functional foods. Recently, there has been a lot of interest in using bananas as a component in functional foods. This is especially true given that bananas' easy digestion makes their starch and non-starch carbohydrates an ideal meal element. The annual production of bananas is around 103 million tonnes. Roughly 35 million tonnes of banana peel are generated each year, and since the peel makes up about 35% of the weight of the entire fruit, there is plenty of it to be used in other ways. In tropical and subtropical areas, like Asia, Latin America and Africa, bananas are an extensively grown and consumed fruit crop [13].

Typically, the peels from consumed bananas are either tossed or used as organic feed for animals. The disposal of these peels could have negative environmental effects. Banana peel possess different phytochemicals, primarily antioxidants. Dessert banana peels are reported to contain pyridoxine as well as secondary metabolites, like catecholamines, phenolics and carotenoid chemicals. In addition to acting as natural antioxidants, carotenoids also help keep food from spoiling. Several researchers have discovered that carotenoids are typically more abundant in F&V peels compared to that of pulp. Depending upon the varieties (plantain or dessert), banana peel contains significantly higher amount of carotenoid levels, out of which trans-β-carotene largely constitute the pro-vitamin A carotenoid group [8]. The principal carotenoids found in banana peel are trans-β-carotene (168.82 ± 6.78 mg/g dry weight), trans-α-carotene (167.68 ± 9.48 mg/g dry weight) and cis-β-carotene (93.32 ± 6.22 mg/g dry weight). Regarding the carotenoid content of both the peel and the pulp, banana peel is a type of significant raw material that could be further utilized [14]. Extensive research into the antihyperglycaemic effects of banana fruit has also been sparked by recent studies on the fruit's antidiabetic potential. Studies have indicated that increasing dietary fibre intake lowers both the total cholesterol and low-density lipoprotein (LDL) levels. The findings also revealed that insoluble dietary fibre from banana peel at measured levels of 60–240 mg showed a lower cholesterol-absorbing potential than the soluble dietary fibre. LDL that has been oxidized causes the formation of free radicals, which in turn causes oxidized LDL to be destroyed. Oxidized LDL causes an increase in the pro-inflammatory genes, which leads to the recruitment of monocytes into the vascular endothelial cells of a dysfunctional blood vessel wall. As such, LDL oxidation must be inhibited if atherosclerosis and cardiovascular diseases are to be managed effectively [15]. Arun et al. (2017) demonstrated that LDL oxidation can be effectively inhibited by ethyl acetate (IC_{50}: 215.55 g/mL) and methanol (IC50: 170.57 g/mL) extracts from banana peels in a dose-dependent manner [16]. A powerful vasoconstrictor, angiotensin-converting enzyme (ACE) converts the angiogenic hormone angiotensin I to angiotensin II, which is essential for maintaining blood pressure. Angiotensin II is a key inducer of insulin resistance and has been linked to the development of vascular problems of diabetes. The study found that banana peel extracts showed an inhibitory impact on ACE similar to that of captopril (100–200 g/mL), the positive control. Such residues have come under increased scrutiny in recent years as a potential source of antioxidant chemicals. Comparing banana peels to other plant residual biomasses, a key characteristic of banana peel as a source of constituents capable of scavenging free radicals as an antioxidant mechanism of action have been observed. One may anticipate a direct correlation between those factors when linking antioxidant activity to the phenolic content of biomass samples. Although vitamins and carotenoids such as beta-carotene are not phenolic components, they may help to increase the antioxidant activity of banana peel extracts [17]. Someya et al. (2002) investigated the antioxidants present in commercial banana peel from Musa cavendishii L., and detection of gallocatechin (one of the antioxidants) was reported [18]. Since the antioxidant capacity of the extract prepared from banana peel against lipid auto-oxidation was higher compared to banana pulp extract, gallocatechin was, in fact, more plentiful in the peel (160 mg/100 g dw) than in the pulp (30.5 mg/100 g dw) in Musa cavendishii L. genotypes. This outcome was in line with the gallocatechin analysis, and the superior antioxidant benefits could be attributed to the gallocatechin's higher concentration. Banana peels could therefore be used in a variety of ways as a natural antioxidant source in meals. Leucocyanidin, a flavonoid found in unripe banana peels, promotes cell proliferation by increasing the addition of thymidine into cellular DNA, thereby hastening the wound-healing process. Unripe banana peel and pulp have been used to heal human gastric ulcers and damaged nipples. Unripe bananas have been demonstrated to be effective in the treatment and prevention of peptic ulcers in studies using rats. Curiously, the active ingredient in unripe bananas is water soluble and loses its potency when the bananas are mature [19]. The most abundant naturally occurring phenolic compound with antioxidant potential found in plant leaves, seeds, bark and flowers is called gallocatechin. The wound-healing ability of a Musa spp. peel gallocatechin-rich extract (GE:106.6 mg/mL) was investigated. The lesions were healed in nine days thanks to gastroenteritis (GE) treatment, which was also able to raise the hydroxyproline level over the course of the treatment. The lesions' potential for GE healing was established by a histological study that revealed fibroblast proliferation and the induction of the re-epithelialization process. Besides, reactive oxygen species (ROS) are essential for cell signalling and effective defence against invasive pathogens [20]. Low amounts of ROS are also required for cell signalling, particularly angiogenesis, even in the absence of infection. Thus, a stronger connection between ROS generation and detoxification is essential for the normal healing of an injury. This investigation indicated that banana peel extract was able to avoid oxidative damage to cellular structures in the lesion bed throughout the experimental period. This ability to control ROS levels was particularly crucial in the early stages of the healing process. This healing process rapidly and efficiently came to a stop in the damaged region that had been trained to support cellular proliferation [8].

Ethanol extracts of banana peel showed a wide range of antibacterial activity against the studied microbes, with notably significant inhibitory effectiveness against P. vulgaris and S. paratyphi, according to a study by Krishna et al. (2013) [21]. The existence of biologically active substances such as tannins, glycosides, terpenoids and flavonoids was discovered through phytochemical research. Clinical pathogenic microorganisms may be treated with these compounds. In research conducted by Dahham et al. (2015), the extract of banana peel made from hexane solvent had the highest level of toxicity towards the human colon cancer cell line HCT-116, inhibiting cell multiplication by 64.02% [22]. The aqueous methanol extract of the banana peel from Nendran had been shown by Durgadevi et al. (2019) to have considerable cytotoxic activity against MCF-7 breast cell lines in a different investigation [23]. Furthermore, ferulic acid, which is highly identified in sucrier banana peel, may contribute to the development of melanogenesis by regulating the expression of the growth factor for vascular endothelium, starting nitric oxide synthase and functioning as a tumour suppressor gene. In research conducted by Arun et al. (2015), banana peel flour from the Nendran type was used to make functional cookies in ratios of 5%, 10% and 15% without changing the amount of wheat flour in the recipe overall [24]. The prepared cookies had high levels of overall dietary fibre, ash and moisture. With an increase in the percentage of banana peel flour, parameters such as cookie index, spread ratios and breaking intensity all decreased. By gradually adding banana peel flour, the phenolic content rose from 4.36 mg gallic acid equivalent (GAE) to 5.28 mg GAE when compared to the control cookie (3.21 mg GAE). Cookies made with 10% banana peel flour are preferable in terms of sensory acceptance because they have better colour, flavour and texture compared to other formulations. A banana peel-based edible food wrapper's increased tensile strength was shown by Santhoskumar et al. (2019) [25]. Because banana peels include inorganic nutrients and fibre, their tensile characteristics are comparable to those of polyethylene. The mechanical characteristics are improved by the inorganic nutrient. These films are also readily available and biodegradable. The banana film dissolves completely in 45 days after being subjected to biotic bacteria in accordance with ASTMD 5338 under controlled composting circumstances. High tensile strength was noted in the bioplastic film made of two biopolymers produced from corn starch and banana peel in another investigation by Sultan and Johari (2017) [26]. In addition to having a high tensile strength, the polyvinyl alcohol matrix's elongation at break reduced as the amount of green banana peel flour it contained rose to 20%. According to earlier research, edible films made from banana peels can help with the growing industry efficiency. Additionally, they enhanced the overall economy by creating a range of goods that utilized plastic wrapping during the manufacturing process. The potential uses of banana peel in food-based applications are shown in Table 4.1.

4.2.2 Mango peel

The mango (Mangifera indica L.) fruit is grown all over the world, but is most popular in tropical nations. It is a member of the family Anacardiaceae in the order Sapindales. Mangoes come in more than a thousand different types. Only a few of the types that are available are produced and traded on a large basis. Currently, mango is grown on over 3.7 million acres of land worldwide. Mango fruit ranks second in terms of production and acreage used among tropical crops, behind only bananas. The importance of mango fruits as a source of minerals, vitamins and other compounds has been well established. Every component of a mango tree, including the leaves, flowers, bark, fruit, pulp, seeds and peel, is full of valuable nutrients [7].

The mango peel has a broad range of valuable bioactive compounds. The moisture content of mango peels varied from 65% to 85%. Essential nutrients such as protein, carbohydrates, vitamins, fat and ash along with bioactive compounds constitute the dry matter of mango peels. The

TABLE 4.1
Utilization of banana peels in food-based applications

Food Products	Banana Peel Cultivar	Composition in Food	Aim	Applications
Cookies	M. paradisiaca	5%, 10% and 15% / flour basis	Development of cookies containing plantain peel flour for use in the prevention of harmful diseases.	High amount of moisture content, dietary fibre, phenolic and ash content were found in the produced cookies. Also, helps in improving texture, taste and colour of cookies.
Ground chicken patties	M. paradisiaca	2% / aqueous extract	Evaluation of antioxidant properties in ground chicken patties aerobically stored at 4°C for 7 days.	The free radical scavenging activity was considerably reduced by aqueous extract of banana peel.
Chicken sausage	M. balbisiana	2%, 4% and 6% / flour basis	Improvement in the dietary fibre content.	TDF and ash content substantially increases. It also deccelerates lipid oxidation.
Egyptian balady flatbread	M. balbisiana	5% and 10% / flour basis	Determining the impact of banana peel flour on the sensory quality and physicochemical characteristics of flatbread.	There was an increase in the fat, ash and protein content. Nevertheless, the carbohydrate content was reduced.
Edible food wrapper	M. paradisiaca	—	Utilization of banana peel to produce edible film wrapper.	The mechanical properties get enhanced due to the inorganic nutrients. Besides, such films easily biodegrade.
Xylitol	—	Flour with xylitol and sucrose in the ratio: 25:75, 0:100, 75:25, 50:50 and 100:0	Utilization of banana peel waste to develop xylitol for testing its impact on the physicochemical qualities of rusks.	The water activity of rusk was significantly decreased by the produced xylitol, leading to improved product self-stability.
Chapatti	M. balbisiana	5%, 10%, 15% and 20% / flour basis	Development of functional chapatti via use of banana peel flour.	The antioxidant activity of the banana peel-based chapatti was 67.5%. Moreover, dough incorporated with banana peel flour resulted in soft chapatti.

Modiyfied with permission from Zaini et al. (2022).

soluble, insoluble and total fibre contents in mango peels were 12.4–27 g/100 g, 23.5–50.3 g/100 g and 35.5–77.6 g/100 g of dry weight (dw), respectively [27]. Ajila and Rao (2013) examined the fibre content of mature and raw peels from Badami and Raspuri mango varieties [28]. The findings demonstrated that the ranges for the total sugar content of soluble and insoluble fibre, respectively, were 72.5–82.7 and 66.3–73.8 g/100 g of fibre dw. Uronic acid concentrations in soluble and insoluble fibres were, respectively, 9.8–21 and 15.2–24 g/100 g of fibre dw. The three principal sugars reported in the mango peels are glucose, galactose and arabinose. The presence of pectic polysaccharides and arabinogalactan was linked to the high levels of arabinose and galactose. In turn, the presence of β-glucans and cellulosic polysaccharides was linked to increased glucose levels. To add more fibre to processed food products today, the food industry still mostly uses cereal bran. However, mango peel provides an alternate source of fibre with a number of nutritional and physiological benefits, such as low energy value and phytic acid concentration, high soluble/insoluble fibre ratio and improved capabilities for water and oil absorption. Mango peel fibre also had a significant concentration of bound phenolic constituents, giving it antioxidant capabilities. The phenolic content of mango peels was considerably high (14.77–131.5 mg/g of mango peels dw). Flavonoids, benzophenones, gallotannins, gallic acid, cinnamic acids, gallates, ellagic acids and xanthones, are eight bioactive families that can be divided depending upon their chemical structures [29]. The phenolic constituents from the Ataulfo variety of mango peels were examined by Pacheco-Ordaz et al. (2018) [30]. To do this, they created three distinct extracts viz. acid and alkaline hydrolysis of bound phenolic constituents and organic extraction of free phenolic constituents. The most prevalent free phenolic compounds detected in terms of the phenolic compounds profile were mangiferin and hexagalloyl glucose. Gallic acid and digallic acid were the bound phenolic compounds that were found in higher concentrations. Ancos et al. (2018) also investigated the phenolic composition of mango peels from the Ataulfo variety [31]. To extract the phenolic constituents from the mango peel, they performed an organic extraction. No particular technique was used to assess bound phenolic constituents. The amounts of phenolic constituents identified in the mango peel were 45.35 ± 2.55 mg/g dw. Besides, the phenolic constituents of lyophilized Ataulfo mango peels were assessed by Blancas-Benitez et al. (2015) [32]. However, two different techniques were used in this research to remove the phenolic chemicals from the food matrix. In the first technique, they performed an acid hydrolysis after an organic aqueous extraction. In the second procedure, soluble and insoluble fibre remnants were subjected to enzymatic hydrolysis, followed by acid hydrolysis. The first approach yielded 72.2 18 mg of GAE/g dw of free phenolic content and 56.39 mg of GAE/g dw of bound phenolic content. However, in the second approach, the enzymatic hydrolysis resulted in the release of 93.7 ± 19.8 mg of GAEs/g dw, whereas the soluble and insoluble fibres were hydrolyzed to produce 33.5 ± 0.7 and 44.8 ± 2.9 mg of GAE/g dw, respectively. Regarding the profile of phenolic compounds, chlorogenic acid and hydroxycinnamic acid were the primary free phenolic constituent and bound phenolic constituent identified using the first approach. Also, the primary phenolic constituents associated with insoluble and soluble fibre were hydroxycinnamic and caffeic acid, respectively, while the primary phenolic constituent during enzymatic hydrolysis was ellagic acid. A number of variables, including genetic variation within and among different mango varieties, hydric stress, soil composition, maturation stage, climate and postharvest preservation affect the phenolic component profile of mango peels [33]. Raspuri and Badami mango cultivars' levels of free phenolic compounds drastically dropped as they ripened, according to Ajila et al. (2007) [34]. In Raspuri and Badami varieties, the reduction observed was 108.85 mg GAE/g dw to 101.10 mg GAE/g dw and from 88.24 mg GAE/g dw to 55.79 mg GAE/g dw, respectively. The bioaccessibility and bioavailability of phenolic compounds influence their potential health benefits. The intestinal permeability of gallic acid and mangiferin found in the mango peel extracts of Ataulfo variety was assessed by Pacheco-Ordaz et al. (2018) [30]. A Caco-2/HT-29 monolayer permeability experiment was carried out to achieve this. Gallic acid and mangiferin recovered basolaterally at rates of 42.39% and 28.84%, respectively. Mango peels are generally regarded as good sources of these bioactive substances, and carotenoids are one of the most significant pigments associated with the colour of the mango peels. The carotenoid level can, however, vary greatly since mango peels from different cultivars can range in hue. Mango cultivars are divided into three categories based on the hue of their peels: red, yellow and green. The carotenoids profile of fresh peels from yellow (Lazzat Baksh, Arka Anmol, Banganapalli and Peach), red (Gulabi, Lalmuni, Tommy Atkins and Janardhan Pasand) and green (Bombay No. 1, Hamlet, Amrapali and Langra) mangoes were analyzed by Ranganath et al. (2018) [35]. In general, green mango peel (3.22–13.95 μg/g fw) and yellow mango peel (1.61–30.97 μg/g fw) revealed higher levels of carotenoid than the red variety (0.82–6.32 g/g fw) across the examined cultivars. Seven carotenoids (zeaxanthin, violaxanthin, β-carotene, trans-violaxanthin butyrate, lutein, luteoxanthin and cis-β-carotene) were found in the studied peels when it came to the carotenoids profile. In green, yellow and red varieties, the most prevalent carotenoid detected was β-carotene along with its isomer cis β-carotene. Along with β-carotene, green- and red-coloured peels displayed high levels of trans-violaxanthin butyrate, whereas yellow peels contained larger levels of violaxanthin. In contrast to what was observed in the aforementioned mango cultivars, lutein was reported to be the most prevalent carotenoid in various mango peels [36]. The bioavailability and bioaccessibility of carotenoid content largely determine the health-promoting aspects. Through an in vitro digestion, Mercado-Mercado et al. (2018) assessed the bioaccessibility of beta-cryptoxanthin, lutein and alpha-carotene, from dried mango peel [37]. They discovered that ultrasonic therapy considerably improved these three carotenoids' bioaccessibility. The bioaccessibility of β-cryptoxanthin, lutein and β-carotene in mango peels subjected to ultrasound treatment was found to be 47.88%, 36.19% and 31.58%, respectively.

Mango peels are a by-product with a great potential for application in the creation of functional foods, nutraceuticals and medicinal therapies since they possess a number of bioactive qualities that have been well-documented. In comparison to wheat flour, Umbreen et al. (2015) examined the functional characteristics, antioxidant activity, as well as chemical composition of powders prepared from ripe and raw mango peels [38]. Mango peel powders have been studied as a substitute for maize flour, wheat flour, mango pulp, millet and wheat semolina in bakery goods. Mango peel powders have also been investigated as a replacement for sugar cane or wheat flour in bakery goods. Jalgaonkar et al. (2018) demonstrated that the substitution of around 6% wheat semolina and millet as well as 11% wheat flour with mango peel powders in both pastas and bakery goods maintains the sensory acceptability and considerably improves the antioxidant activity and bioactive composition of such food products [39]. Nevertheless, when greater amounts of mango peel powders were incorporated, it resulted in hard texture, more reddish and yellowish colour as well as harsh taste due to the high level of phenolic constituents. The production of air bubbles during fermentation and a decrease in gluten concentration, which interfered with the creation of 3-D network structure may have contributed to changes in the texture of bakery items enhanced with mango peel powders. According to Ramirez-Maganda et al. (2015), powders created from mango peels and pasta can substitute 73% of wheat flour and sugarcane in muffins while also enhancing their antioxidant activity, chemical composition and sensory qualities [40]. In vitro starch digestibility of bread produced with wheat flour and powdered mango peel was investigated by Chen et al. (2019) [41]. They discovered an apparent inverse relationship between the rate of starch digestion and the quantity of mango peel powder added to bread (6%–22%). For example, the rate of starch digestion and the amount of digested starch at the conclusion of digestion were both reduced by 20% mango peel powder, from 0.0374 minute and 100% to 0.0169 minute and 80%, respectively. In order to preserve chicken mince, Kanatt and Chawla (2018) created a biodegradable and non-edible primary packing material and assessed its effectiveness [42]. Three solutions viz. cyclodextrin (6%), gelatine (3%) and poly vinyl alcohol (6%), as well as ethanol extract of Langra mango peel variety, which accounts for 5% (vol:vol) of the final solution—were combined and dried to produce the packing material. The studied packaging material considerably reduced the formation of thiobarbituric acid reactive substances (TBARS) and the microbiological load throughout the refrigerated storage of 12 days when compared with control (packing film without mango peel extract), prolonging the shelf life of minced chicken meat from 3 to 10 days.

4.2.3 Pitaya peel

The pitaya (Hylocereus spp.), also known as the dragon fruit or pitahaya, is a fruit that grows on some strains of cacti and is native to tropical areas of Mexico, Central America and South America. As one of the most significant fruit crops, it is widely grown in tropical or subtropical regions all over the world. Pitayas can often be divided into three categories based on the colour of their peel and flesh: white flesh/yellow peel pitaya (YP), red peel/white pulp pitaya (RW) and red peel/red pulp pitaya (RR). The three species Hylocereus megalanthus (YP), Hylocereus undatus (WP) and

Hylocereus polyrhizus (RP) are the ones that are most frequently farmed and eaten. The principal by-products of pitaya processing are pitaya peels, which make up around one-third of whole fruits. Pitaya peels, however, have the potential to be recycled as excellent sources of nutrients and bioactive substances [6].

Pitaya has a large concentration of phytochemicals with bioactivities, including terpenoids, betacyanins, polysaccharides and phenolic compounds. These phytochemicals have profound pharmacological benefits, and they have been shown to help people manage type 2 diabetes, obesity and cancer, among others with very less toxicological side effects. Among the various phytochemicals, pitaya is considered to be one of the major sources of betalains, betacyanins and betaxanthins. Three enzymes, viz. tyrosinase, 4,5-DOPA-extradiol-dioxygenase and betanidin-glucosyltransferase, contribute significantly to the generation of betalains. Citramalic acid, in particular, may be closely related to betalains generation when the pitaya ripens since it has a lower level of amino acids in the peel and flesh while having a higher content of sugars and organic acids [43]. The vivid red colour is largely due to the betacyanins that are plentiful in pitaya peels. Betacyanins have greater potential for use in food applications than anthocyanins because it may maintain an optimum pH range of 3–8 and exhibit better antioxidant properties. The primary betacyanins found in pitaya peels are isophyllocactin, phyllocatin, isobetanin and hylocerenin. The stability of betacyanins may be increased by using certain metal ions and disaccharide solutions. The betacyanins extracted from pitaya peels can be used as either natural colourants or functional ingredients in different food products, thereby increasing the value-added index of the F&Vs while having a very less adverse impact on the environment. Betacyanins have a variety of bioactivities and health-promoting effects. Atherosclerosis, obesity, coronary heart disease, hyperlipidaemia, type 2 diabetes and neurodegenerative diseases are just a few of the illnesses that are currently thought to be fuelled in part by oxidative stress. Many researchers have considered the dim prospects of using phytochemicals from fruit and vegetable waste to combat oxidative stress. In this context, phenolic constituents with excellent antioxidant characteristics show enormous potential. Pitaya peels possess a variety of phenolic constituents in addition to betacyanins, which have recently piqued the curiosity of many researchers [44]. In research carried out by Suleria et al. (2020), the total phenol content (TPC), total flavonoid content (TFC) and total tannin content (TTC) of the pitaya peels were found to be 0.45 ± 0.12 mg/g dw, 0.03 ± 0.01 mg/g dw and 0.03 ± 0.01 mg/g dw, respectively [45]. The major flavonoids detected were quercetin 3-O-glucuronide, kaempferol 3-O-glucoside, quercetin 3-O-galactoside and catechin with quantified values of 1.7 ± 0.4 mg/g, 2.4 ± 0.7 mg/g, 4.5 ± 0.7 mg/g and 7.5 ± 0.9 mg/g dw, respectively, were identified. The principal phenolic acids detected were ferulic acid, coumaric acid, syringic acid, caftaric acid and chlorogenic acid with quantified values of 2.7 ± 0.8 mg/g, 2.8 ± 0.1 mg/g, 3.1 ± 0.9 mg/g, 3.5 ± 0.7 mg/g and 4.4 ± 0.9 mg/g dw, respectively. Additionally, other phenolic substances such as flavones, anthocyanins and flavanols, were also discovered. In a different study, microwave technology was used to extract bioactive compounds from pitaya peels. Here, 17 flavonoid molecules were effectively discovered, including kaempferol 3-O-rutinoside, quercetin 3-O-rutinoside and isorhamnetin 3-O-glucoside [46]. In another study, the TPC of the peel and flesh of red peel/white pulp (RW) and red peel/red pulp (RR) pitaya varieties was determined to be 15.85 ± 0.77 and 4.11 ± 0.83 mg GAE/g dw, respectively. In contrast, the peel and flesh of RW had TFC of 14.33 mg/g dw and 3.52 mg/g dw per gramme, respectively, whereas RR showed 18.16 mg/g dw and 9.56 mg/g dw per gramme, respectively [47]. Based on the diversity of species, it is possible to infer that the TPC of pitaya peels may vary. In line with the findings of several other studies, it was discovered in this investigation that the peel contained more phenolic chemicals than the pulp. The study also looked at the association between TPC and antioxidant capabilities, and it found a favourable one. Before being tested, phenolic compounds are typically extracted using specific solvents, therefore the results may differ depending on the extraction methods used as well as the types and origins of the compounds. The impact of extraction techniques on phytochemicals recovered from pitaya peels has been studied in recent research. The extraction techniques have an impact on the quantity of phenolic compounds that were recovered

[48]. Researchers examined the TFC and TPC of pitaya peel extracts made using conventional and enzyme-assisted extraction techniques. It was found that the application of enzyme treatment resulted in extracts having higher levels of TFC (1290.25 mg CE/100 g dw) and TPC (1050.22 mg GAE/100 g dw) compared to the conventional extraction methods which showed a TFC of 218.31 mg CE/100 g fw and TPC of 552.66 mg GAE/100 g fw. Another class of bioactive compounds, viz. terpenoids, are predominantly found in pitaya peels. They have a number of health-promoting characteristics including the reduction of hyperlipidaemia, oxidative stress, cardiovascular disease and cancer risk. Squalene, a polyunsaturated hydrocarbon, is the major terpenoid molecule found in the peels of white pitayas [49]. Two pitaya species, viz. H. polyrhizus and H. undatus, had their peel extracts prepared using a supercritical carbon dioxide process and gas chromatography-mass spectrometry was utilized to assess their chemical composition. The principal terpenoid compounds identified in the peel extracts of H. polyrhizus and H. undatus were composed of squalene, campesterol, γ-sitosterol, stigmasterol, α and β-amyrin and β-sitosterol. In addition to vitamin B, amino acids and proteins, alkaloids are a type of nitrogenous substances found in plant tissues [50]. A recent study evaluated the different types of metabolites found in the green and red pitaya peels [51]. In this study, methanol was utilized for the recovery of phytochemicals from pitaya peels, and the metabolites were then analyzed via liquid chromatography-mass spectrometry (LC-MS). This research showed that the main alkaloids detected in the pitaya peels were dopamine hydrochloride, choline, spermine, N-benzylmethylene isomethylamine, 6-deoxyfagomine, trigonelline and serotonin. However, choline and N-benzylmethylene isomethylamine were found in the highest concentration in the pitaya peels. In contrast to the green pitaya peels, the red cultivars showed a larger accumulation of alkaloids, which may be due to age difference and variety. Because of their safety and wide antibacterial spectrums, there have been several attempts to research the antimicrobial characteristics of natural bioactives. Pitaya peel extract exhibits a broad antimicrobial spectrum against a variety of fungi and bacteria, including gram-positive bacteria like Bacillus cereus, Enterococcus faecalis and Listeria monocytogenes and gram-negative bacteria, like Salmonella typhi, Enterobacter aerogenes and Pseudomonas aeruginosa, as well as yeasts, like Candida albicans (Aspergillus flavus, Botrytis cinereal and Cladosorium herbarum) [44]. In a further investigation, extracts of the white and red pitaya peels were made by treating them with 95% ethanol, n-hexane and chloroform, and their antibacterial properties were then assessed. All food-borne bacteria were studied, and it was discovered that every fraction could stop their development in the range of 1.30–10 mg/mL. In this investigation, red pitaya peels had a stronger inhibitory impact than white pitaya peels, and n-hexane and chloroform fractions had a greater inhibitory effect than 95% ethanol [52]. Pitaya peels have been shown to have anticancer properties in some trials. Researchers evaluated the peel and flesh of red and white pitayas for their capability to inhibit the proliferation of cancer cells in a study. With IC_{50} (50% inhibitory concentration) values of 453 and 451μg/mL as well as 419 and 481 μg/mL, respectively, the study discovered that red and white pitaya peel extract could successfully inhibit the growth of MCF-7 (human breast cancer cells) and AGS (human gastric adenocarcinoma cells) [47]. Furthermore, extracts from red and white pitaya peels designed as red peel extract (RPE) and white peel extract (WPE) were synthesized via supercritical CO_2 by Luo et al., (2014) [50]. The study showed that WPE (0.8 mg/mL) could considerably inhibit the growth of MGC-803 (human gastric cancer cells), Bcap-37 (human breast cancer cells) and PC3 (human prostate cancer cells). The rate of inhibition obtained by WPE were 55.2%, 62.4% and 60.7%, respectively, whereas RPE (0.8 mg/mL) showed an inhibitory rate of 78.9%, 62.7% and 67.3%, respectively. Here, the concentration of extracts was favourably connected with the inhibitory actions against such cancer cells, and the primary active ingredients may be α and β-amyrin and γ-sitosterol. A study used 2% pitaya peel powder (PPP) as a natural antibacterial and antioxidant agent to increase the time limit for storing pasteurized milk [53]. The incorporation of PPP could offer specific antioxidant properties and maintain a normal pH of milk that had been kept at an ambient temperature for 11 hours. This may be because PPP's bioactives have an inhibitory effect on the growth of organisms that cause spoiling. Some researchers analyzed the impact of pitaya peel

extracts on the physicochemical composition of fermented milk. They discovered that the fermentation process of fermented milk was strengthened by the lactic acid bacteria and that the 62% of pitaya peel extract could boost the fat and total flavonoid levels of the milk. Noodles now contain PPP, a partial flour substitute that has been added to raise product quality. In comparison to noodles without PPP, dried and cooked noodles with PPP showed higher betacyanins and polyphenols, indicating better antioxidant activity. PPP significantly increased the appearance of redness and the colour property remained constant while being stored. High levels of PPP were introduced, which reduced elasticity and extensibility while enhancing tensile strength and cutting force. According to the results of the sensory evaluation, adding 3% or 6% of PPP in place of a certain flour promised to create noodles that are enriched with nutrients while causing little harm to the goods' sensory qualities [54]. A study showed that PPP could also be combined with regular wheat flour to make cookies [55]. PPP was used in this case to partially replace wheat flour in varying quantities (5, 10 and 15%), resulting in cookies that were higher in fibre and carbohydrates than their non-PPP-added counterparts. Additionally, using PPP in place of some of the wheat flour resulted in increased diameter and spread ratio while lowering crumb height when compared to the control. Furthermore, the PPP inclusion did not cause the food items' sensory appeal to suffer. As such, PPP has a potential to be used in the production of cookies as a wheat flour alternative and nutritional enhancer. The potential uses of pitaya peel in food-based applications are shown in Table 4.2.

4.2.4 Pomegranate peel

Pomegranate (Punica granatum L.), often known as a seeded or granular apple, is a tasty fruit that is eaten all over the world. The fruit is indigenous to China, India, Afghanistan and Iran. The pomegranate fruit is well known for having significant concentrations of bioactive phenolic and flavonoid chemicals. The peel, seeds and pomace of pomegranates have all traditionally been regarded as agricultural waste. However, the peels include a network of internal membranes that make up nearly 26–30% of the weight of the entire fruit and are distinguished by significant levels of phenolic chemicals, such as hydrolyzable tannins and flavonoids. Pomegranate peel (PoP) and juice contain high concentrations of these chemicals, which are responsible for 92% of the fruit's antioxidant action [5].

The primary class of phytochemicals found in pomegranate peels are phenolic compounds. The latter are abundant in phenolic acids, flavonoids and high molecular weight hydrolyzable tannins (ellagitannins). PoP's phenolic compounds, which include hydroxybenzoic acids, anthocyanins, gallagyl esters, gallotannins, dihydroflavonol, ellagitannins and hydroxycinnamic acids are linked to the antioxidant activity of the fruit. However, ellagitannins, which are represented by punicalagin, gallic acid and ellagic acid are the fruit's predominant phenolic compounds. The most potent antioxidants found in PoP's tannins are its hydrolyzable polyphenols, particularly its ellagitannins. These substances—gallagic acid, punicalin, punicalagin and ellagic acid—have been demonstrated to possess enhanced biological activities that are pleiotropic and antioxidant, particularly acting synergistically with one another [56]. In addition, punicalagin (16.67–245.47 mg/g dry matter), an ellagitannin having a molecular weight (MW) of 1,084 Da, is the main phenolic component found in pomegranate peels. Along with having anti-free radical capabilities, gallic acid, ellagic acid and punicalagin also have antibacterial effects on intestinal flora, particularly enteric pathogens like Escherichia coli, Salmonella spp., Shigella spp. and Vibrio cholera. Pomegranate peel extract (PoPx) has been used to treat a number of common illnesses, including inflammation, diarrhoea, intestinal worms, cough and infertility [57]. Among the most popular fruits, pomegranates contain the highest punicalagin content. Punicalagin possesses antioxidant, antifungal and antibacterial effects, according to studies. Punicalagin hydrolyzes into smaller polyphenolic compounds in the small intestine under physiologically normal circumstances because it is water soluble. Punicalagin is made up of 11–20 g/kg of ellagitannins from the powdered peel. Punicalagin, a significant polyphenol antioxidant found in pomegranate peel, also possesses antiproliferative action, suppressing

TABLE 4.2
Application of pitaya peel in food products

Cultivar and Origin	Sample	Analyses	Results	Functions	Application
Red pitaya peel (China)	Peel powder	Determination of sensory analysis, betacyanin content, texture analysis and phenolic content.	Improvement in hardness, redness, along with phenolic and betacyanin content. Reduction in elasticity, springiness and cohesiveness.	Antioxidant, colourant	Steamed bread
White pitaya peel (Malaysia)	Peel powder	Determination of sensory analysis, nutritional analysis, along with physical property.	Improvement in spread ratio, along with carbohydrate, fibre and ash content. Reduction in protein and moisture content.	Fortifies nutrition and can substitute wheat flour	Cookie
Red pitaya peel (China)	Peel powder	Evaluation of sensory and textural analysis, phenolic and betacyanin content, along with antioxidant activity.	Enhances the tensile strength, antioxidant activity, redness, along with phenolic and betacyanin content. Decreases both elasticity and extensibility.	Antioxidant, colourant	Noodles
White pitaya peel (India)	Peel powder	Evaluation of sensory, texture, nutritional and microbiological analysis, along with total phenolic content.	Increase in the emulsion stability, dietary fibre, ash, protein, cooking yield, total phenolics. Reduces odour, hardness, chewiness and gumminess.	Fortifies nutrition and acts as antibacterial and antioxidant agents	Chicken nuggets
Red pitaya peel (Brazil)	Peel powder	Determination of pH along with antioxidant property.	The quality can be maintained when stored for 12 h.	Antioxidant, antibacterial agent	Pasteurized milk
Red pitaya peel (Brazil)	Distilled water extract	Evaluation of pH and nutritional properties.	The activity of lactic acid bacteria increases throughout fermentation. Protein and carbohydrate content decreases, while flavonoid and lipid content increases after fermentation.	Fermenting promoter	Fermented milk
Red pitaya peel (Brazil)	Microencapsulated 40% ethanol extract	Antioxidant property, colourant.	Decrease in the redness, protein and lipid oxidation throughout five days of cold storage.	Antioxidant	Pork patty treated by UV-C

Modified with permission from Jiang et al. (2021) [6].

proliferation against all cell lines by 30% to 100%. Punicalagin has been linked to a number of biological processes, including anti-inflammatory, anticancer, antimicrobial and antioxidant [58]. Punicalagin has 16 hydroxyls per molecule, which makes it more antioxidant-active than hydrolyzable tannins such anthocyanins and ellagic acid by a factor of 1.5, 6.5 and 20 correspondingly, according to Gil et al. (2000) [59]. Gallic acid, catechin and ellagic acid, which have relative contents of 12.58–25.90, 8.68–12.65 and 0.44–3.04 mg/g dry matter, are additional polyphenol components found in pomegranate peels. Ellagic acid can be found both unbound and bound (EA-glycosides and ellagitannins). It is well established in the literature that ellagic acid is effective in treating a number of low to mild chronic illnesses with very slow rates of progression. Potential as a chemopreventive drug for the treatment of cancer has also been demonstrated. Ellagic acid has been shown to lessen white fat deposits and triglycerides levels accumulated in the body following regular ingestion of high-fat meals, in addition to its other well-known ethnopharmacological effects. The cytoprotective properties of ellagic acid from PoPx on oxidatively damaged live cells, oxidative DNA damage and depletion of the non-protein sulphhydryl pool have been proven in numerous investigations. The antioxidant activity of PoPx is closely correlated with higher ellagic acid concentrations [60]. According to reports, ellagic acid concentrations in fruit peel and juice are 10–50 mg/100g and 1.38–2.38 mg/100ml, respectively [61]. Pomegranate peels also include caffeic acid, cyanide and p-coumaric acid. Pomegranate peels have a total phenol concentration that has been estimated to be between 18 mg/g dry matter and 510 mg/g dry matter. This large range can be explained by the fact that both the total and individual phenolic content of PoPx depends on a number of variables, including the cultivar, sample pre-treatment, extraction solvent and technique, etc. Different solid/solvent ratios (1:25 g/mL to 1:35 g/mL) have been employed thus far, along with a variety of solvents including water, ethanol, methanol, acetone and ethyl acetate. Furthermore, because it can promote the growth of Bifidobacterium spp. and Lactobacillus spp., pomegranate peel extracts have the potential to be employed as prebiotics [1, 62]. PoPx has been used in prior research as an edible packaging component to increase the shelf life of fresh fruits, vegetables, meat and meat products, seafood, dairy and bakery food items. Pomegranate peel extract has demonstrated antioxidant and antibacterial capabilities in this situation by lowering the rate of microbial degradation during storage, avoiding moisture loss, managing gas exchange and minimizing lipid oxidation. The effects of edible packaging enhanced with PoPx on capsicum and guava fruit over storage periods of 20 and 25 days at 10°C were examined in a study by Nair et al. (2018) [63]. The results showed that the food products' overall sensory and postharvest quality had improved, as well as their shelf life. Additionally, pomegranate peel enriched edible packaging has been utilized on a variety of food items to increase shelf life by preserving postharvest qualities while in storage. PoPx in edible packaging typically demonstrates a variety of functionalities, including oxygen and mechanical permeability, water loss prevention, shelf life and tensile strength, among others along with health-related effects such as lipolysis, oxidation and antimicrobial properties. It has been shown that adding pomegranate peel extract as a natural reinforcing agent boosts the polymeric matrix's capacity to scavenge free radicals, increasing the potential of the material for use in biodegradable food packaging [56].

PoPx has been demonstrated to cause human breast cancer cells to undergo apoptosis (MCF-7). In earlier research, using PoPx with genistein to treat breast cancer cells resulted in considerably stronger MCF-7 inhibitory and cytotoxic effects. PoPx also showed the ability to prevent the expression of angiogenesis markers, phosphorylation of p38 and C-Jun mitogen-activated protein kinases and activation of pro-survival signalling pathways, in addition to the inhibition of cell proliferation. Additionally, it has been demonstrated that PoPx reduces the expression of reporter genes that are nuclear factor kappa B (NF-kB)-dependent and linked to aggressive breast cancer phenotypes' proliferative, invasive and motile behaviours [64]. PoPx with anticancer capabilities was effectively evaluated at a dose of 300 mg/ml in conjunction with 1 mM tamoxifen to sensitize and increase the activity of the latter and decrease the proliferation of resistant MCF-7 cells. The possibility of using polyphenols, flavonoids and condensed and hydrolyzable tannins—which are derived from fruits, vegetables, herbs and spices—to treat or prevent a variety of illnesses has been investigated. The antibacterial actions of phenolic compounds involve the interaction of phenolics with protein sulphhydryl groups and/or membrane proteins of microorganisms, which results in bacterial mortality by precipitation of membrane proteins and inhibition of enzymes like glycosyltransferases [65]. On the Indian subcontinent, PoPx is traditionally used to treat food-borne illnesses and urinary tract infections. PoP ellagitannins, punicalagin, ellagic acid and gallic acid, however, have been widely used as natural antimicrobial agents against Staphylococcus aureus and haemorrhagic Escherichia coli because of their capacity to precipitate membrane proteins and inhibit enzymes like glycosyltransferases, resulting in cell lysis. An 80% methanolic extract of PoP applied in vivo and on-site had an inhibitory impact on Listeria monocytogenes, Staphylococcus aureus, Escherichia coli and Yersinia enterocolitica. The lowest bactericidal concentration for Listeria monocytogenes, however, has been shown to be higher dosages of PoPx (24.7 mg/ml) [56]. Additionally, Sandhya and Kumar (2018) examined the effects of an ethanolic PoPx encapsulated with skim milk powder and whey protein concentrate on the functional characteristics and storage stability of curd at concentrations of 0.5–1.5% [66]. The total phenolic content and antioxidant activity increased along with PoPx concentration, whereas sensory qualities decreased. In instance, flavour scores dropped as the level of PoPx addition increased concurrently. The authors hypothesized that this might be caused by the PoPx's astringency and bitterness. Additionally, it was discovered that the texture, colour and appearance were equivalent to the control samples up to a level of addition of 1%, but further addition led to a considerable decline in the corresponding scores. Based on these findings, the authors decided that the maximum degree of addition for future research would be 1%. In comparison to the control sample, the fortified product had more stable titratable acidity, pH, whey syneresis, microbiological count and sensory characteristics throughout storage. Incorporating 0.2% aqueous PoPx into chicken products increased their shelf life by 2–3 weeks without affecting their sensory qualities, as claimed by Kanatt et al. (2010) [67]. PoPx significantly inhibited Staphylococcus aureus growth, demonstrating its antimicrobial activity. Additionally, the observed oxidative rancidity of the control sample was higher over the course of storage than it was in the enriched chicken product. The antibacterial efficacy of pomegranate peel ethanolic extract against Salmonella Kentucky in chicken flesh was recently examined by Wafa et al. (2017) [68]. During 23 days of storage at 4°C, the incorporation of extract at doses of 0.8 and 1.6 mg/g chicken breast caused the growth rate of Salmonella to decrease. However, the addition of the extract also caused the colour of the flesh to shift from pink to yellow. Commercial strawberry-infused orange juice and tomato juice were given a boost in antioxidant activity by adding dried PoPx. Participants in the sensory analysis gave the juices with higher extract concentrations the lowest ratings due to the astringent taste of pomegranate peels. Therefore, a 0.5% maximum extract concentration was suggested. Recently, untreated, high-pressure and thermally processed carrot juice were combined with dry PoPx (2.5 mg/mL) by Trigo et al. (2019) [69]. While the extract had no impact on the sensory quality of the juices, the fortified juices had higher antioxidant activity, lower counts of psychrophiles and mesophiles and higher total phenolic and hydrolyzable tannin levels than the control samples.

4.3 RECOVERY OF BIOACTIVE COMPOUNDS FROM VEGETABLE PEEL

4.3.1 Potato peel

After rice, wheat and maize, the potato (Solanum tuberosum L.), which is a staple food for 1.3 billion people, is the most significant food crop. In 2018, more than 368 million tonnes of potatoes were produced worldwide. Since there are more than 5,000 recognized potato varieties, it is thought that among farmed species, potatoes have the highest genetic diversity. Potato processing typically starts with peeling the tubers, for which techniques like abrasion, lye peeling and steam peeling have been recorded. Large volumes of peel waste are also produced by the potato starch, flour and canning sectors and its disposal causes serious environmental problems. The majority of the trash produced by potato processing businesses is either dumped in landfills, which has negative environmental effects, or it is used as animal feed with no added value. Nevertheless, this waste could be used to recover useful substances including protein from processing wastewaters, starch from tuber flesh and phytochemicals and critical minerals from peels [10].

The variety in size, shape, texture and colour, as well as in terms of chemical composition and utilization, among others, are the factors that distinguishes potato species from one another in terms of phenotypic variety. Because of the presence of high levels of phenolic constituents with known health-promoting effects, viz. antioxidant activity, potato peel (PP) can be considered a rich source of phytochemicals. These constituents can be used in different food applications to increase the food product sustainability. The phenolic chemicals found in PP are also said to be a significant source of benefits for human health, including antioxidant and antibacterial characteristics. Up to ten times more phenolic chemicals are present in PP than in potato flesh [70]. Additionally, numerous phenolic constituents found in tuber peel that are associated with the defence mechanism of the plant against pathogens, including p-coumaric acid, coniferyl aldehyde, caffeic acid, ferulic acid, coniferyl alcohol and vanillic acid are only occasionally found in potato flesh. PP comes in many different types, each of which has a different total amount of phenolic chemicals from the others. Nevertheless, caffeic acid and its isomers are shown to be the secondary significant molecules in the majority of investigations, followed by chlorogenic acid and its isomers [71]. As per Riciputi et al. (2018), the total phenolic constituents were made up of 49.3–61% chlorogenic acid [72]. Additionally, caffeic acid can partially degrade into chlorogenic acid when exposed to direct light while being stored at ambient temperature. Thus, caffeic acid emerges to be the dominant phenolic constituent in PP waste generated from industries, as they are typically kept outdoors, unprotected from sunlight. Rutin, quercetin, catechin, ferulic and p-coumaric acids, as well as other phenolic substances, have also been linked to PP. In PP aqueous extracts, Hsieh et al. (2016) determined that the concentration of total phenolic acid was 85.8 mg GAE/100 g dw, while the flavonoid content was 27.5 mg GAE/100 g dw, with the most common components being ferulic acid, caffeic acid, chlorogenic acid and p-coumaric acid [73]. The utilization of various extraction settings, extraction solvents and extraction techniques can be partly blamed for the concentration variation in the phenolic constituents among the existing research. For example, Silva-Beltr'an et al. (2017) used maceration with continuous stirring for 72 hours to examine the composition of phenolic constituents in PP via acidified ethanol and aqueous extracts [74]. The total flavonoids (3.4 mg quercetin equivalent/g dw) and total phenolic contents (13.5 mg GAE/g dw) in the acidified ethanolic extract were higher than those in the aqueous extract (1.2 mg quercetin equivalent/g dw and 4.6 mg GAE/g dw, respectively). Similar observations were made by Albishi et al. (2013) during the evaluation of the total concentration of bound, esterified and free phenolic constituents in PP of four distinct potato varieties (Russet, Purple, Yellow and Innovator) [75]. The solvent and the process used were methanol/acetone/water (7:7:6; v/v/v) and solid–liquid extraction, respectively. The content of bound phenolics was higher in the varieties viz. Russet and Innovator (5.1 and 5.7 mg GAE/g dw, respectively) compared to the concentration in free phenolic contents (2.9 mg GAE/g dw and 3.1 mg GAE/g dw, respectively), which emphasizes the significance of bound phenolic content calculation in analytical protocols for determining the content of all phenolic constituents more precisely. In addition to the conventional solid–liquid extraction method, methods like ultrasound- and microwave-assisted treatments can further increase the effectiveness of extracting phenolic chemicals from polypropylene. Kumari et al. (2017) observed a greater rate of extraction for phenolic constituents using an ultrasound-assisted extraction approach rather than a solid–liquid extraction procedure [76]. The findings showed that the optimum solvent combination for the recovery of phenolic constituents from PP was 80% methanol/water (v/v) and that 33 kHz ultrasonic treatment was superior to higher frequency treatment (42 kHz) for recovering polyphenols. Furthermore, Singh et al. (2011) used a response surface technique study in conjunction with microwave-assisted extraction to optimize variables like extraction duration, solvent concentration and microwave power [77]. The greatest amount of total phenolics (3.88 mg/g dw) was achieved with a recovery period of 14 min, a solvent composition of 68.23%

methanol/water and a microwave power of 15.67%. The optimum concentration of ferulic acids (0.50 mg/g dw) and caffeic acids (1.33 mg/g dw) were obtained at 100% methanol, 15 min extraction time and 15% microwave power. On the other hand, the highest concentration of chlorogenic acid (1.40 mg/g dw) was achieved at a shorter recovery period of 5 min, while keeping the operating conditions of other extraction parameters same. The study concluded that microwave-assisted extraction was found to be more effective than the conventional methods for the recovery of phenolic constituents from PP. Anthocyanins, which give purple and red fleshed potato types their colours, are a significant class of chemicals that can be produced from PP. Normally, the potato skin is of the same colour as that of the flesh. Nevertheless, there are some varieties in which the colour of the flesh significantly varies from the skin [10]. Four distinct varieties, out of which three had yellow-coloured flesh and one had purple-coloured flesh, were compared for their phenolic compound content in PP by Albishi et al. (2013) [75]. Of the four types under study, the purple-fleshed cultivar's PP had the highest concentrations of esterified and free phenolic constituents. Additionally, the purple potato cultivar accumulated most of the pigments in the outer cortices and skin of the tubers, with the peel's anthocyanin concentration being 10.75 times greater than that of the flesh. In order to determine the polyphenol and anthocyanin content in the skin and flesh of potato tubers from 56 coloured potato varieties, Oertel et al. (2017) utilized Ultra-performance liquid chromatography in combination with UV and mass spectrometry detection [71]. They discovered that the anthocyanin accumulation patterns of the potato genotypes might be used to classify them. A total of six sets of anthocyanin from the red variety and 12 groups from the purple and blue varieties were selected, for a total of 18 anthocyanin profiles. The primary pigments of red tubers, specifically pelargonid in-3-p-coumaroylrutinoside-5-glucoside, were verified to be acylated pelargonidin glycosides. The two most prevalent anthocyanins in purple and blue tubers were malvidin 3-feruloylrutinoside-5-glucoside and petunidin 3-p-coumaroylrutinoside-5-glucoside, with the former being much more prevalent in the peel than in the flesh. In general, the flesh of blue and purple cultivars had a lower concentration of malvidin derivatives than the peels. Malvidin's antibacterial action has been connected to the defence mechanism of potato against harmful bacteria, which may explain why it is more abundant in the peel, which serves as the outer defence of tubers against aggressors from the outside. The glycoalkaloids, another class of substances that predominately exist in PP, primarily consist of two molecules, viz. α-chaconine and α-solanine. Glycoalkaloids can be hazardous to humans to a certain extent, although at doses of roughly 1–2 mg/kg of human weight, they are safe to consume [78]. The most prevalent glycoalkaloid in PP is often α-chaconine, which is said to be around five times more bioactive than α-solanine. However, these substances have bioactivities that are advantageous to human health, including anti-inflammatory, anticancer, antibacterial and anti-obesity properties [70].

PP extracts have proven to have antibacterial efficacy against Salmonella Typhimurium and Eschericia coli. The PP acidified ethanol extract showed antiviral activity against human enteric viruses in research by Silva-Beltr'an et al. (2017) [74]. The human intestinal viral surrogates of the MS2 and Av-05 bacteriophages were successfully suppressed by the PP extracts. Plaque-forming units per unit of volume (PFU/mL) of MS2 and Av-05 were decreased by 3.5 log 10 and 2.9 log10, respectively, after a 3 h incubation with PP extracts at a dosage of 5 mg/mL. This reduction was dose-dependent. According to the findings, the extracts from PP has the capacity to be a powerful weapon against human enteric viruses. Also, PP have been investigated for their ability to fight diabetes and obesity [78]. For example, adding PP powder to the diet prevented mice (fed with a high-fat diet) from gaining weight, indicating its potential as an anti-obesity functional food. In this study, 10–20% PP powders were added to mice's adipogenic high-fat meals (27% fat by weight) for 21 days. The peel-containing diet resulted in considerable weight gain reduction in mice of up to 73% when compared to the control diet. Given that the mice's weight gains were inversely connected with the diet's contents of α-chaconine and α-solanine, the effectiveness of PP powder as an anti-obesity treatment was probably due to its high glycoalkaloid content. More recently, PP fibre in bread (0.5 g fibre/100 g flour) was used by Curti et al. (2016) to explore its capacity to decrease bread staling [79]. The addition of potato fibre to bread, according to the authors, increased the amount of frozen water in the product and made the bread crumb softer after seven days of storage. At the conclusion of the storage duration, the crumb hardness for the formulation with added PP were lower than those for the control formulation (3.8 ± 0.7 N and 4.4 ± 0.6 N). Crawford et al. (2019), on the other hand, added PP to quinoa flatbreads in an effort to lower the amount of acrylamide, a potentially harmful substance created during the thermal processing of grains [80]. When PP powder from the Russet cultivar was added to quinoa flour at 6%, the cooked flatbreads' acrylamide level significantly decreased when compared to the control formulation (from 479 to 371 g/kg). According to the findings, quinoa flatbreads with PP supplements may one day act as a healthy, low-acrylamide and gluten-free diet food. The antioxidant impact was measured via anisidine and peroxide values in addition to sensory analysis using water and ethanolic extracts of PP from Bintje and Sava cultivars of the Danish potato. The results revealed that the Sava variety of PP ethanolic extract at a concentration of 2,350 mg/kg was 5.8 times more efficient than BHT in lowering the peroxide value and 8 times more efficient in lowering the anisidine value, however, the aqueous extracts did not exhibit antioxidant action. According to the sensory assessment findings, oil enriched with water extracts had more rancidity than oil supplemented with ethanolic PP extracts [81]. The growth inhibition of Bacillus cereus (B. cereus) in cooked rice by various F&V peel powders, including PP, was investigated by Juneja et al. (2018) [82]. Rice containing four strains of B. cereus was combined with the powders at a 10% (w/w) concentration. The amount of B. cereus increased by 1.93, 2.82, 3.83, or 3.58 log cfu/g when aliquots (5 g) of control cooked rice were cooled from 55.8 °C to 7.1°C in 12, 15, 18, or 21 hours, respectively. Under similar settings as the control rice sample, the results for the rice samples supplemented with organic gold PP powder for the rise in B. cereus levels were 0.25, 0.49, 1.42 and 2 log cfu/g. These findings showed how effective PP can be at slowing the spread of B. cereus in food products. The potential uses of potato peels in various food-based applications are shown in Table 4.3.

4.3.2 Onion peel

The onion (Allium cepa), one of the most widely grown and consumed crops worldwide, is the item most frequently utilized in Indian cuisine. The top five countries that contribute to the world's total production of onions are China (23.85 MT), India (19.64 MT), Egypt (3.22 MT), the US (3.15 MT) and Iran (2.20 MT). But whether done in a commercial setting or a home kitchen, processing onions produces a lot of waste, including the non-edible components of the onion as well as the outer skin and peel. Additionally, improper disposal of this waste could have an adverse influence on the ecosystem because it can neither be utilized to prepare fodder due to the strong onion smell nor as a fertilizer. Because of this, onion wastes, particularly onion peels, go unused while being a rich source of bioactive substances such phenols, flavonoids and flavanols. Therefore, using onion peels and their extracts to make products for the biomedical and pharmaceutical industries could be a good way to lessen environmental damage [9].

During industrial processing, large volumes of onion waste are produced, including the skin/peel and exterior fleshy leaves. These wastes are highly concentrated in phytochemicals and bioactive components. Phytochemicals such as, phenolics, flavonoids, flavanols, anthocyanins, tannins, vanillic acid and ferulic acid are present in the peel. The onion peel and skin had the highest levels of phenolics and flavonoids when compared to the edible onion bulb and meat. Based on dry matter, the top-to-bottom outer scales of an onion had higher levels of total flavanols (19.38, 15.32, 7.91 mg/g), total flavonoids (19.48, 25.88, 43.21 QE/g) and total phenolics (19.68, 30.48, 52.66 mg GAE/g) than the inner scales (9.39 mg GAE/g, 7.15 mg QE/g and 6.21 mg/g), as well as the entire onion (18.28 mg GAE/g; 11.28 mg QE/g; 8.91 mg/g). Compared to edible flesh, onion peel contains a higher concentration of phytochemicals, with quercetin being the main phenolic component [83]. Moreover, protocatecoyl quercetin, quercetin-3-glycoside, quercetin dimer hexoside, quercetin-4'-glucoside, quercetin dimer, isorhamnetin-4'-glycoside, protocatechuic acid, quercetin trimer and

TABLE 4.3
Application of potato peel in food industry

Application	Type of extract	Analyses	Results
Bread	Fibre extracted from potato peel	Texture and shelf life	The bread was blended with potato fibre (0.5 g/100 g flour) which decreases the hardness after a duration of 6 days when compared against the control.
Flatbread	PP powder	Acrylamide content	The flour of flatbread formulation was incorporated with 5% potato peel powder which significantly reduces the acrylamide content (responsible for toxicity).
Cake	PP powder	Sensory and textural evaluation	The cakes were incorporated with potato peel powders having higher concentration of protein and dietary fibre which enhances the texture and increases the nutritional content of the product.
Soybean oil	Ethanol extracts of potato peel	Oxidative stability	The soybean oil was stabilized by the use of potato peel extracts at 60°C for 13 days which reduces the oxidative indices such as totox, peroxide and p-anisidine.
Soybean and sunflower oils	Methanol extracts of potato peel	Oxidative stability	The soybean and sunflower oils were incorporated with methanol extracts of potato peels at 65°C for three days. Improved antioxidant property of the potato peel extract was reported for both the vegetable oils when compared to artificial antioxidants, viz. BHA and BHT.
Salmon (cooked)	Water-Methanol-acetone extracts of potato peel	Antioxidant activity	The formation of oxidative rancidity in fish was retarded by the addition of phenolic compound-enriched extracts. The oxidation of cooked salmon was effectively inhibited after a storage period of 6 days when compared against the control.
Cooked rice	PP powder	Antimicrobial activity	Rice samples were incorporated with potato peel powder (10% w/w) which inhibits B. cereus throughout the cooling period w(55.1°C–7.1°C) at 12 h, 15h, 18h and 21 h.
Soybean oil	Petroleum ether extract of potato peel	Oxidative stability over shelf life	The antioxidant capacity of the potato peel extract was reported to be very high, similar to that of artificial antioxidants, viz. BHT and BHA. This assisted in deccelerating the formation of oxidation products in soybean oil throughout the storage period of 60 days.

Modified with permission from Sampaio et al. (2020) [10].

quercetin-7,4'-diglycoside among others have been detected in onion peel [84]. All of these bioactive substances give onion peel a range of therapeutic benefits, including protection against cancer, diabetes, obesity, erectile dysfunction and neurodegenerative and cardiovascular diseases. Additionally, the onion peel's outer scale contains 12 distinct fatty acids, of which half were unsaturated and the other half were saturated. The fatty acid composition of onion peel includes stearic acid (8.81%), myristic acid (1.28%), palmitic acid (9.80%), lauric acid (0.94%), arachidic acid (0.59%), erucic acid (0.63%), oleic acid (18.65%), lignoceric acid (0.54%) and linoleic acid (52.87%), among others, suggesting that the total unsaturated fatty acids was 76.79%, and total saturated fatty acids was 21.42%. The total phenol concentration (mg GAE/g) of whole yellow onions grown in southern Sweden's farms ranged from 27.8 to 51.1 and 54.7 to 70.5 depending upon its dry weight, respectively [85]. On the other hand, yellow onion peels from Galaţi, Romania, contained somewhat more total polyphenolic components (98.33 mg GAE/g dw). These onion peels were said to contain large amounts of flavonoids (56.31 mg QE/g dw). Yellow onions grown in Korea showed a lot of phenolic and flavonoid components in their ethanolic onion peel extracts, including epicatechin (275 μg/g dm), p-coumaric acid (579.3 μg/g dm), morin (158.7 μg/g dm) and vanillic acid (255 μg/g dm). In comparison to other coloured onion skins, including pearl, red, yellow and white, phenolic compounds (free, esterified and bound) showed that pearl onion skin had the highest amount, followed by red onion skin, yellow onion skin and white onion skin. Nevertheless, the red variety contained more flavonoids (20.22 mg/g of free flavonoids), followed by the pearl variety (19.64 mg/g), and the white variety (0.08 mg/g) had the least. In addition to these varietal and regional variations, the extraction techniques used also had an impact on the amounts of phenolics and flavonoids [86]. The ethanol extraction at 110°C showed the highest concentration of total phenolics and flavonoids (369.48 mg GAE/g extract; 179.88 mg QE/g extract), followed by subcritical water extraction (SWE) (221.67 mg GAE/g extract; 121.48 mg QE/g extract) and hot water extraction (119.55 mg GAE/g extract; 55.52 mg QE/g extract). However, at 165°C, SWE showed the lowest concentration (57.74 mg GAE/g extract; 28.15 mg QE/g extract). The most abundant flavonoids, including kaempferol, quercetin and quercetin 3,4'-diglucoside were found in higher concentration in pearls, followed by red onion skin, yellow onion skin and white onion skin [9]. Furthermore, Lee et al. (2014) characterized the quercetin content in hot water (24.81 mg/g dw), ethanol (63.41 mg/g dw), SWE– 165°C (13.16 mg/g dw) and SWE– 110°C (45.35 mg/g dw) extracts [87]. It was reported that apart from quercetin, the peel extracts (methanolic, ethanolic and acetonic) consist of compounds such as resveratrol, epicatechin, morin, catechin, naringenin, quercitrin, and epigallocatechin gallate. In a high-fat diet, onion peel extract enhanced the mRNA levels of adiponectin from adipose tissue, notably from mesenteric fats, which had a positive impact on controlling the inflammation brought on by obesity and its associated diseases. According to Moon et al. (2013), extract of onion peel (25–100 μg/mL) suppressed the triglyceride content and lipid accumulation in a 3T3-L1 cell culture model [88]. The peel extracts inhibited adipogenesis via downregulating adipocyte protein (AP)-2 markers in adipocytes and upregulating the fatty acid binding protein (FABP)-4 by improving the free fatty acid transport to β-oxidation. When compared to the control (methanol), SWE extracts of onion peel shown superior antibacterial activity against various strains of the pathogenic bacterium Staphylococcus aureus by lowering cell growth by 0.6–1.2 log CFU/mL. But compared to quercetin, the SWE extract of onion peel exhibited slightly less efficacy, because SWE extracts include smaller amounts of quercetin [89]. The structures of some of the important bioactive compounds present in the onion peel are shown in Figure 4.1.

Onion peel quercetin disrupts energy metabolism, cytoplasmic membrane operations and nucleic acid production in order to operate as an antibacterial agent. Another study showed that onion peel extract SWE: 110°C at 0.6 mg/mL concentration had bactericidal effects against Bacillus cereus, while SWE: 160°C had a bacteriostatic effect at 1.2 mg/mL because it included antimicrobial components such quercetin and quercetin oxidation products. By rupturing the cell wall, producing cell lysis and allowing antimicrobial chemicals to enter the cell while leaking cell contents, these substances demonstrated antioxidant capability. Thus, a reaction with bacterial DNA ultimately leads to cell death [90]. Also, inhibition of the Acetylcholinesterase (AChE) and butyrylcholinesterase (BuChE) enzymes was proposed to be able to restore acetylcholine levels and the cholinergic activity of the brain, thereby demonstrating therapeutic capacity in the treatment of Alzheimer's disease [91]. Besides, AChE and BuChE can potentially be inhibited by extracts of onion solid waste (OSW). The management and regulation of Alzheimer's disease is indicated by the AChE and BuChE inhibitory action of ethyl acetate (70.8; 74.6), ethanol (64.5; 68.2) and methanolic (58.2; 60.5) extracts of OSW. The anticancer activity of onion peel extracts against breast cancer was reported by Choe et al. (2020) [92]. Cancerous cells exhibit uncontrolled cell division and abnormalities in apoptotic processes. Natural extracts typically stop cell division or trigger the apoptotic machinery in malignant cells to target cell proliferation. Through inhibition of the phosphatidylinositol 3 kinase (PI3K)-protein kinase B (Akt) signalling pathway in the MDA-MB-231 cell, subfraction 1 (F1) of red onion peel crude ethanolic extract decreased the cell proliferation and enhanced the cell cycle arrest in MDA-MB-231 (triple-negative breast cancer cell line), and this anticancer effect was superior to the natural antineoplastic agent viz. camptothecin.

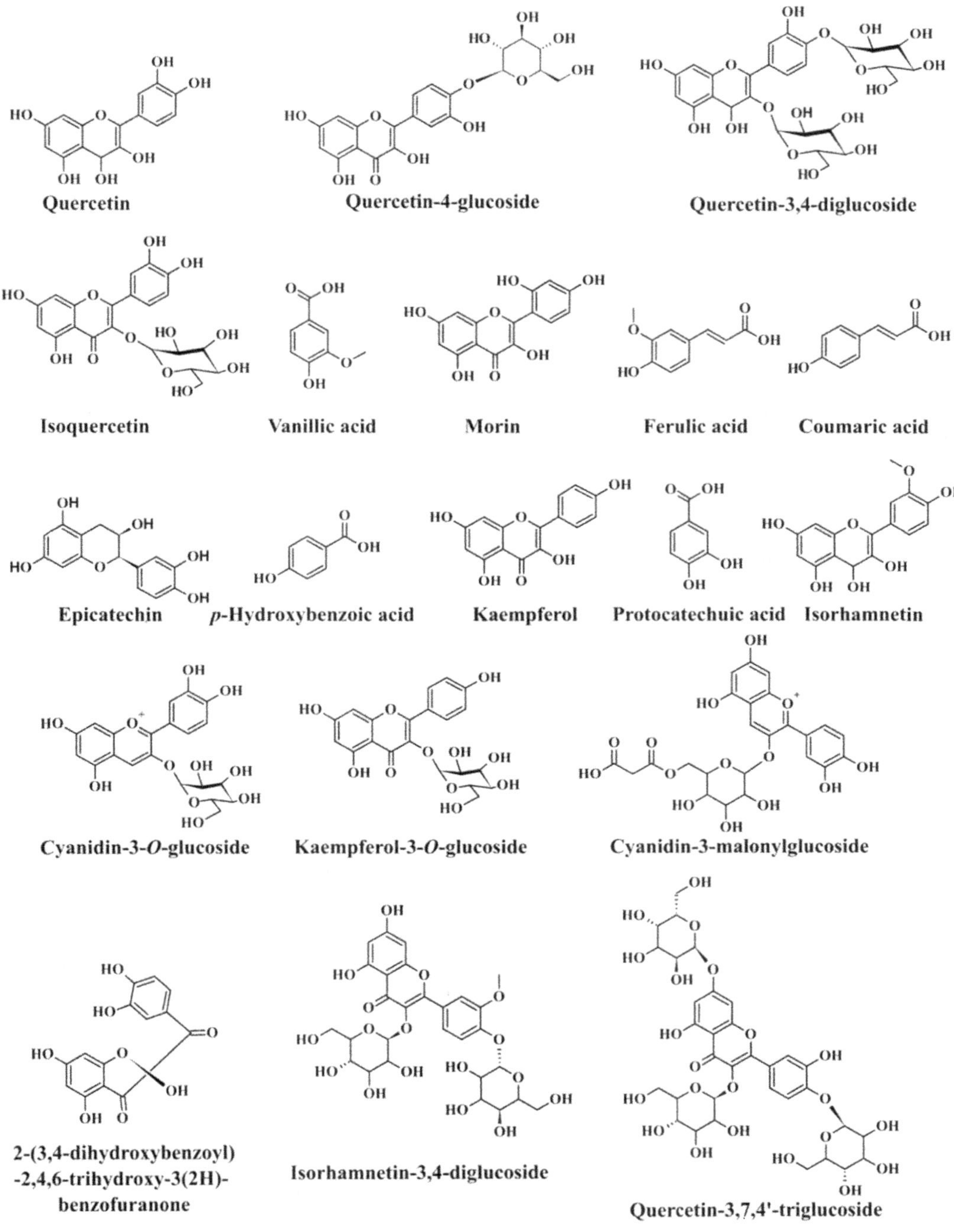

FIGURE 4.1 Structures of bioactive compounds derived from onion peels (Reproduced with permission from Kumar et al. (2022) [9]).

4.3.3 Papaya peel

Papaya (C. papaya L.) is a member of the Caricaceae family, which also includes 35 species and six genera. One of the 35 species (Vasconcellea monoica) is monoecious, two (papaya and V. cundimarsensis) are trioecious and the remaining 32 are dioecious. The majority of papaya production occurs in tropical and subtropical nations, with India, Brazil, Indonesia, Nigeria and Mexico being the top papaya producers. Papaya is used for many different things, but doing so produces a lot of trash, especially during and after harvest. Papaya peels (PaP), which make up around 12% of the weight of the fruit, are the main by-products of papaya processing. Even though papaya peels have medicinal value, they are typically discarded during the fruit preparation process. If PaP are not properly disposed of, they may cause environmental damage. Nevertheless, PaP can be used as a valuable source of bioactive substances, dietary additives, or nutritional supplements. High interest has been sparked by this concept for recovering bioactive chemicals from papaya peels [93].

Several bioactive substances, including carotenoids, phenolic compounds, vitamins A, C and E; pantothenic acid; minerals (potassium and magnesium); folate and fibre are abundant in papaya peel. These substances have a variety of positive health effects on our bodies that are attributed to their antioxidant properties. A source of the digestive enzyme papain is papaya peel. Additionally, papaya peel has valuable bioactive components that can be exploited to generate new foods, pharmaceuticals, nutraceutical supplements and dietary additives. The primary phenolic acids found in papaya peel included ferulic acid (186.63–277.49 mg/100 g), p-coumaric acid (135.64–229.59 mg/100 g) and caffeic acid (112.89–175.51 mg/100 g). Only traces of these acids were observed in the fresh fruit at various stages of ripeness [94]. According to Rivera-Pastrana et al. (2010), papaya pulp only includes traces of protocatechuic acid, gallic acid and caffeic acid, but papaya peel contains high concentrations of quercetin-3-O-rutinoside, caffeic acid, myricetin, protocatechuic acid, quercetin, gallic acid and kaempferol [95]. The primary carotenoids identified in papaya genotypes, viz. yellow and red varieties as well as papaya cv. Pococi, are β-cryptoxanthin myristate (80–268 μg/100 g), β-carotene (218–541 g/100 g), lycopene (10–4266 μg/100 g), β-cryptoxanthin (135–487 g/100 g), β-cryptoxanthin laurate (635–1221 g/199 g) and β-cryptoxanthin caprate (120–536 g/100 g). Furthermore, Ali et al. (2012) reported carotenoid contents in fresh papaya peel, such as lutein (95–321 μg/100 g), β-carotene (127–724 μg/100 g), zeaxanthin (20–28 μg/100 g), α-carotene (17–32 μg/100 g), β-cryptoxanthin (119–3801 μg/100 g) and lycopene (109–4141 μg/100 g) [96]. Dietary fibres are well known for their ability to decrease cholesterol and eliminate toxins from the digestive system, among other health benefits. Papaya peels are made up of 68–87% moisture, 7–20% protein, 0.20–2% fat and 3–12% ash. These peels might be thought of as a source of both macro- and trace minerals. The papaya peels are abundant in minerals (K, Ca, P and Mg), according to studies. Minerals are essential for the healthy operation of the body's physiological and metabolic systems. In fact, the amount of these nutrients decreases as the fruit ripens, making unripe papaya a good source of carbs, proteins and vitamins. The protein, lipid and carbohydrate contents of PaP have been seen to drastically decrease with ripening [93]. Final testing of PaP reveals that carbon (38.10%) and nitrogen (1.49%) are present. When the C-to-N ratio of the raw materials is between 25 and 30:1, microbes can develop and flourish appropriately. The outcome demonstrates that PaP has a healthy C-to-N ratio (26:1), indicating that microbial interactions can take place without difficulty on the surface of PaP. At different stages, papaya is a strong source of vitamin A and has a relatively high vitamin C content. With maturity of the fruit, the amounts of non-nutritive components, such as phenol, alkaloid, flavonoids, tannin and saponin, in PaP diminishes. The saponin in PaP provides unripe fruit its bitter flavour and benefits the upper digestive tract. Papaya peel from both unripe and ripe fruits had 37.9–75.7 mg GAE/g of total phenolics in aqueous and ethanolic extracts [97]. According to Annegowda et al. (2014), the amounts of total flavonoids in aqueous and ethanolic extracts of unripe and ripe papaya were 2.24–3.10 and 1.48–2.86 mg CE/g, respectively [98]. In unripe and ripe papaya peels, the same authors reported total tannins values of approximately 1.25 and 3.75 mg/g, respectively. Alkaloid, which is present in PaP, is advantageous for the heart's general health. Tannin provides a number of health advantages despite being present in extremely small amounts. It was observed that the fruit peel consists of quercetin-3-O-glycoside, feruloyl quinic acid, feryloyl-O-hexoside, chlorogenic acid, caffeoylhexose deoxyhesoxide, 5-hydroxycaffeoyl-O-glycoside and p-coumaroyl quinic acid derivative. Besides, PaP also contain volatile fatty acids, viz. valeric acid (0.5 ± 0.1 g/kg DM), acetate (23.7 ± 1.8 g/kg DM), total volatile fatty acids (36.0 ± 2.2 g/kg DM), propionate (7.6 ± 0.3 g/kg DM), digestible organic matter (716 ± 9 g/kg DM) and butyrate (4.3 ± 0.6 g/ kg DM) [99]. The individual antioxidant characteristics of the three primary phenolic components (ferulic acid, p-coumaric acid and caffeic acid) discovered in papaya peel showed a decrease in the contributions of the caffeic and ferulic acids from 14.98% to 8.09% and from 6.92% to 6.22%, respectively. The p-coumaric acid, on the other hand, went up from 0.86% to 0.94%. These findings suggest that each compound's antioxidant capabilities depend on its structure and concentration within the food matrix [100]. Numerous biochemical changes that impact all the cell components characterize the various ripening stages. The volatile chemicals that produce a distinctive scent and colour and changes in the metabolism of organic acids may be the cause of the distinctive aroma in fruit peels. As they are respired or converted to sugars, the amounts of organic acids that are present in the unripe stage typically decrease during the ripening stage. PaP can be used as an alternative food because it is a source of important minerals and nutrients. Additionally, the peel and leaf have higher mineral levels than the pulp. Polyphenols, fibres, carbs, proteins, fatty acids, tannins and minerals are just a few of the many beneficial substances found in PaP. Due to the presence of these chemicals, PaP is a raw material for numerous goods with added value. Due to its accessibility, polyphenol is suited for use in pharmaceutical applications (e.g., cosmetics) [97]. Soluble polyphenols had organic and conventional levels of 0.55 ± 0.02 and 0.37 ± 0.13, respectively, while hydrolyzable polyphenols have organic and conventional contents of 2.03 ± 0.22 and 0.53 ± 0.06 (mg GAE mL VE-1). PaP also contains flavonoids, such as catechin (740.22 mg/g crude extract) and epicatechin (9.75 mg/g crude extract) [101]. Maximum phenolic and flavonoid extraction can be carried out in the presence of a combination of aqueous and organic solvent, according to Sultana et al. (2009), who noted that the maximum total flavonoid content in 80% ethanolic extract of papaya peel suggested as much [102]. The total amount of polyphenol in PaP extracted using various solvents is as follows, according to Tafese Bezuneh (2015): ethanol, 241.63 ± 3.54; methanol, 207.20 ± 0.08; acetone, 231.90 ± 0.07; and water, 149.58 ± 0.14 (mg GAE/g, dry weight) [103]. In order to extract polyphenols from PaP, the choice of solvent is crucial. An alternative method to extract bioactive compounds from PaP uses pulsed electric fields and high-voltage electrical discharge-assisted extraction, which has advantages, such as shorter extraction times, lower temperature and solvent requirements, improved extraction yields and lower energy consumption when compared to traditional methods [104]. Additionally, the combination of carbon and nitrogen makes it an appropriate substrate for microbes; as a result, PaP is suited for the fermentation-based production of useful goods including enzymes, biogas and methane. Pectin also has a crucial part in the human diet. Consuming pectin aids in lowering blood cholesterol levels. PaP can be removed of pectin in a variety of ways. Pectin can be made through chemical extraction using citric and hydrochloric acids under various time, temperature and pH conditions. Pectin was taken out of PaP by Altaf et al. (2015), who then investigated its physicochemical properties [105]. Pectin was extracted in their investigation using citric acid and hydrochloric acid at various times, temperatures and pH levels to provide total yields that ranged from 2.8–16% and 1.9–9.9%, respectively. On a weight basis, the pectin extracted from PaP using the ethanol and aluminium chloride precipitation extraction procedures are 2.23 and 5.84%, respectively. Compared to ethanol-precipitated pectin (48.39%), aluminium-precipitated pectin (72.43%) contains much more galacturonic acid. While only a little quantity of high methyl-esterified, branching water-soluble pectin could be recovered from diverse kinds of PaP, the different PaP produced low methyl-esterified, linear Ca^{2+}-cross-linked homogalacturonan with a high molar mass [106].

PaP has a high nutritional value and is a good source of vitamins and minerals. The combination of nutrients found in PaP, including minerals, vitamins, phenolic antioxidants and dietary fibre, may be the cause of its protective effects on pathological and physiological abnormalities such cardiovascular disease, inflammation and ageing. According to the findings of epidemiological studies, eating fruits regularly lowers the chance of developing chronic diseases like ageing, cardiovascular diseases and cancer, especially colon cancer [12]. Using the disc diffusion method, the antibacterial activity of several papaya peel extracts was examined against four bacterial species. The study's findings demonstrated that every papaya peel extract was effective against all tested bacteria (E. coli, P. aeruginsa, B. subtilis, S. aureus). Compared to other studied extracts, the absolute ethanolic extract of papaya peel had the highest antibacterial activity; this may be because the active ingredients are well soluble in organic solvents like ethanol [107]. According to Prabhu et al. (2017), the chloroform extract of raw C. papaya peel (25 mg/mL) and the acetone extract of ripe C. papaya peel (25–0.39 mg/mL) both demonstrated vital antidiarrheal action against the gut pathogens [108]. Ripe C. papaya peel extract had significant antidiarrheal action against Plesiomonas shigelloides, with concentrations ranging from 50 mg/mL to 0.39 mg/mL. According to Akindele et al. (2011), the herbal combination DAS-77, which combines young Mangifera indica bark and dried root of the papaya plant, is effective in treating diarrhoea [109]. When DAS-77 was tested on mice, the results revealed that it has antidiarrheal properties. Alkaloids, tannins, flavonoids and saponins are just a few of the many secondary metabolites found in papaya peels. These compounds have been demonstrated to have a notable anti-inflammatory effect. Papain and chymopapain, two proteolytic enzymes found in papaya peels, have been shown to have anti-inflammatory and immunomodulatory properties. Trypsin, chymotrypsin and papain, together with other proteolytic enzymes, lower TGF-1 levels in rheumatoid arthritis, herpes zoster and osteomyelofibrosis [110]. Additionally, the enzyme's (papain) proteolytic activity in dried PaP was around ten times lower than that in latex. To create meat tenderizers, spice batches were combined with 35% and 45% PaP depending on the enzyme's activity. After that, the cooked samples were examined. Because of its good flavour and soft texture, the meat that had been marinated with 25% green PaP at an ambient temperature received the most favourable reviews after two hours. This demonstrates that the green PaP has a lot of promise for usage as a low-cost meat tenderizer when dried, crushed and combined with other spices [97]. Papaya peels are also used to make jam when combined with other ingredients like jelly, pickles, candied fruit, puree, blended beverages, canned slices and chunks, concentrate and spray-dried enzyme, to name a few.

4.3.4 Pumpkin peel

Pumpkin (Cucurbita maxima), of the Cucurbitaceae family, is widely utilized as a functional food and herbal remedy. There are at least 12 different species of pumpkin in the genus Cucurbita, three of which are particularly well-known: Cucurbita pepo, Cucurbita moschata and Cucurbita maxima. Pumpkin is a significant crop in the Cucurbitaceae family. The pharmaceutical, agriculture, food processing and feed industries have recently given greater attention to pumpkin and pumpkin waste due to the availability of nutritive and health-protective phytochemicals in fruit waste, particularly in pumpkin peels. Proteins, carbohydrates, fibre and fats are all crucial nutrients found in pumpkin by-products, along with vitamins and other necessary bioactive substances. As a result, even though these by-products are frequently underutilized and destroyed, it should be highlighted that their use could be advantageous to both health and economic sustainability [111].

Compared to the pulp that is usually ingested, pumpkin peels include a substantial quantity of fibre, protein and minerals like calcium and magnesium, while having lower amounts of carbs, lipids and potassium. Carotenoids, the pigment agents with the ability to improve health, are abundant in pumpkin skin. Pumpkin peel is a good source of biologically active substance. Among other things, carotenoids, pectin, tocopherol, polysaccharides, cucurmosin and polyphenols are among the most significant bioactive substances found in pumpkin peels. The physicochemical, sensory and cooking qualities of beef burger were improved by adding pumpkin peel flour, which has significant mineral and dietary fibre contents. Pumpkin peels have the potential to be used as a manufacturing substrate for oxidative enzymes produced by indigenous fungus strains [11]. During the phytochemical screening of pumpkin peel, it was discovered that both methanolic and ethanolic extracts contained flavonoids and phenolic substances. The total phenolic content of pumpkin peel extract was reported to be 22.88 mg of gallic acid in 1 g of dried extract by Bahramsoltani et al. (2017) [112]. Also, for 2,2-diphenyl-1-picrylhydrazyl (DPPH) radical scavenging activity, the IC 50 value was observed to be 4.010 mg/ ml, while the ferric reduction potential was found to be 143.57 mmol Fe^{2+}/g. Three crown pumpkin waste stream fractions, viz. peel, pulp and seeds were combined to create pumpkin flour by Norfezah et al. (2011) [113]. They divided the pumpkin's fractional makeup into four categories: skin (10–12%), pulp (3–4%), edible flesh (79–82%) and seeds (4–6%). According to research by Mala and Kurian (2016), the phytochemicals present in the pumpkin peels are ascorbic acid (19.15 mg/100 g), β-carotene (12.05 mg/100 g), iron (43.10 mg/100 g) and phosphorus (320.21 mg/100 g) [114]. Pumpkin peel has a TPC value of 5.19 mg GAE/g. Both the TFC and TPC content in the pumpkin peel extract at various methanolic concentration were calculated by Asif et al. (2017) [115]. It was reported that utilization of 80% methanol extract results in the TFC and TPC values of 0.38 mg CE/100 g and 1.77 mg GAE/100 g respectively. The study found that extracts from pumpkin peel demonstrated good antibacterial potential against several bacterial strains when evaluated using the disc diffusion method, with inhibition zone values ranging from 10 mm to 15 mm. Pumpkin peel's significant antibacterial and antioxidant properties suggest that functional foods high in antioxidants may be made with this fruit's peel. The chemical makeup and biological content of powdered pumpkin peel were analyzed by Badr et al. (2011) [116]. The amount of copper, calcium, zinc and iron in pumpkin peel powder was determined, and the corresponding values were found to be 12.91, 5,571, 42.92 and 247.33 mg/100 g dw, respectively. Also, the pumpkin peel contained a β-carotene concentration of 748.85 μg/100 g. When the amino acid content of powdered pumpkin peel was examined, aspartic acid (2.64%) emerged as the amino acid with the highest concentration, followed by glutamic acid (2.48%) and leucine (1.17%). Besides, pumpkin peel contained the highest proportion of tryptophan amongst the various pumpkin by-products. When pumpkin peel extract's antimicrobial activity was examined using the agar disc method, inhibition zones measuring 17 mm in diameter against the gram-positive bacterium Streptomyces viridochromogenes and 14 mm in diameter against the fungus Mucor meihi were found. Also, the dietary fibre present in pumpkin peel, viz. pectin, has been shown to decelerate the digestion of starch, which aids in the treatment of diet-related disorders like diabetes. Pumpkin peel has been found to contain a wide variety of amino acids, viz. phenylalanine, glutamic acid, valine, alanine, serine, isoleucine, arginine, threonine, aspartic acid, tyrosine, histidine, methionine and leucine among others [112]. Song et al. (2018) enhanced the ultrasound-assisted extraction (UAE) in a study using pumpkin peel after identifying the ideal solvent for the traditional extraction of carotenoids, which turned out to be a blend of ethanol and petroleum ether [117]. These scientists assessed the impact of the primary process variables (ultrasonic power, extraction duration and liquid–solid ratio) on carotenoid yield in order to achieve this goal. UAE produced a total carotenoid content of 363 g/g under ideal conditions (205 W, 25 min and a solvent to material ratio of 30 mL/g), increasing yields by 92% above traditional extraction. Furthermore, pectic oligosaccharides were isolated from pumpkin peel and divided into three fractions (water, EDTA and alkali soluble oligosaccharides), with the alkali soluble fraction having the greatest total sugar and uronic acid levels (25 and 46 g/100 g, respectively) [118]. In a different study by Lalnunthari et al. (2019), pectin extraction from pumpkin peel was optimized and the resulting pectin and protein from pumpkin seeds were utilized to create biodegradable films [119]. Additionally, the phenolic content and antioxidant activity of pumpkin peel and seed fractions were examined. According to data published by Saavedra et al. (2015), in the best operational conditions assessed, peels had a higher phenolic content than seeds, with values of 11 and 6.1 mg GAE/g dw, respectively [120]. Peel has been found to have a positive association between the two variables, while seeds do not; this difference may be due to high tocopherol and beta-sitosterol levels in the seeds. The yield of 1,069 mg GAE/100 g dw was obtained under the experimental conditions of temperature 70°C, operating time 0.5 hr and solute/

solvent 1:25 w/v. The study evaluated the effect of various extraction solvents (70% ethanol, 70% methanol, 70% acetone, 100% dichloromethane and ultra-pure water) on the content of phenolic compounds in pumpkin peels. Ascorbic acid, chlorophyll a and chlorophyll b were measured by Blanco-Diaz et al. (2015) in a different investigation [121]. They found that the first one had values between 0.42 and 1.2 mg/g dry weight. The same study also demonstrated that pulp's chlorophyll level was about 21 times lower than that of the peels.

The nutritional profile of pumpkin has been discovered to decelerate dietary-related disorders like diabetes because of the presence of sufficient levels of carbohydrate and dietary fibres, particularly pectin. Research by Rahayu et al. (2020) on diabetic rats produced by alloxan showed that pumpkin peel flour exhibited hypolipidemic and hypoglycaemic effects [122]. One of these widely grown medicinal plants for treating ailments like diabetes is the pumpkin, whose fruits have long been used by people. Pumpkin pulp extracts showed promise as antioxidants and antimicrobials. Pumpkin extracts containing high levels of carotenoids, known as phytocomplexes, are a strong source of bioactive chemicals and slow the proliferation of cells in a human chronic lymphocytic leukaemia cell line. In the disc diffusion method, the ethanol and methanol extracts of pumpkin peel have antibacterial activity with an inhibitory zone of 6–10 mm [123]. Jun et al. (2006) isolated pectic polysaccharides from pumpkin peel and evaluated their impact on the development of human gut microbes [118]. While growth-retarding activity was seen for bad intestinal bacteria like E. coli and C. perfringins, the pectic polysaccharide fractions showed growth-promoting activity for good intestinal bacteria including B. bifidium, B. longum and L. brevis. They came to the conclusion that pumpkin peel's pectic polysaccharides have growth-promoting effects on beneficial gut bacteria and also possess glucose and bile acid-lowering actions, making them suitable as a functional dietary element. Pumpkin peel has also been used to treat burn wounds, gastrointestinal haemorrhage, hepatic problems and peptic ulcers. Pumpkin fruit peel is helpful for hot and dry disorders like burn wounds because of its cool and wet nature. Some formulations of pumpkin peel are advised for the healing of burn wounds [111]. Additionally, pumpkin peel has been tested on mice for its anti-fatigue potential and contains carotenoids and γ-tocopherol. In order to create biscuits with superior physical and sensory qualities, pumpkin peel flour was also evaluated for its immediate composition and functional features. It was then used at various replacement amounts with wheat flour [124]. The flavourful flesh of pumpkin fruits is widely valued as an ingredient in a variety of food products for people of all ages. The most popular products created from processed pumpkin fruit include dried goods, pomade, juice and pickles. Through the process of freeze-drying, pumpkin peel and pulp flours were obtained to use in the creation of cakes at various replacement levels. The evaluation of the sensory and chemical characteristics of pumpkin cakes produced positive findings [125]. Pumpkin peel has begun to be processed into a variety of food products, including bread goods, soups, sauces, extruded goods and spices due to the presence of nutritious components.

4.4 CONCLUSIONS

In order to meet the growing daily demand of consumers, the modern food processing industry is creating and developing a variety of food products, which inevitably leads to the production of food by-products. More efforts should be made to return these by-products to the food supply chain or allied industries with a view to sustainability and the circular economy. Peel waste produced by F&V processing contains a large number of added-value bioactive compounds with considerable potential for use in the food industry and for improving human health. This section explains how valuable phytochemical compounds with desirable properties, primarily consisting of phenolic constituents, carotenoids, dietary fibres, flavonoids and hydrolyzable tannins, are present in F&V peels like banana peel, mango peel, pitaya peel, pomegranate peel, potato peel, onion peel, papaya peel and pumpkin peel. Vitamin C, pectin and fibre are particularly abundant in these peels. The composition of food (especially in terms of fibre, phenolic compounds and carotenoid content) and health benefits were improved by the addition of F&V peel powders (increased prebiotic and antioxidant activity and decreased the glycaemic index). Additionally, it has been shown that F&V peels possess strong nutraceutical properties, including, antioxidant, antimicrobial, antidiabetic, anti-obesogenic, antibacterial, anticancer and antiproliferative properties, among others. Such F&V peels also demonstrated significant promise in the food business (a) as functional food additives with nutritional, antioxidant and antibacterial qualities and (b) film matrixes or active ingredients for food packaging or edible coatings. Nevertheless, despite ongoing research into the phytochemical content and health benefits of F&V peels, more thorough research is needed before such waste products may be turned into a phytomedicine or even an allopathic medicine.

ACKNOWLEDGEMENT

The study is supported by the Indian National Academy of Engineering (INAE/121/AKF/22), Gurgaon, India. The authors are solely responsible for all of the opinions, results and conclusions expressed in this study; INAE's viewpoints are not necessarily reflected in any of these aspects.

REFERENCES

1. K. Kaderides, A. Kyriakoudi, I. Mourtzinos, A.M. Goula, Potential of pomegranate peel extract as a natural additive in foods, *Trends Food Sci. Technol.* 115 (2021) 380–390. https://doi.org/10.1016/j.tifs.2021.06.050.
2. D.R. Berdahl, R.I. Nahas, J.P. Barren, Synthetic and natural antioxidant additives in food stabilization: Current applications and future research, *Oxid. Foods Beverages Antioxid. Appl.* (2010) 272–320. https://doi.org/10.1533/9780857090447.2.272.
3. D. Panwar, A. Saini, P.S. Panesar, H.K. Chopra, Unraveling the scientific perspectives of citrus by-products utilization: Progress towards circular economy, *Trends Food Sci. Technol.* 111 (2021) 549–562. https://doi.org/10.1016/j.tifs.2021.03.018.
4. Z.E. Martins, O. Pinho, I.M.P.L.V.O. Ferreira, Food industry by-products used as functional ingredients of bakery products, *Trends Food Sci. Technol.* 67 (2017) 106–128. https://doi.org/10.1016/j.tifs.2017.07.003.
5. N. Kumar, D. Daniloski, Pratibha, Neeraj, N.M. D'Cunha, N. Naumovski, A.T. Petkoska, Pomegranate peel extract – A natural bioactive addition to novel active edible packaging, *Food Res. Int.* 156 (2022) 111378. https://doi.org/10.1016/j.foodres.2022.111378.
6. H. Jiang, W. Zhang, X. Li, C. Shu, W. Jiang, J. Cao, Nutrition, phytochemical profile, bioactivities and applications in food industry of pitaya (Hylocereus spp.) peels: A comprehensive review, *Trends Food Sci. Technol.* 116 (2021) 199–217. https://doi.org/10.1016/j.tifs.2021.06.040.
7. M.H.A. Jahurul, I.S.M. Zaidul, K. Ghafoor, F.Y. Al-Juhaimi, K.L. Nyam, N.A.N. Norulaini, F. Sahena, A.K.M. Omar, Mango (Mangifera indica L.) by-products and their valuable components: A review, *Food Chem.* 183 (2015) 173–180. https://doi.org/10.1016/j.foodchem.2015.03.046.
8. A. Pereira, M. Maraschin, Banana (Musa spp) from peel to pulp: Ethnopharmacology, source of bioactive compounds and its relevance for human health, *J. Ethnopharmacol.* 160 (2015) 149–163. https://doi.org/10.1016/j.jep.2014.11.008.
9. M. Kumar, M.D. Barbhai, M. Hasan, S. Punia, S. Dhumal, Radha, N. Rais, D. Chandran, R. Pandiselvam, A. Kothakota, M. Tomar, V. Satankar, M. Senapathy, T. Anitha, A. Dey, A.A.S. Sayed, F.M. Gadallah, R. Amarowicz, M. Mekhemar, Onion (Allium cepa L.) peels: A review on bioactive compounds and biomedical activities, *Biomed. Pharmacother.* 146 (2022) 112498. https://doi.org/10.1016/j.biopha.2021.112498.
10. S.L. Sampaio, S.A. Petropoulos, A. Alexopoulos, S.A. Heleno, C. Santos-Buelga, L. Barros, I.C.F.R. Ferreira, Potato peels as sources of functional compounds for the food industry: A review, *Trends Food Sci. Technol.* 103 (2020) 118–129. https://doi.org/10.1016/j.tifs.2020.07.015.
11. X. Rico, B. Gullón, J.L. Alonso, R. Yáñez, Recovery of high value-added compounds from pineapple, melon, watermelon and pumpkin processing by-products: An overview, *Food Res. Int.* 132 (2020) 109086. https://doi.org/10.1016/j.foodres.2020.109086.
12. S. Siddique, S. Nawaz, F. Muhammad, B. Akhtar, B. Aslam, Phytochemical screening and in-vitro evaluation of pharmacological activities of peels of Musa sapientum and Carica papaya fruit, *Nat. Prod. Res.* 32 (2018) 1333–1336. https://doi.org/10.1080/14786419.2017.1342089.

13. H.T. Vu, C.J. Scarlett, Q. V Vuong, Optimization of ultrasound-assisted extraction conditions for recovery of phenolic compounds and antioxidant capacity from banana (Musa cavendish) peel, *J. Food Process. Preserv.* 41 (2017) 13148. https://doi.org/10.1111/jfpp.13148.
14. M.W. Davey, J. Keulemans, R. Swennen, Methods for the efficient quantification of fruit provitamin A contents, *J. Chromatogr. A.* 1136 (2006) 176–184. https://doi.org/10.1016/j.chroma.2006.09.077.
15. H.M. Zaini, J. Roslan, S. Saallah, E. Munsu, N.S. Sulaiman, W. Pindi, Banana peels as a bioactive ingredient and its potential application in the food industry, *J. Funct. Foods.* 92 (2022) 105054. https://doi.org/10.1016/j.jff.2022.105054.
16. K.B. Arun, S. Thomas, T.R. Reshmitha, G.C. Akhil, P. Nisha, Dietary fibre and phenolic-rich extracts from Musa paradisiaca inflorescence ameliorates type 2 diabetes and associated cardiovascular risks, *J. Funct. Foods.* 31 (2017) 198–207. https://doi.org/10.1016/j.jff.2017.02.001.
17. H.T. Vu, C.J. Scarlett, Q.V. Vuong, Phenolic compounds within banana peel and their potential uses: A review, *J. Funct. Foods.* 40 (2018) 238–248. https://doi.org/10.1016/j.jff.2017.11.006.
18. S. Someya, Y. Yoshiki, K. Okubo, Antioxidant compounds from bananas (Musa Cavendish), *Food Chem.* 79 (2002) 351–354. https://doi.org/10.1016/S0308-8146(02)00186-3.
19. F.R. Novak, J.A. Guerra de Almeida, R. De Souza e Silva, Banana peel: A possible source of infection in the treatment of nipple fissures, *J. Pediatr. (Rio. J).* 79 (2003) 221–226. https://doi.org/10.2223/jped.1023.
20. K.E. Heim, A.R. Tagliaferro, D.J. Bobilya, Flavonoid antioxidants: Chemistry, metabolism and structure-activity relationships, *J. Nutr. Biochem.* 13 (2002) 572–584. https://doi.org/10.1016/S0955-2863(02)00208-5.
21. Venkatesh Krishna; Girish Kumar, Antibacterial activity of ethanol extract of Musa Paradisiaca Cv. Puttabale and Musa Cv. Grand Naine, *Asian J. Pharm. Clin. Res.* 6 (2013) 4–7. https://doi.org/10. 1008/s91130-108-0052-3.
22. S.S. Dahham, T.A. Mohamad, Y.M. Tabana, A. Majid, Antioxidant activities and anticancer screening of extracts from banana fruit (Musa sapientum), *Acad. J Cancer Res.* 8 (2015) 28–34. https://doi.org/10.5829/idosi.ajcr.2015.8.2.95162.
23. S. Durgadevi, A. Saravanan, S. Uma, Antioxidant potential and antitumour activities of Nendran banana peels in breast cancer cell line, *Indian J. Pharm. Sci.* 81 (2019) 464–473. https://doi.org/10.36468/pharmaceutical-sciences.531.
24. K.B. Arun, F. Persia, P.S. Aswathy, J. Chandran, M.S. Sajeev, P. Jayamurthy, P. Nisha, Plantain peel - A potential source of antioxidant dietary fibre for developing functional cookies, *J. Food Sci. Technol.* 52 (2015) 6355–6364. https://doi.org/10.1007/s13197-015-1727-1.
25. A.U. Santhoskumar, R. Vaishnavi, T. Karunakaran, N. Jaya Chitra, Studies on mechanical properties and biodegradation of edible food wrapper from banana peel, *Asian J. Adv. Basic Sci.* 07 (2019) 1–4. https://doi.org/10.33980/ajabs.2019.v07i02.001.
26. N.F.K. Sultan, W.L.W. Johari, The development of banana peel/corn starch bioplastic film: A preliminary study, *Bioremediation Sci. Technol. Res.* 5 (2017) 12–17. https://doi.org/10.54987/bstr.v5i1.352.
27. S. Marçal, M. Pintado, Mango peels as food ingredient/additive: Nutritional value, processing, safety and applications, *Trends Food Sci. Technol.* 114 (2021) 472–489. https://doi.org/10.1016/j.tifs.2021.06.012.
28. C.M. Ajila, U.J.S. Prasada Rao, Mango peel dietary fibre: Composition and associated bound phenolics, *J. Funct. Foods.* 5 (2013) 444–450. https://doi.org/10.1016/j.jff.2012.11.017.
29. A.M. Gómez-Caravaca, A. López-Cobo, V. Verardo, A. Segura-Carretero, A. Fernández-Gutiérrez, HPLC-DAD-q-TOF-MS as a powerful platform for the determination of phenolic and other polar compounds in the edible part of mango and its by-products (peel, seed, and seed husk), *Electrophoresis.* 37 (2016) 1072–1084. https://doi.org/10.1002/elps.201500439.
30. R. Pacheco-Ordaz, M. Antunes-Ricardo, J.A. Gutiérrez-Uribe, G.A. González-Aguilar, Intestinal permeability and cellular antioxidant activity of phenolic compounds from mango (Mangifera indica cv. ataulfo) peels, *Int. J. Mol. Sci.* 19 (2018) 21–28. https://doi.org/10.3390/ijms19020514.
31. B. de Ancos, C. Sánchez-Moreno, L. Zacarías, M.J. Rodrigo, S. Sáyago Ayerdí, F.J. Blancas Benítez, J.A. Domínguez Avila, G.A. González-Aguilar, Effects of two different drying methods (freeze-drying and hot air-drying) on the phenolic and carotenoid profile of 'Ataulfo' mango by-products, *J. Food Meas. Charact.* 12 (2018) 2145–2157. https://doi.org/10.1007/s11694-018-9830-4.
32. F.J. Blancas-benitez, G. Mercado-mercado, A.E. Quirós-sauceda, E. Montalvo-gonzález, G.A. González-aguilar, S.G. Sáyago-ayerdi, Food & function dietary fi ber and in vitro kinetics release of polyphenols in Mexican ' Ataulfo ' mango (Mangifera indica L.) by-products, *Food Funct.* 6 (2015) 859–868. https://doi.org/10.1039/c4fo00982g.

33. F.J. Blancas-Benitez, R. de Jesús Avena-Bustillos, E. Montalvo-González, S.G. Sáyago-Ayerdi, T.H. McHugh, Addition of dried 'Ataulfo' mango (Mangifera indica L) by-products as a source of dietary fiber and polyphenols in starch molded mango snacks, *J. Food Sci. Technol.* 52 (2015) 7393–7400. https://doi.org/10.1007/s13197-015-1855-7.
34. C.M. Ajila, S.G. Bhat, U.J.S. Prasada Rao, Valuable components of raw and ripe peels from two Indian mango varieties, *Food Chem.* 102 (2007) 1006–1011. https://doi.org/10.1016/j.foodchem.2006.06.036.
35. K.G. Ranganath, K.S. Shivashankara, T.K. Roy, M.R. Dinesh, G.A. Geetha, K.C.G. Pavithra, K.V. Ravishankar, Profiling of anthocyanins and carotenoids in fruit peel of different colored mango cultivars, *J. Food Sci. Technol.* 55 (2018) 4566–4577. https://doi.org/10.1007/s13197-018-3392-7.
36. J. Ruales, N. Baenas, D.A. Moreno, C.M. Stinco, A.J. Meléndez-Martínez, A. García-Ruiz, Biological active ecuadorian mango 'tommy atkins' ingredients: An opportunity to reduce agrowaste, *Nutrients.* 10 (2018) 1–14. https://doi.org/10.3390/nu10091138.
37. G. Mercado-Mercado, E. Montalvo-González, G.A. González-Aguilar, E. Alvarez-Parrilla, S.G. Sáyago-Ayerdi, Ultrasound-assisted extraction of carotenoids from mango (Mangifera indica L. 'Ataulfo') by-products on in vitro bioaccessibility, *Food Biosci.* 21 (2018) 125–131. https://doi.org/10.1016/j.fbio.2017.12.012.
38. H. Umbreen, M.U. Arshad, F. Saeed, N. Bhatty, A.I. Hussain, Probing the functional potential of agro-industrial wastes in dietary interventions, *J. Food Process. Preserv.* 39 (2015) 1665–1671. https://doi.org/10.1111/jfpp.12396.
39. K. Jalgaonkar, S.K. Jha, M.K. Mahawar, Influence of incorporating defatted soy flour, carrot powder, mango peel powder, and moringa leaves powder on quality characteristics of wheat semolina-pearl millet pasta, *J. Food Process. Preserv.* 42 (2018) 13575. https://doi.org/10.1111/jfpp.13575.
40. J. Ramírez-Maganda, F.J. Blancas-Benítez, V.M. Zamora-Gasga, M. de L. García-Magaña, L.A. Bello-Pérez, J. Tovar, S.G. Sáyago-Ayerdi, Nutritional properties and phenolic content of a bakery product substituted with a mango (Mangifera indica) "Ataulfo" processing by-product, *Food Res. Int.* 73 (2015) 117–123. https://doi.org/10.1016/j.foodres.2015.03.004.
41. Y. Chen, L. Zhao, T. He, Z. Ou, Z. Hu, K. Wang, Effects of mango peel powder on starch digestion and quality characteristics of bread, *Int. J. Biol. Macromol.* 140 (2019) 647–652. https://doi.org/10.1016/j.ijbiomac.2019.08.188.
42. S.R. Kanatt, S.P. Chawla, Shelf life extension of chicken packed in active film developed with mango peel extract, *J. Food Saf.* 38 (2018) e12385. https://doi.org/10.1111/jfs.12385.
43. Y. Wu, J. Xu, Y. He, M. Shi, X. Han, W. Li, X. Zhang, X. Wen, Metabolic profiling of pitaya (hylocereus polyrhizus) *during fruit development and maturation, Molecules.* 24 (2019) 1–16. https://doi.org/10.3390/molecules24061114.
44. G.C. Tenore, E. Novellino, A. Basile, Nutraceutical potential and antioxidant benefits of red pitaya (Hylocereus polyrhizus) extracts, *J. Funct. Foods.* 4 (2012) 129–136. https://doi.org/10.1016/j.jff.2011.09.003.
45. H.A.R. Suleria, C.J. Barrow, F.R. Dunshea, Screening and characterization of phenolic compounds and their antioxidant capacity in different fruit peels, *Foods.* 9 (2020) 13–19. https://doi.org/10.3390/foods9091206.
46. F. Ferreres, C. Grosso, A. Gil-Izquierdo, P. Valentão, A.T. Mota, P.B. Andrade, Optimization of the recovery of high-value compounds from pitaya fruit by-products using microwave-assisted extraction, *Food Chem.* 230 (2017) 463–474. https://doi.org/10.1016/j.foodchem.2017.03.061.
47. H. Kim, H.-K. Choi, J.Y. Moon, Y.S. Kim, A. Mosaddik, S.K. Cho, Comparative antioxidant and antiproliferative activities of red and white pitayas and their correlation with flavonoid and polyphenol content, *J. Food Sci.* 76 (2011) 38-C45. https://doi.org/10.1111/j.1750-3841.2010.01908.x.
48. F. Fathordoobady, H. Mirhosseini, J. Selamat, M.Y.A. Manap, Effect of solvent type and ratio on betacyanins and antioxidant activity of extracts from Hylocereus polyrhizus flesh and peel by supercritical fluid extraction and solvent extraction, *Food Chem.* 202 (2016) 70–80. https://doi.org/10.1016/j.foodchem.2016.01.121.
49. G.V.S. Bhagya Raj, K.K. Dash, Ultrasound-assisted extraction of phytocompounds from dragon fruit peel: Optimization, kinetics and thermodynamic studies, *Ultrason. Sonochem.* 68 (2020) 105180. https://doi.org/10.1016/j.ultsonch.2020.105180.
50. H. Luo, G. Xiong, K. Ren, S.R. Raman, Z. Liu, Q. Li, C. Ma, D. Li, Y. Wan, Air DBD plasma treatment on three-dimensional braided carbon fiber-reinforced PEEK composites for enhancement of in vitro bioactivity, *Surf. Coatings Technol.* 242 (2014) 1–7. https://doi.org/10.1016/j.surfcoat.2013.12.069.
51. X. Lin, H. Gao, Z. Ding, R. Zhan, Z. Zhou, J. Ming, Comparative metabolic profiling in pulp and peel of green and red pitayas (*Hylocereus polyrhizus* and *Hylocereus undatus*) reveals potential valorization in the pharmaceutical and food industries, *Biomed Res. Int.* 2021 (2021) 6546170. https://doi.org/10.1155/2021/6546170.

52. M.M. Nurmahani, A. Osman, A. Abdul Hamid, F. Mohamad Ghazali, M.S. Pak Dek, Short communication antibacterial property of hylocereus polyrhizus and hylocereus undatus peel extracts, *Int. Food Res. J.* 19 (2012) 77–84. https://doi.org/10.5240/sh10191151.
53. R. Faridah, A. Mangalisu, F. Maruddin, Antioxidant effectiveness and pH value of red dragon fruit skin powder (Hylocereus polyrhizus) on pasteurized milk with different storage times, *IOP Conf. Ser. Earth Environ. Sci.* 492 (2020) 25–32. https://doi.org/10.1088/1755-1315/492/1/012051.
54. S.-Y. Shiau, G.-H. Li, W.-C. Pan, C. Xiong, Effect of pitaya peel powder addition on the phytochemical and textural properties and sensory acceptability of dried and cooked noodles, *J. Food Process. Preserv.* 44 (2020) 14491. https://doi.org/10.1111/jfpp.14491.
55. L.H. Ho, N.W. Binti Abdul Latif, Nutritional composition, physical properties, and sensory evaluation of cookies prepared from wheat flour and pitaya (Hylocereus undatus) peel flour blends, *Cogent Food Agric.* 2 (2016) 12–19. https://doi.org/10.1080/23311932.2015.1136369.
56. T. Ismail, P. Sestili, S. Akhtar, Pomegranate peel and fruit extracts: A review of potential anti-inflammatory and anti-infective effects, *J. Ethnopharmacol.* 143 (2012) 397–405. https://doi.org/10.1016/j.jep.2012.07.004.
57. M. Aviram, N. Volkova, R. Coleman, M. Dreher, M.K. Reddy, D. Ferreira, M. Rosenblat, Pomegranate phenolics from the peels, arils, and flowers are antiatherogenic: Studies in vivo in atherosclerotic apolipoprotein E-deficient (E0) mice and in vitro in cultured macrophages and lipoproteins, *J. Agric. Food Chem.* 56 (2008) 1148–1157. https://doi.org/10.1021/jf071811q.
58. K. Kaderides, I. Mourtzinos, A.M. Goula, Stability of pomegranate peel polyphenols encapsulated in orange juice industry by-product and their incorporation in cookies, *Food Chem.* 310 (2020) 125849. https://doi.org/10.1016/j.foodchem.2019.125849.
59. M.I. Gil, F.A. Tomas-Barberan, B. Hess-Pierce, D.M. Holcroft, A.A. Kader, Antioxidant activity of pomegranate juice and its relationship with phenolic composition and processing, *J. Agric. Food Chem.* 48 (2000) 4581–4589. https://doi.org/10.1021/jf000404a.
60. G.J. Kelloff, C.W. Boone, J.A. Crowell, V.E. Steele, R. Lubet, C.C. Sigman, Chemopreventive drug development: Perspectives and progress, *Cancer Epidemiol. Biomarkers Prev.* 3 (1994) 85–98. https://doi.org/10.1053/jf010406b.
61. N.P. Seeram, R. Lee, D. Heber, Bioavailability of ellagic acid in human plasma after consumption of ellagitannins from pomegranate (Punica granatum L.) juice, *Clin. Chim. Acta.* 348 (2004) 63–68. https://doi.org/10.1016/j.cccn.2004.04.029.
62. P. Kandylis, E. Kokkinomagoulos, Food applications and potential health benefits of pomegranate and its derivatives, *Foods.* 9 (2020) 31–37. https://doi.org/10.3390/foods9020122.
63. M.S. Nair, A. Saxena, Characterization and antifungal activity of pomegranate peel extract and its use in polysaccharide-based edible coatings to extend the shelf-life of capsicum (Capsicum annuum L.), *Food Bioprocess Technol.* 11 (2018) 1317–1327. https://doi.org/doi.org/10.1007/s11947-018-2101-x.
64. M.A.L. Jeune, J. Kumi-Diaka, J. Brown, Anticancer activities of pomegranate extracts and genistein in human breast cancer cells, *J. Med. Food.* 8 (2005) 469–475. https://doi.org/10.1089/jmf.2005.8.469.
65. S. Naz, R. Siddiqi, S. Ahmad, S.A. Rasool, S.A. Sayeed, Antibacterial activity directed isolation of compounds from Punica granatum, *J. Food Sci.* 72 (2007) M341–M345. https://doi.org/10.1111/j.1750-3841.2007.00533.x.
66. S. Sandhya, K. Khamrui, W. Prasad, M.C.T. Kumar, Preparation of pomegranate peel extract powder and evaluation of its effect on functional properties and shelf life of curd, *Lwt-Food Sci. Technol.* 92 (2018) 416–421. https://doi.org/10.1016/j.lwt.2018.02.057.
67. S.R. Kanatt, R. Chander, A. Sharma, Antioxidant and antimicrobial activity of pomegranate peel extract improves the shelf life of chicken products, *Int. J. Food Sci. Technol.* 45 (2010) 216–222. https://doi.org/10.1111/j.1365-2621.2009.02124.x.
68. B.A. Wafa, M. Makni, S. Ammar, L. Khannous, A. Ben Hassana, M. Bouaziz, N.E. Es-Safi, R. Gdoura, Antimicrobial effect of the Tunisian Nana variety Punica granatum L. extracts against Salmonella enterica (serovars Kentucky and Enteritidis) isolated from chicken meat and phenolic composition of its peel extract, *Int. J. Food Microbiol.* 241 (2017) 123–131. https://doi.org/10.1016/j.ijfoodmicro.2016.10.007.
69. J.P. Trigo, E.M.C. Alexandre, A. Oliveira, J.A. Saraiva, M. Pintado, Fortification of carrot juice with a high-pressure-obtained pomegranate peel extract: Chemical, safety and sensorial aspects, *Int. J. Food Sci. Technol.* 55 (2020) 1599–1605. https://doi.org/10.1111/ijfs.14386.
70. M. Friedman, V. Huang, Q. Quiambao, S. Noritake, J. Liu, O. Kwon, S. Chintalapati, J. Young, C.E. Levin, C. Tam, L.W. Cheng, K.M. Land, Potato peels and their bioactive glycoalkaloids and phenolic compounds inhibit the growth of pathogenic trichomonads, *J. Agric. Food Chem.* 66 (2018) 7942–7947. https://doi.org/10.1021/acs.jafc.8b01726.

71. A. Oertel, A. Matros, A. Hartmann, P. Arapitsas, K.J. Dehmer, S. Martens, H.P. Mock, Metabolite profiling of red and blue potatoes revealed cultivar and tissue specific patterns for anthocyanins and other polyphenols, *Planta.* 246 (2017) 281–297. https://doi.org/10.1007/s00425-017-2718-4.
72. Y. Riciputi, E. Diaz-de-Cerio, H. Akyol, E. Capanoglu, L. Cerretani, M.F. Caboni, V. Verardo, Establishment of ultrasound-assisted extraction of phenolic compounds from industrial potato by-products using response surface methodology, *Food Chem.* 269 (2018) 258–263. https://doi.org/10.1016/j.foodchem.2018.06.154.
73. Y.L. Hsieh, Y.H. Yeh, Y.T. Lee, C.Y. Huang, Dietary potato peel extract reduces the toxicity of cholesterol oxidation products in rats, *J. Funct. Foods.* 27 (2016) 461–471. https://doi.org/10.1016/j.jff.2016.09.019.
74. N.P. Silva-Beltrán, Phenolic compounds of potato peel extracts: Their antioxidant activity and protection against human enteric viruses, *J. Microbiol. Biotechnol.* 27 (2017) 234–241. https://doi.org/10.4014/jmb.1606.06007.
75. T. Albishi, J.A. John, A.S. Al-Khalifa, F. Shahidi, Phenolic content and antioxidant activities of selected potato varieties and their processing by-products, *J. Funct. Foods.* 5 (2013) 590–600. https://doi.org/10.1016/j.jff.2012.11.019.
76. B. Kumari, B.K. Tiwari, M.B. Hossain, D.K. Rai, N.P. Brunton, Ultrasound-assisted extraction of polyphenols from potato peels: Profiling and kinetic modelling, *Int. J. Food Sci. Technol.* 52 (2017) 1432–1439. https://doi.org/10.1111/ijfs.13404.
77. A. Singh, K. Sabally, S. Kubow, D.J. Donnelly, Y. Gariepy, V. Orsat, G.S.V. Raghavan, Microwave-assisted extraction of phenolic antioxidants from potato peels, *Molecules.* 16 (2011) 2218–2232. https://doi.org/10.3390/molecules16032218.
78. S. Elkahoui, G.E. Bartley, W.H. Yokoyama, M. Friedman, Dietary supplementation of potato peel powders prepared from conventional and organic russet and non-organic gold and red potatoes reduces weight gain in mice on a high-fat diet, *J. Agric. Food Chem.* 66 (2018) 6064–6072. https://doi.org/10.1021/acs.jafc.8b01987.
79. E. Curti, E. Carini, A. Diantom, E. Vittadini, The use of potato fibre to improve bread physico-chemical properties during storage, *Food Chem.* 195 (2016) 64–70. https://doi.org/10.1016/j.foodchem.2015.03.092.
80. L.M. Crawford, T.S. Kahlon, S.C. Wang, M. Friedman, Acrylamide content of experimental flatbreads prepared from potato, quinoa, and wheat flours with added fruit and vegetable peels and mushroom powders, *Foods.* 8 (2019) 5–7. https://doi.org/10.3390/foods8070228.
81. S.F. Koduvayur Habeebullah, N.S. Nielsen, C. Jacobsen, Antioxidant activity of potato peel extracts in a fish-rapeseed oil mixture and in oil-in-water emulsions, *J. Am. Oil Chem. Soc.* 87 (2010) 1319–1332. https://doi.org/10.1007/s11746-010-1611-0.
82. V.K. Juneja, M. Friedman, T.B. Mohr, M. Silverman, S. Mukhopadhyay, Control of Bacillus cereus spore germination and outgrowth in cooked rice during chilling by nonorganic and organic apple, orange, and potato peel powders, *J. Food Process. Preserv.* 42 (2018) 13558. https://doi.org/doi.org/10.1111/jfpp.13558.
83. N. Marefati, V. Ghorani, F. Shakeri, M. Boskabady, F. Kianian, R. Rezaee, M.H. Boskabady, A review of anti-inflammatory, antioxidant, and immunomodulatory effects of Allium cepa and its main constituents, *Pharm. Biol.* 59 (2021) 287–302. https://doi.org/10.1080/13880209.2021.1874028.
84. L. Campone, R. Celano, A.L. Piccinelli, I. Pagano, S. Carabetta, R. Di Sanzo, M. Russo, E. Ibañez, A. Cifuentes, L. Rastrelli, Response surface methodology to optimize supercritical carbon dioxide/co-solvent extraction of brown onion skin by-product as source of nutraceutical compounds, *Food Chem.* 269 (2018) 495–502. https://doi.org/10.1016/j.foodchem.2018.07.042.
85. S.C.M. Burri, A. Ekholm, Å. Håkansson, E. Tornberg, K. Rumpunen, Antioxidant capacity and major phenol compounds of horticultural plant materials not usually used, *J. Funct. Foods.* 38 (2017) 119–127. https://doi.org/10.1016/j.jff.2017.09.003.
86. T. Albishi, J.A. John, A.S. Al-Khalifa, F. Shahidi, Antioxidative phenolic constituents of skins of onion varieties and their activities, *J. Funct. Foods.* 5 (2013) 1191–1203. https://doi.org/10.1016/j.jff.2013.04.002.
87. K.A. Lee, K.T. Kim, H.J. Kim, M.S. Chung, P.S. Chang, H. Park, H.D. Pai, Antioxidant activities of onion (Allium cepa L.) peel extracts produced by ethanol, hot water, and subcritical water extraction, *Food Sci. Biotechnol.* 23 (2014) 615–621. https://doi.org/10.1007/s10068-014-0084-6.
88. J. Moon, H.J. Do, O.Y. Kim, M.J. Shin, Antiobesity effects of quercetin-rich onion peel extract on the differentiation of 3T3-L1 preadipocytes and the adipogenesis in high fat-fed rats, *Food Chem. Toxicol.* 58 (2013) 347–354. https://doi.org/10.1016/j.fct.2013.05.006.

89. K.A. Lee, K.T. Kim, S.Y. Nah, M.S. Chung, S.W. Cho, H.D. Paik, Antimicrobial and antioxidative effects of onion peel extracted by the subcritical water, *Food Sci. Biotechnol.* 20 (2011) 543–548. https://doi.org/10.1007/s10068-011-0076-8.
90. W.J. Kim, K.A. Lee, K.T. Kim, M.S. Chung, S.W. Cho, H.D. Paik, Antimicrobial effects of onion (Allium cepa L.) peel extracts produced via subcritical water extraction against Bacillus cereus strains as compared with ethanolic and hot water extraction, *Food Sci. Biotechnol.* 20 (2011) 1101–1106. https://doi.org/10.1007/s10068-011-0149-8.
91. N.H. Greig, D.K. Lahiri, K. Sambamurti, Butyrylcholinesterase: An important new target in Alzheimer's disease therapy, *Int. Psychogeriatrics.* 14 Supplement 1 (2002) 77–91. https://doi.org/10.1017/S1041610203008676.
92. Sin Pei Choe; Poh Chiew Siah, The cellular activities of the subfraction of red onion peel crude ethanolic extract in MDA-MB-231 cells, *Pharmacognosy Res.* 10 (2020) 24–30. https://doi.org/10.4103/pr.pr_20_20.
93. J.M. Dotto, S.A. Abihudi, Nutraceutical value of Carica papaya: A review, *Sci. African.* 13 (2021) e00933. https://doi.org/10.1016/j.sciaf.2021.e00933.
94. L.E. Gayosso-García Sancho, E.M. Yahia, G.A. González-Aguilar, Identification and quantification of phenols, carotenoids, and vitamin C from papaya (Carica papaya L., cv. Maradol) fruit determined by HPLC-DAD-MS/MS-ESI, *Food Res. Int.* 44 (2011) 1284–1291. https://doi.org/10.1016/j.foodres.2010.12.001.
95. D.M. Rivera-Pastrana, E.M. Yahia, G.A. González-Aguilar, Phenolic and carotenoid profiles of papaya fruit (Carica papaya L.) and their contents under low temperature storage, *J. Sci. Food Agric.* 90 (2010) 2358–2365. https://doi.org/10.1002/jsfa.4092.
96. A. Ali, S. Devarajan, M.I. Waly, M.M. Essa, M.S. Rahman, Nutritional and medicinal values of papaya (Carica papaya L.), *Nat. Prod. Their Act. Compd. Dis. Prev.* 3 (2012) 307–324. https://doi.org/10.20959/wjpps20178-9947.
97. P.D. Pathak, S.A. Mandavgane, B.D. Kulkarni, Waste to wealth: A case study of papaya peel, *Waste Biomass Valorization.* 10 (2019) 1755–1766. https://doi.org/10.1007/s12649-017-0181-x.
98. H.V. Annegowda, R. Bhat, K.J. Yeong, M.T. Liong, A.A. Karim, S.M. Mansor, Influence of drying treatments on polyphenolic contents and antioxidant properties of raw and ripe papaya (Carica papaya L.), *Int. J. Food Prop.* 17 (2014) 283–292. https://doi.org/10.1080/10942912.2011.631248.
99. T. Negesse, Nutritional composition, volatile fatty acids production, digestible organic matter and anti-nutritional factors of some agro-industrial by-products of ethiopia, Ethiop, *J. Sci.* 32 (2009) 149–156. https://doi.org/10.8870/3018-0696316.
100. L.E.G.-G. Sancho, E.M. Yahia, G.A. González-Aguilar, Contribution of major hydrophilic and lipophilic antioxidants from papaya fruit to total antioxidant capacity, *Food Nutr. Sci.* 04 (2013) 93–100. https://doi.org/10.4236/fns.2013.48a012.
101. A.L.K. Faller, E. Fialho, Polyphenol content and antioxidant capacity in organic and conventional plant foods, *J. Food Compos. Anal.* 23 (2010) 561–568. https://doi.org/10.1016/j.jfca.2010.01.003.
102. B. Sultana, F. Anwar, M. Ashraf, Effect of extraction solvent/technique on the antioxidant activity of selected medicinal plant extracts, *Molecules.* 14 (2009) 2167–2180. https://doi.org/10.3390/molecules14062167.
103. T. Tafese Bezuneh, UV - Visible spectrophotometric quantification of total polyphenol in selected fruits, *Int. J. Nutr. Food Sci.* 4 (2015) 397. https://doi.org/10.11648/j.ijnfs.20150403.28.
104. O. Parniakov, F.J. Barba, N. Grimi, N. Lebovka, E. Vorobiev, Impact of pulsed electric fields and high voltage electrical discharges on extraction of high-added value compounds from papaya peels, *Food Res. Int.* 65 (2014) 337–343. https://doi.org/10.1016/j.foodres.2014.09.015.
105. U. Altaf, G. Immanuel, F. Iftikhar, Extraction and characterization of pectin derived from papaya (Carica Papaya Linn.) peel, in: *Int. J. Sci. Eng. Technol.*, 2015: pp. 15–23. https://doi.org/10.4236/dlt.2015.48a303.
106. D. Boonrod, K. Reanma, H. Niamsup, Extraction and physicochemical characteristics of acid-soluble pectin from raw papaya (Carica papaya) peel, *Chiang Mai J. Sci.* 33 (2006) 129–135. https://doi.org/10.1016/j.chngscf.2021.e01723.
107. H.J. De Boer, A. Kool, A. Broberg, W.R. Mziray, I. Hedberg, J.J. Levenfors, Anti-fungal and anti-bacterial activity of some herbal remedies from Tanzania, *J. Ethnopharmacol.* 96 (2005) 461–469. https://doi.org/10.1016/j.jep.2004.09.035.
108. A.K. Prabhu, S.M. Devadas, R. Lobo, P. Udupa, K. Chawla, M. Ballal, Antidiarrheal activity and phytochemical analysis of Carica papaya fruit extract, *J. Pharm. Sci. Res.* 9 (2017) 1151–1155. https://doi.org/10.4276/jpse.2017.48a052.

109. A.J. Akindele, O. Awodele, A.A. Alagbaoso, O.O. Adeyemi, Antidiarrhoeal activity of DAS-77 (a herbal preparation), Nig. Q. J. *Hosp. Med.* 21 (2011) 317–323. https://doi.org/10.5950/3025-069552.
110. L. Desser, D. Holomanova, E. Zavadova, K. Pavelka, T. Mohr, I. Herbacek, Oral therapy with proteolytic enzymes decreases excessive TGF-β levels in human blood, *Cancer Chemother. Pharmacol. Suppl.* 47 Supplement (2001) 10–15. https://doi.org/10.1007/s002800170003.
111. A. Hussain, T. Kausar, S. Sehar, A. Sarwar, A. Haseeb, M. Abdullah, S. Noreen, A. Rafique, K. Iftikhar, M. Yousaf, J. Aslam, M. Abid, A Comprehensive review of functional ingredients , especially bioactive compounds present in pumpkin peel , flesh and seeds , and their health benefits, *Food Chem. Adv.* 1 (2022) 100067. https://doi.org/10.1016/j.focha.2022.100067.
112. R. Bahramsoltani, M.H. Farzaei, A.H. Abdolghaffari, R. Rahimi, N. Samadi, M. Heidari, M. Esfandyari, M. Baeeri, G. Hassanzadeh, M. Abdollahi, S. Soltani, A. Pourvaziri, G. Amin, Evaluation of phytochemicals, antioxidant and burn wound healing activities of Cucurbita moschata Duchesne fruit peel, *Iran. J. Basic Med. Sci.* 20 (2017) 799–806. https://doi.org/10.22038/ijbms.2017.9015.
113. M.N. Norfezah, A. Hardacre, C.S. Brennan, Comparison of waste pumpkin material and its potential use in extruded snack foods, *Food Sci. Technol. Int.* 17 (2011) 367–373. https://doi.org/10.1177/1082013210382484.
114. K.S. Mala, A.E. Kurian, Nutritional composition and antioxidant activity of pumpkin wastes, *Int. J. Pharma. Chem. Bio. Sci.* 6 (2016) 336–344. https://doi.org/10.1016/j.ijpcbs.2016.02.052.
115. M. Asif, S.A.R. Naqvi, T.A. Sherazi, M. Ahmad, A.F. Zahoor, S.A. Shahzad, Z. Hussain, H. Mahmood, N. Mahmood, Antioxidant, antibacterial & antiproliferative activities of pumpkin (cucurbit) peel & puree extracts -An in vitro study, *Pak. J. Pharm. Sci.* 30 (2017) 1327–1334. https://doi.org/10.1041/jps010231a.
116. S.E.A. Badr, M. Shaaban, Y.M. Elkholy, M.H. Helal, A.S. Hamza, M.S. Masoud, M.M. El Safty, Chemical composition and biological activity of ripe pumpkin fruits (Cucurbita pepo L.) cultivated in Egyptian habitats, *Nat. Prod. Res.* 25 (2011) 1524–1539. https://doi.org/10.1080/14786410903312991.
117. J. Song, Q. Yang, W. Huang, Y. Xiao, D. Li, C. Liu, Optimization of trans lutein from pumpkin (Cucurbita moschata) peel by ultrasound-assisted extraction, *Food Bioprod. Process.* 107 (2018) 104–112. https://doi.org/10.1016/j.fbp.2017.10.008.
118. H. Il Jun, C.H. Lee, G.S. Song, Y.S. Kim, Characterization of the pectic polysaccharides from pumpkin peel, *LWT - Food Sci. Technol.* 39 (2006) 554–561. https://doi.org/10.1016/j.lwt.2005.03.004.
119. C. Lalnunthari, L.M. Devi, L.S. Badwaik, Extraction of protein and pectin from pumpkin industry by-products and their utilization for developing edible film, *J. Food Sci. Technol.* 57 (2020) 1807–1816. https://doi.org/10.1007/s13197-019-04214-6.
120. M.J. Saavedra, A. Aires, C. Dias, J.A. Almeida, M.C.B.M. De Vasconcelos, P. Santos, E.A. Rosa, Evaluation of the potential of squash pumpkin by-products (seeds and shell) as sources of antioxidant and bioactive compounds, *J. Food Sci. Technol.* 52 (2015) 1008–1015. https://doi.org/10.1007/s13197-013-1089-5.
121. M.T. Blanco-Díaz, R. Font, D. Martínez-Valdivieso, M. Del Río-Celestino, Diversity of natural pigments and phytochemical compounds from exocarp and mesocarp of 27 Cucurbita pepo accessions, *Sci. Hortic.* 197 (2015) 357–365. https://doi.org/10.1016/j.scienta.2015.09.064.
122. M. Rahayu, M. Kasiyati, A. Martsiningsih, B. Setiawan, F. Khasanah, Hypoglicemic and antioxidant activity of yellow pumpkin (Curcubitamoschata) in diabetic rats, *Indian J. Public Heal. Res. & Dev.* 11 (2020) 1300–1304. https://doi.org/doi.org/10.37506/v11%2Fi1%2F2020%2Fijphrd%2F194022.
123. T. Xia, Q. Wang, Hypoglycaemic role of Cucurbita ficifolia (Cucurbitaceae) fruit extract in streptozotocin-induced diabetic rats, *J. Sci. Food Agric.* 87 (2007) 1753–1757. https://doi.org/doi.org/10.1002/jsfa.2916.
124. S. Abdulaali, S. George, Preparation of Pumpkin Pulp and Peel Flour and Study Their Impact in the Biscuit Industry, *J. Biol. Agric. Healthc.* 10 (2020) 111–117. https://doi.org/10.7176/jbah/10-6-05.
125. A. Nawirska, A. Figiel, A.Z. Kucharska, A. Sokół-Łetowska, A. Biesiada, Drying kinetics and quality parameters of pumpkin slices dehydrated using different methods, *J. Food Eng.* 94 (2009) 14–20. https://doi.org/10.1016/j.jfoodeng.2009.02.025.

5 Utilization of seeds for the synthesis of bioactive compounds

5.1 INTRODUCTION

By-products of the food industry are a significant source of pollution on a worldwide scale. Food losses and waste, which account for over 1.3 billion tonnes of food annually—or 16% of the world's food supply—occur throughout the food supply chain. Food losses of 20–40% for fruits and vegetables start with the beginning agricultural output and continue through all stages of production, all the way to the ultimate consumer. Fruit and vegetable processing generates a broad variety of by-products, notably in the juice business, viz. peels, leaves, undesirable seeds, stones, pulp and cull fruits. The majority of this material is ultimately thrown away [1]. By raising the demand for post-harvest and agro-industrial by-products that may be employed as sources of bioactive compounds with antioxidant qualities and applied in the generation of functional goods, the transformation of such by-products into value-added food items might offset losses. Numerous investigations on fruit and vegetable by-products (FVBP) have reported a broad range of phytochemicals, viz. phenolics, flavonoids, anthocyanins, carotenoids and vitamin C. FVBP can take the form of peels, seeds, flowers, leaves, stems and pomace, among others. These substances may also be utilized in pharmaceutical and medical goods as nutraceuticals. Some FVBP may be used as natural additives in the pharmaceutical, biotechnology and food industrial sectors owing to their bioactive chemical content and biological activity in humans [2, 3].

Despite the fact that other plant components contain bioactive substances, this chapter will concentrate on the seed extract because there is not much research on the topic. The use of seeds would greatly decrease issues with ultimate disposal into the environment because seeds are considered to be the highest phytochemical containing components in fruit and vegetables and are typically regarded to be trash in the agro-industrial sectors. Depending on the kind of seed, the extraction processes used to obtain initial crude extracts include a complex blend of different plant metabolites, viz. flavonoids, alkaloids, fatty acids and phenolic acids [4]. The processing of tomato-based goods produces tomato (Solanum lycopersicum L.) seed waste, which is composed of 5–10% tomato pomace and is a rich source of nutrients such as proteins, amino acids, fatty acids, fibre and functional compounds with remarkable nutraceutical qualities [5]. Rich in minerals, grape seeds are valuable both nutritionally and medicinally. Fifteen phenolic compounds, including flavonols, proanthocyanidin, (–)-epicatechin, (+)-catechin and kaempferol, among others, are the primary components of grape seed extract (GSE). The biological properties of the active components in grape seed extract include neuroprotection, antioxidant, anticancer, anti-inflammation and lowering of hypertension and cholesterol [6]. Date palm seeds are high in phenolics, phytosterols, carotenoids, dietary fibre, anthocyanin and tocopherols, among other phytochemicals. Date seed can be used to make decaffeinated coffee or as an ingredient in animal feed. Date seeds are a good source of dietary fibre and polyphenols, which help to maintain the health of the digestive system [7]. Pomegranate seeds are regarded as a good source of phenolic compounds and high-quality oil (10–25%). Conjugated linolenic acids, particularly phospholipids, tocopherols, triterpenes, punicic acid and phytosterols are all abundant in pomegranate seed oil. The pomegranate seed's secondary and primary metabolites, such as polyphenols, which include fatty acids, hydrolyzable tannins and lipids, were thought

DOI: 10.1201/9781003315469-5

to be responsible for the majority of the therapeutic benefits [8]. Moisture, ash, protein, nitrogen-free extract, starch and fat are the main components of avocado seeds (lower fat content than the pulp). Neutral lipids make up 77–80% of the lipids in seeds, whereas glycolipids and phospholipids account for 7.4% and 10.9%, respectively. Avocado seeds also include dietary fibre, a range of polyphenols such as epicatechin, epigallocatechin, kaempferol, procyanidins, quercetin and cinnamic acid, among others [9]. Around 30% of the oil in papaya seeds is made up of mostly palmitic, stearic, oleic and linoleic acids, as well as tocopherols and carotenoids with beneficial nutritional and functional qualities. This residue has been utilized as a raw material for a number of processes, including the manufacturing of biodiesel, nutritional supplements and yeast substrate. Other phytochemicals/bioactive compounds found in papaya seeds include anzymemyosin, benzylisothyocyanate (BITC), carpaine, caricin (sinigrin) and glucotropacolin [10].

The importance and benefits of various bioactive compounds derived from fruit and vegetable (F&V) seeds, such as grape seed, date seed, tomato seed, papaya seed, avocado seed and pomegranate seed, are highlighted in this chapter. Evaluation of the characteristics and nutrient loss in F&V waste has also been demonstrated. Additionally, the health-promoting benefits and utilization of bioactive compounds in various applications recovered from F&V seeds are elaborately discussed and summarized.

5.2 CHARACTERISTICS AND NUTRIENT LOSS IN FRUIT AND VEGETABLE WASTE

The physical, chemical and biological features of fruit and vegetable waste streams have been widely evaluated to establish their potential in a variety of applications including feedstock. For the by-products to be utilized in various processes, the examination of waste composition is essential [11]. Volatile solids, total solids, ash and moisture content are among the physical qualities that are most frequently studied. Biological features (organisms and pathogens) and chemical characteristics (toxic metals, protein, phosphate, hemicellulose, cellulose, nitrogen, starch, pH, total organic carbon and biological oxygen requirement) are less investigated despite their importance. F&V waste are plentiful in biodegradable organic compounds, have a high moisture content (80–90%) and are a great source of minerals, carbohydrates, organic acids, soluble sugars and starch [12]. According to Diaz et al. (2017), carbohydrates make up 70–90% of the dry weight of discarded fruits and vegetables, making them the most prevalent component [13]. Nutrients such as dietary fibres, minerals, vitamins, lipids and proteins are abundant in F&V waste. They may also be significant providers of nutrients and bioactive substances, such as prebiotics and oligosaccharides, as well as phenolic compounds (phenolic acids, flavonoids and tannins). F&V waste is a source of natural food components, additions or supplements with high nutritional value since it includes many types of phytochemicals. Nutrients including minerals, vitamins and polyphenols in seeds and peel are frequently abundant in agricultural waste and loss. In the US, a study on the nutrients lost through food waste at the retailer and consumer level found that the waste generated from fruits alone was responsible for 23% of the total fibre loss, 11% of the magnesium lost, 16% of the potassium lost, 11% of the vitamin B6 lost and 52% of the vitamin C lost. Moreover, the waste generated from vegetables was responsible for 35% of the total loss of fibre, 10% of the total loss of calcium, 14% of the total loss of iron, 20% of the total loss of magnesium, 29% of the total loss of potassium, 18% of the total loss of sodium, 11% of the total loss of folate, 25% of the total loss of thiamine, 25% of the total loss of vitamin B6, 10% of the total loss of potassium and 45% of the total loss of vitamin C. Microbial activity offers an environmentally beneficial method of producing value-added compounds such as polypeptides, polysaccharides, prebiotics and phytochemicals [14, 15]. The phases of harvest, postharvest and consumption imparts the most micronutrient losses along the F&V supply chain, according to Food and Agriculture Organization (FAO), albeit this varies between nations. If part of this edible trash is recovered, it could reduce the challenges of vulnerable communities' poor nutritional intakes. Dietary recommendations increasingly take sustainability into account. Techniques for managing traditional and alternative/emerging F&V waste are divided into two groups (waste disposal method and waste valorisation method). Composting, burning, incineration and open-field dumping are examples of traditional techniques. The downsides of conventional waste disposal systems have been thoroughly researched in the literature which include leachates, hazardous gas emissions, slow degradation rates, eutrophication, water pollution and health and environmental risks. Traditional treatment methods are unsuccessful for managing F&V waste because of traits including high amounts of organic content, moisture content and slow biodegradability [16]. A number of studies are looking at establishing other F&V waste management strategies that are more effective, affordable and ecologically friendly in order to get around the drawbacks of the current F&V waste treatment systems. Waste valorisation, also known as the process of converting waste items into high-value products including fuels, materials and chemicals has drawn a lot of interest in recent years. Studies have demonstrated that, compared to the edible sections of fruit and vegetables, the post-processing and non-edible parts have greater levels of phytochemicals. F&V waste has a great deal of potential for transformation into value-added products including fuels, biopesticides, enzymes, biocolourants, chemicals, animal feed and perfumes due to its chemical properties [17].

5.3 PRESENCE OF BIOACTIVE COMPOUNDS IN FRUIT AND VEGETABLE SEEDS

5.3.1 Tomato seed

One of the most commonly grown vegetables globally is the tomato. In 2018, there were 182 million tonnes of tomatoes produced worldwide, according to FAOSTAT. In addition to being eaten raw, tomatoes are often made into sauces, purees and many other products. Tomato by-products, viz. seeds, peels and pulp, are produced in sizeable numbers (between 1.5% and 5%) during these operations. Nutraceutical and bioactive components are reported in such products. Tomato processing generates a substantial amount of waste, thereby making it challenging to dispose of them and posing serious environmental risks. Bioactive substances like lycopene and ß-carotene, which have been demonstrated to contain high antioxidant activity in several investigations, are abundant in tomato waste [18]. Since tomato seeds have a diverse array of phytochemicals, they are thought to be a potential antioxidant source. They serve as a source of carotenoids and bioactive phenolic compounds, which are essential for controlling a variety of physiological and metabolic processes in the body. Since seeds are primarily composed of protein and oil, they have a very distinct composition from the other parts. Other compounds, such as phenols and lycopene, are also present in seeds, albeit considerably less so than in the peel fractions [19]. Tomato seeds are especially rich in (i) nucleosides, viz. adenosine, inosine and guanidine, (ii) carotenoids, viz. lycopene and β-carotene and (iii) phenolic compounds (PCs), viz. phenolic acids and flavonoids. PCs, which are made up of various parts, have a variety of bioactivities, including antiplatelet, antibacterial, anti-neurodegeneration, anticancer, anti-inflammatory, antimutagenic and cardioprotective qualities. They efficiently carry out these functions on their own or in collaboration with other tomato seed components. By modifying several cellular signal transduction pathways and activating natural defensive systems, PCs demonstrate potential protective measures. The potential of dietary phytochemicals as functional food and medicinal components is best exemplified by their bioactivity [20]. PCs are a class of small bioactive molecules united by the presence of one or more hydroxyl moieties and at least one aromatic ring in their chemical structure. PCs have the ability to modify food's sensory qualities by altering their flavour, taste, colour and astringency. PCs found in food are predominantly flavonoids and phenolic acids. Fourteen flavonoids, including variants of quercetin, isorhamnetin and kaempferol, were identified by phytochemical evaluation of an aqueous tomato seed extract. The phenolic concentration of tomato seed extract was measured to be 20,657 mg/100 g [21]. Of the overall flavonoid content, derivatives of quercetin made up about 37%. PCs, viz. kaempferol-3-O-sophoroside-7-O--rhamnoside, quercetin-3-O-sophoroside-7-O-glucoside, isorhamnetin-3-O-sophoroside-7-O-rhamnoside, quercetin-3-O-sophoroside-7-O-rhamnoside, quercetin-3-O-sophoroside, kaempferol-3-sophoroside-7-O-glucoside, quercetin-3-O-(2-pentosyl, quercetin-3-O-gentiobioside-7-O-glucoside, 6-rhamnosyl) glucoside, kaempfer-ol-3-O-(2-sophorosyl) glucoside and kaempferol-3-O-(2-pentosyl) glucoside were detected for the first time in tomato seeds via high-performance liquid chromatography/photodiode array detector/electrospray ionization-mass spectrometry [5]. Also, tomato peel includes a variety of flavonoids, including rutin and naringenin, which are beneficial to human health. Naringenin (83.03 mg/kg dw) was reported to be the most prevalent flavonoid, succeeded by caffeic acid (25.50 mg/kg dw). In a separate investigation, Valdez-Morales et al. (2014) analyzed the presence of bioactive compounds along with their antioxidant capabilities in the seeds of four distinct tomato varieties [22]. Using high-performance liquid chromatography with photodiode-array detection (HPLC-DAD), the authors were able to identify and measure a total of 16 components, including eight phenolic acids and eight flavonoids. The PC levels of various cultivars in methanolic tea seed oil extract (TSE) were determined to be 66.2–120.6 mg eq of GA (gallic acid) and 436.2–680.5 mg eq of CAT (catechin)/100 g of sample. According to reports, tartaric esters, flavonoids and phenolic subgroups each contained 52–68.5 mg eq of QUER (quercetin)/100 g and 3.3–6.4 mg eq of CA/100 g, respectively. β-coumaric acid, caffeic acid, ferulic acid and sinapic acid, made up the majority of the phenolic acids in TSE; on the other hand, gallic acid, trans-cinnamic acid, vanillic acid and chlorogenic acid were only present in trace levels. Compared to myricetin, naringenin, quercetin, isorhamnetin, kaempferol and flavonoids, it was shown that quercetin-3-O-glucoside and rutin were present in greater amounts. Total phenolic content (TPC) levels in tomato pulp are much lower than in the seeds and peel [23]. Using high performance liquid chromatography mass spectrometry (HPLC-MS), 21 different bioactive compounds, including 15 flavonoids (Coumaric acid, phloridzin, apigenin-7-O-glucoside, epicatechin, kaempferol, gallic acid, genistein, ferulic acid, quercetin, phloretin, daidzein, luteolin-7-O-glucoside, procyanidin B2, rutin and kaempferol-3-O-glucoside), 2 carotenoids (lycopene, β-carotene) and 3 nucleotides (adenosine, guanosine and inosine) were identified in aqueous TSE. Results revealed that aqueous extracts (3.37% m/m) had greater extraction yields than ethanolic extracts (1.38% m/m), which was most likely caused by the polarity and viscosity of the extraction solvents. Because of their role in powerful platelet antiaggregant action, the nucleotides found in TSE also function as bioactive substances [19]. Carotenoids are a different group of bioactive substances found in tomato seeds in addition to phenolics. Carotenoids are one of the most functionally and chemically varied classes of bioactive compounds on the planet. They are extensively present pigments. Because of their capacity to scavenge free radicals, carotenoids possess a tremendous antioxidant capacity. Consuming carotenoids has been linked to a lower risk of developing some malignancies, cardiovascular illnesses, cataracts and macular degeneration. Using HPLC, a carotenoid profile of tomato seed oil revealed the presence of β-carotene and the isoforms of lycopene (lycopene cis 1, lycopene cis 2 and lycopene cis 3, lycopene all trans). In comparison to supercritical CO_2 extraction and ethanol accelerated solvent extraction (ASE), the study observed that hexane-ASE treatment had the greatest levels of β-carotene and lycopene. The seed fraction's lycopene content was calculated to be 122.87 g/g [24]. After lycopene, the second-most prevalent carotenoid found in tomato seeds is β-carotene, a precursor to vitamin A. In spite of their significant presence in tomato seeds, there are currently no direct studies evaluating the bio-absorption and bioavailability of tomato seed compounds, with the majority of studies being conducted on whole tomatoes or their individual bioactive components. For example, research that examined the levels of β-carotene and lycopene in human tissues and serum revealed that β-carotene predominated in the liver, adrenal gland, kidney, ovary and fat, whereas lycopene was shown to be the main carotenoid in the testes [25]. Another investigation found higher levels of phytoene, phytofluene and lycopene in human blood as well as protection from UV-light-induced erythema following consumption of tomato-based products containing roughly 10 mg of lycopene per day [26]. A significant amount of flavonoid present in tomato seeds, called quercetin-3-O-sophoroside, has been shown to be absorbed intact, methylated and sulphated at the jejunum without deglycozylation. The study also discovered that quercetin-3-O-sophoroside was broken down by the cecal microbiota into 11 phenolic acids [27]. When exposed to thrombin and arachidonic acid-mediated platelet aggregation, tomato seed-derived flavonoid derivatives and phenolic conjugates showed greater antiplatelet activity. Tomato seeds include substances such chlorogenic, caffeic, ferulic and p-coumaric acids that may also have comparable effects. Tomato seeds have a fat content ranging of 15–30%, with 80% of those fatty acids being unsaturated ones like palmitic, oleic and linoleic acids. These fatty acids stop human platelet phospholipase A2, which prevents atherogenesis. Additionally, it has been demonstrated that linoleic acid prevents platelet aggregation, tissue factor expression and the progression of arterial thrombosis. By reducing hyperlipidaemia and controlling gut microbiota through their seed oil, tomatoes also have cardioprotective effects [28]. In mice treated with tomato seed oil diets, He et al. (2020) showed that hyperlipidaemia was reduced [29]. The main regulators of the oxidation process are carnitine palmitoyltransferase-I (CPT1A), acyl-CoA dehydrogenase long chain (ACADL) and PPAR-alpha (PPAR-α). The expression of CPT1A, ACADL and PPAR-α was shown to increase oxidation in tomato seed oil, which led to a decrease in plasma lipid levels. Phytosterols found in tomato seed oil reduced the absorption of cholesterol as a result of their structural resemblance to cholesterol. Through sterol regulatory element-binding protein-2 (SREBP-2), tomato seed oil indirectly modifies the activity of HMG-CoA reductase to lower cholesterol levels. To generate protein hydrolyzates from tomato seeds with angiotensin-converting-enzyme (ACE) inhibitory activity, Moayedi et al. (2017) employed Bacillus subtilis [30]. By interacting with the catalytic sites of ACE, the protein fractions rich in hydrophobic amino acids including leucine, phenylalanine, valine, tryptophan, alanine and tyrosine, as well as positively charged residues like arginine and lysine can reduce ACE activity. Moreover, following fermentation with Lactobacillus spp., the ACE inhibitory efficacy of tomato seed rose ten-fold. Additionally, coumarins and other polyphenolic substances found in TSE have ACE2 modulatory properties. According to research conducted both in vitro and in vivo, the use of coumarins prevent inflammation and acute lung damage in mice by blocking the downregulation of Ang (1-7) and ACE2, which reduces the production of pro-inflammatory cytokines [31]. The chemical structures of different bioactive compounds present in tomato seeds are shown in Figure 5.1.

5.3.2 Date seed

The date palm, a member of the Palmaceae family, is a significant fruit-bearing tree that is cultivated widely throughout the world's arid and semiarid regions. Given the present scenario of food supplies worldwide and the predicted demand for food in the future, date palms have the potential to provide an acceptable and effective source of energy and food supply. Dates vary from most fruits in that they go through four phases of botanical development: Kimri, Khalaal, Rutab and Tamr. At the Tamr stage, when dates have a total soluble solids content between 65 and 70 brix, they are considered edible. Both nutritional and therapeutic value may be found in the date fruit and its by-products, including seeds [32]. Similar to the majority of other fruit seeds, date seeds (DS) possess significant levels of bioactive substances such carotenoids, polyphenols, especially phenolic acids and flavonoids. Carotenoids are the main phytochemicals observed in the lipid fractions of DS. By avoiding the development of several chronic illnesses, carotenoids, also known as tetraterpenoids, are known to promote human health. The total carotenoid contents were found to range from 1.38 mg/100 g to 3.53 mg/100 g. These concentrations were less than those found in Tunisian DS that had completely ripened (5.51 mg/100 g) [33]. Similar to this, Habib and Ibrahim (2011) examined several carotenoids in DS from the UAE-obtained Khalas cultivar [34]. β-carotene (3142 g/kg), zeaxanthin (10.8 g/kg), β-cryptoxanthin (20.4 g/kg), lutein (1599 g/kg) and lycopene (19.5 g/kg) were the main carotenoids identified. The variations in carotenoid concentrations among DS types were linked to elements such as degree of ripeness, ambient circumstances, postharvest processing technique and storage conditions. According to Al-Farsi et al. (2005), the utilization of sun-drying as a postharvest processing procedure results in considerable loss of total carotenoids compared to fresh fruit [35].

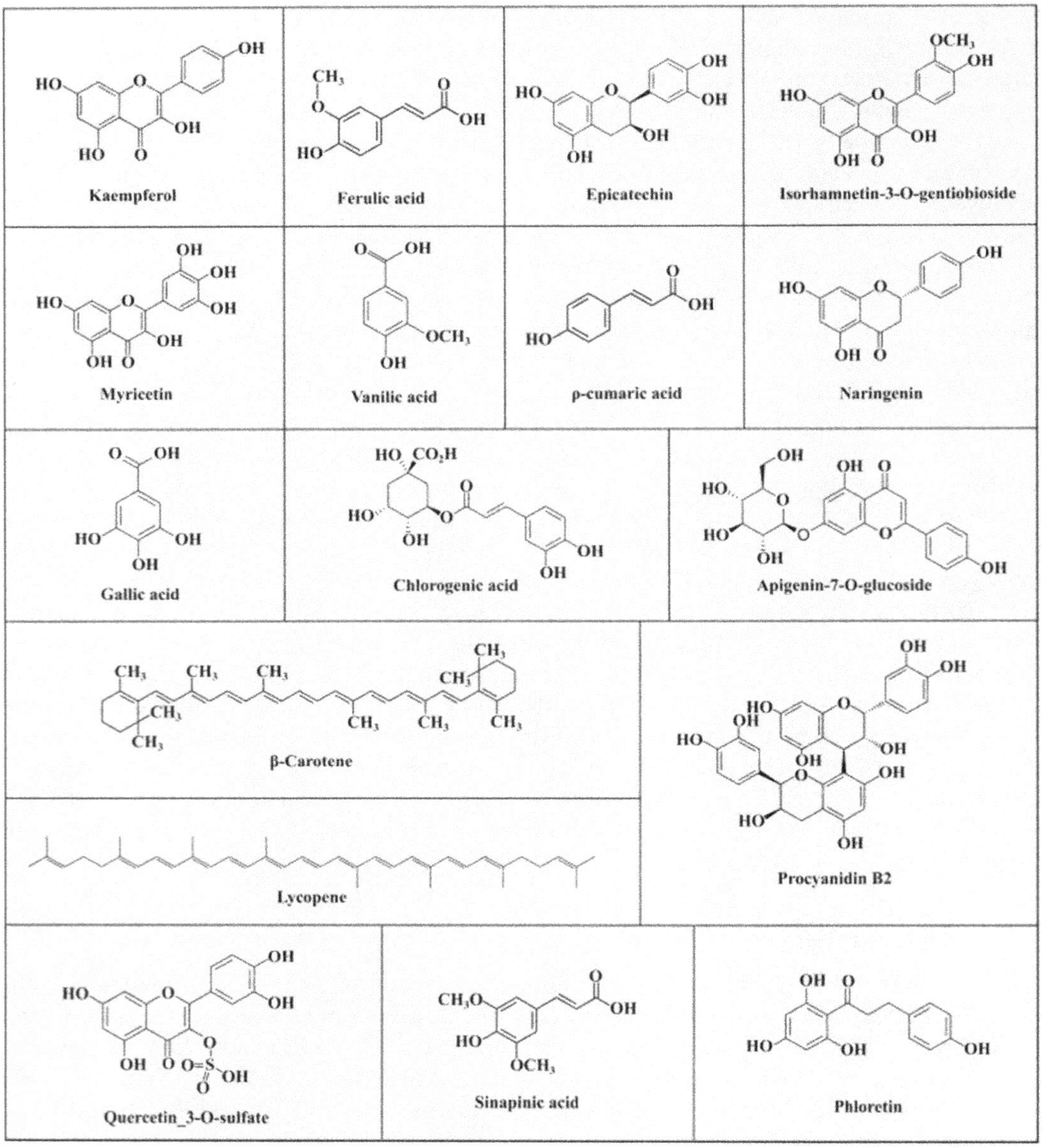

FIGURE 5.1 Chemical structures of different bioactive compounds in tomato seeds [Modified from Kumar et al. (2021) [5]].

The carotenoid pigment's composition is typically regarded as one of the most important aspects of oil quality as it corresponds to the colour formation, which is a fundamental element in the evaluation of oil quality. According to Al-Farsi and Lee (2008), DS often have greater carotenoid levels than dried and fresh date fruit (31.5–3,025 g/kg) [36]. As a result, DS may be thought of as a possible source of carotenoids. In addition, tocopherols and tocotrienols are additional phytochemicals discovered in lipid fractions of the DS. They are part of the vitamin E family and necessary for human health since they have some antioxidant qualities. The quantity and composition of such bioactive compounds, which can stabilize oil owing to the presence of natural antioxidants, vary across different types of vegetable oils. Studies have been conducted to ascertain the tocopherol content of DS cultivars collected from various geographical locations. The three main protocols in Tunisian DS are α-tocotrienol (34.01mg/100 g), γ-tocopherol (11.30mg/100 g) and γ-tocotrienol (4.63mg/100 g). It is interesting to note that the oil of DS contains significant amounts of α-tocopherol acetate, a different type of vitamin E with greater oxygen and light stability than tocopherol [33, 37]. In addition, the analysis of the tocopherols in 12 DS cultivars revealed that β-tocotrienol was the predominate tocopherol in each variety. As a result of the high tocopherol content of DS, it can be utilized as a natural source of antioxidants. The lipid-soluble phytochemicals known as phytosterols have a structure that is similar to that of cholesterol. Two hundred phytosterols are naturally present in plants, with fruits and vegetables having the highest concentrations [38]. Although phytosterols are present in the date fruit's edible component as well, they are mostly stored in the date pit and have been used for ages to treat hormone-related illnesses. DS contains phytosterols such as oestrogen, brassicasterol, esterone and ergasterol. In addition, Δ-sitosterol (75%) is the main sterol portion of oil from DS. Other sterol fractions including Δ5,24-stigmastadienol, Δ5-avenasterol and campesterol were also discovered in significant concentrations, although Δ7-avenasterol, stigmasterol and cholesterol are only detected in minute quantities. Khalas variety of UAE DS has been shown to contain certain phenolic acids, viz. caffeic acid, protocatechuic acid and coumaric acid [7]. Studies comparing the phenolic content of the DS in Mabseeli variety and its concentrates made with various solvents were carried out by Al-Farsi & Lee, (2008) [39]. A total of 9 phenolic acids were detected in the DS, of which protocatechuic acids, p-hydroxybenzoic acid and o-coumaric acid were the most prevalent. These nine phenolic acids included five cinnamic acid derivatives, four vanillic acid hydroxylated derivatives and five derivatives of benzoic acid. It was concluded that the phenolic profile of the fruit may be affected by the type of solvent employed during phenolic extraction.

Plant flavonoids have radical scavenging and antioxidant properties, which can stave off the development of chronic and cardiovascular illnesses. Three Moroccan DS types (Bousthammi, Boufgous and Majhoul) were the subject of recent studies and the results showed that they have significant flavonoid concentrations, ranging from 1224 to 1844 mg of rutin equivalent per 100 g dw. Similar to previous solvent (methanol and ethanol) extracts, water and 50% acetone extract of DS with subsequent fractionation by butanone have been shown to improve the total flavonoid content [40]. Additionally, utilizing UHPLC-DAD ionization-mass spectrometry, a thorough examination on the kind and quantity of flavonoid component found in DS was conducted by Habib et al. (2014) [41]. About 99% of the total polyphenols were found to be flavan-3-ols, which were observed in monomeric and polymeric forms and distributed as epicatechin (47.2 g/kg) and catechin (3.41 g/kg). Another study found 13 different kinds of Saudi Arabian DS to contain the flavonoids rutin and catechin. Red grape seeds have been shown to contain 9,207 mg catechin/100 g dw, white grape seeds have 8,220 mg catechin/100 g dw, watermelon seeds have 969 mg catechin/100 g dw and gooseberry has 801 mg catechin/100 g dw. The TPC values for grape seeds are higher than those found in their polyphenolic extracts, where a TPC value of 5,422 mg gallic acid equivalent (GAE)/100 g dw for date seed of the Bousthammi type and 5,756 mg FAE/1000 g dw for the khalas type were found [7, 42]. Additionally, date seed extract from acetone showed antiviral action against lytic Pseudomonas phage ATCC 14209-B1 starting at a lower dosage of 11 mg/ml, with reduced phage activity and entirely restricting bacterial lysis. The direct binding of the seed extract to phages is probably the mechanism through which this inhibitory effect occurs. Ajwa date seed extract use over time preserves healthy kidney and liver function. Date seeds can be used to make peptides with strong antioxidant activity and ACE inhibitors for use as functional additives in different food items because they have a significant quantity of protein (5.2 g/100 g) in them. To investigate the possible usage of DS as functional food ingredients, research was done employing DS flour (5%) and DS hydrolyzates (2.7%) to create baked goods, viz. muffins [43]. They reported that both DS flour and hydrolyzates significantly increased hydroxyl radicals as well as 2,2-diphenyl-1-picrylhydrazyl (DPPH) radical scavenging activity. Furthermore, compared to the control and other food items made with DS flour, the integration of DS hydrolyzates into baked goods, like muffins, significantly increased the ACE-inhibition activity due to its high antioxidant capacity. Also, because of the presence of high nutritional value, which depends on the dietary fibre content of DS, they can be used to manufacture foods and supplements that are high in fibre. In addition, flavon-3-ols were shown to be the most prevalent polyphenols in bread made with DS, as opposed to other samples where the component was not observed. As previously mentioned, baking may lower the amounts of phenolic compounds, and this may not be due to the compounds being oxidized since yeast uses the oxygen that is now available. Instead, phenolics may be degraded via heat throughout the baking and mixing processes [44]. Numerous studies have also observed increased dietary fibres in baked goods enhanced with various types of date seeds. Additionally, when examined during in vitro and in vivo studies, a broad spectrum of antibacterial characteristics has been documented in several date cultivars. For example, both gram-negative and gram-positive bacteria are inhibited by methanolic and acetone extracts from Ajwa date seeds. Numerous more studies have provided additional evidence for dates' high antibacterial capability after the initial study confirming the fruit extract's antiviral activity. For instance, Egyptian date extracts in ethanol and water demonstrated potent antimicrobial action against different types of harmful bacteria [7, 45]. The chemical structures of different bioactive compounds present in date seeds are shown in Figure 5.2.

5.3.3 Papaya seed

The herbaceous plant Carica papaya Linn is a member of the Caricaceae family. The plants can grow up to 10 metres tall and do well in tropical climates. Papaya production was 13 million metric tonnes in 2017, with India and Brazil accounting for the majority of global output, followed by Mexico, Nigeria and Indonesia. Papain, terpenoids and alkaloids, among others, are present in both ripe and unripe papaya leaves and seeds. Additionally, Carica papaya L. is a natural source of carotenoids, phenolic compounds, vitamin C and vitamin E [46]. Only plants and some types of fungi can produce phenolic chemicals, which have various chemical configurations. These substances, which are a by-product of the plant's secondary metabolism, are linked to many physiological processes. The total phenolic content of the papaya fruit changes as it ripens. According to reports, papaya peel contains between 5 and 28 mg/g of total phenolics per gramme, compared to papaya seed's 0.3–3 mg/g. There were just two phenolic compounds found in papaya seeds: vanillic acid and coumaric acid. While the papaya seed includes glucosinolates, tocopherols, carotenoids and benzyl isothiocyanate, the fruit pulp is abundant in minerals and vitamins. Papaya fruit includes linalool, 4-terpinol and monoterpenoids, and papaya oil is found in the seeds and also contains flavonoids, kaempferol and myricetin [47]. The seeds can serve as an alternate supply to essential oil because they are high in unsaturated fats. Protein, dietary fibre, minerals, antioxidants and phytochemicals are present in both the seeds and the leaves. Regardless of the cultivars, the leaves and seeds consist of around 14–28% protein. The seeds possess a good amount of fat (21–30%) and carbohydrate (8–58%), making them suitable as supplementary energy sources for populations that are undernourished. Essential fatty acids, viz. oleic acid, and lipophilic phytochemicals are the primary constituents of papaya seed oil (21.05–30.07%) [48]. Oleic acid has a well-established reputation for being anti-inflammatory and even provides beneficial impacts on genes linked to tumours. There are also other reports of the presence of stearic acids, palmitic acids, linolenic acids and arachidic acids in papaya seeds. Compared to other oilseeds like corn (4.4–6.2%) and soybean (15–20%), papaya seeds' high lipid content makes them more attractive economically for industrial use. Additionally, papaya seeds and peels are considered an excellent mineral source. Minerals are essential for the healthy operation of the body's metabolic and physiological functions. The human body may receive 70% of Ca, 60% of Cu, 35% of Fe, 80% of P, 15% of K, 80% of Zn, 120% of Mg, 130% of Mn, respectively, from the consumption of 100 g of papaya seeds [49]. Dietary fibres are abundant in papaya seeds and peels as well. Dietary fibres are well known for their ability to decrease cholesterol and eliminate toxins from the digestive system, among other health benefits. Phytochemicals are abundant in papaya seeds. They include beneficial phytochemicals, such as carotenoids, phytosterols, tocopherols and phenolics. In general, phytochemicals have a variety of advantageous features, such as the ability to inhibit the growth of cancer cells and to protect against cellular oxidative damage [50]. The papaya seed includes a variety of bioactive substances,

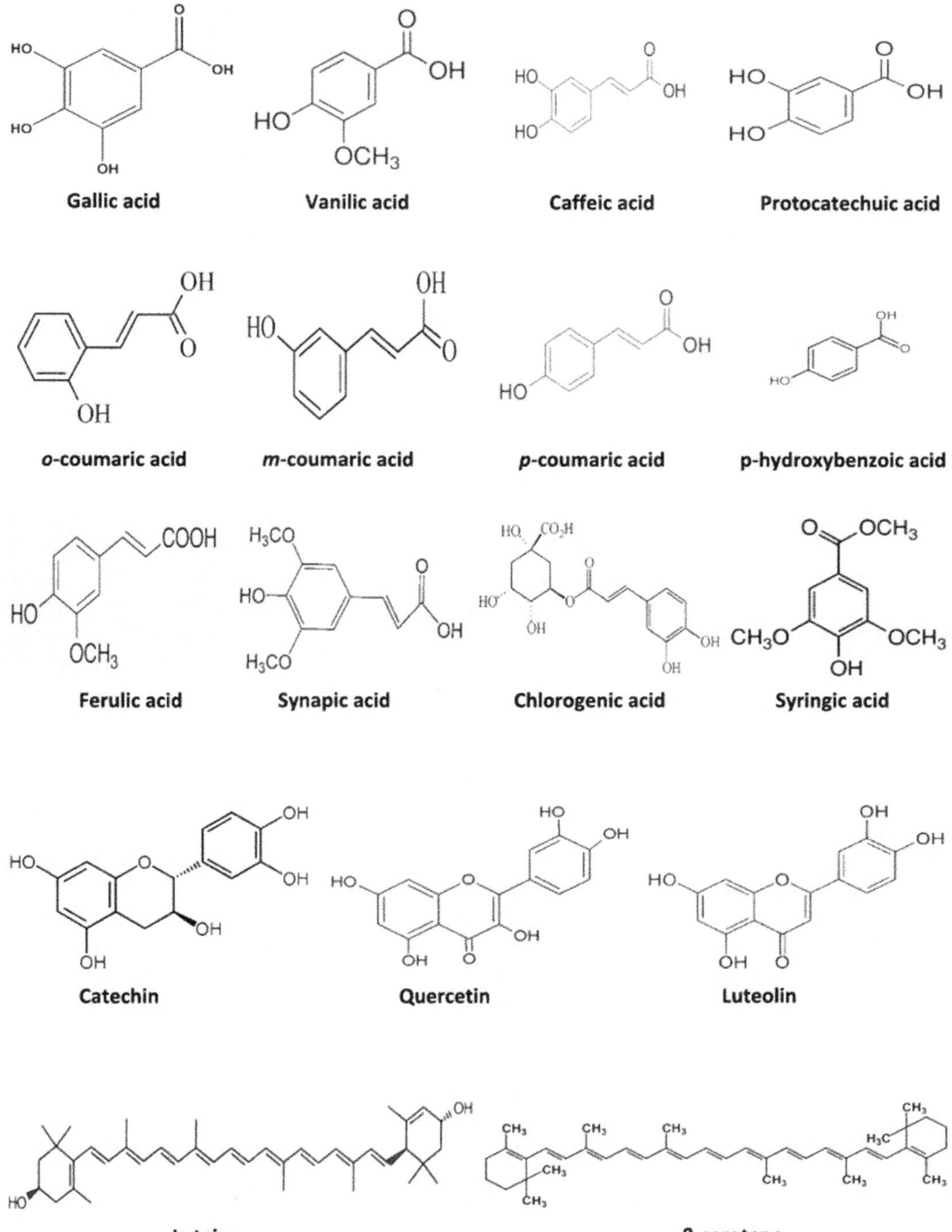

FIGURE 5.2 Chemical structures of different bioactive compounds in date seeds [Modified with permission from Maqsood et al. (2020) [7]].

including (i) phenolics, such as caffeic acid, p-Hydroxybenzoic acid, quercetin-3-galactoside, ferulic acid, p-Coumaric acid and kaempferol-3-glucoside; (ii) phytosterols, including β-Sitosterol, stigmasterol and campesterol; and (iii) other phytochemicals such as terpenoids, strobosteroids, oxalates, phlobatannins and anthraquinones [51, 52]. The seeds are a rich source of 5-hydroxy feruloyl quinic acid, syringic acid hexoside, p-coumaryl trimethyl glycoside, 5-hydroxy caffeic quinic acid, cyanidin-3-O-glucose, acetyl p-coumaryl quinic acid, chlorogenic acid, feruloyl quinic acid and kaempferol-3-O-rhamnoside [47]. The amounts of vitamin C and total β-carotene in papaya seeds range from 0.08 mg/g to 3.37 mg/g and 0.54 g/g to 1.2 g/g, respectively, depending on the varieties and geographic regions. The chemical structures of different bioactive compounds present in papaya seeds are shown in Figure 5.3.

Papaya leaves displayed a higher quantity of phenolics than papaya seeds. This suggests that phenolic acids and flavonoids are more abundant in leaves than in seeds. Carpaine and an anthelmintic alkaloid found in C. papaya's leaves, seeds and fruit are effective at removing worms from the human alimentary canal [54]. Additionally, it is known that papaya seeds and leaves have antifungal properties against Aspergillus flavus, Fusarium spp., Penicillium citrinium, Candida albicans and Colletotrichum gloeosporioides. β-carotene, riboflavin, thiamine, niacin and ascorbic acid concentrations in papaya seed were reported to be 0.03–0.04, 0.03–0.07, 0.05–0.07, 0.2–0.3 and 8.2–12.9 mg/100 g fresh weight, respectively. Other seed spices including oregano, mustard and coriander have similar amounts of other, viz. riboflavin and thiamine. Also, the highest benzyl isothiocyanate (BITC) levels were found in the seed (10–15.5 mmol/kg), followed by the peel (2–6 mmol/kg) and pulp (0.2–0.35 mmol/kg) of the papaya fruit. The young papaya leaf (1.2–610 mg/100 g) and papaya seed (285–294 mg/100 g) were found to contain the highest amounts of BITC, respectively, thereby increasing the possibility of using benzoyl isothiocyanate in cancer chemotherapy [54]. It has been proposed that the ability of isothiocyanates to stimulate phase II enzymes, such as nicotinamide adenine dinucleotide phosphate, glutathione S-transferase and quinine reductase is what accounts for their anticarcinogenic properties. Regardless of cultivar, BITC predominates in papaya seed extracts. The seed's BITC contains anthelmintic characteristics that have been used effectively to treat intestinal parasites [55, 56]. Alkaloids carpaine and carpasemine are primarily responsible for papaya seed's antihelminthic action. Compared to other plant parts, papaya seeds have a high lipid content, with oleic acid (78.5%) as the principle fatty acid. The seed contained tannin as well, with a value of 6.27 g/100 g dw. Papaya seeds, pulp and pericarp have been shown to contain benzyl glucosinolate and benzyl isothiocyanate [57]. Oleic acid was extracted from the seed by Ghosh et al. (2017) [58]. Along with mass spectrometry, FTIR, proton nuclear magnetic resonance (^{1}HNMR) and Carbon-13 (C13) nuclear magnetic resonance (^{13}CNMR) spectroscopy were used to study the structure of the isolated molecule. The antifertility component was isolated from the ethyl acetate extract of C. papaya seeds and its structure was clarified using spectroscopic techniques. Tetrahydropyridin-3-yloctanoate 1,2,3,4-tetrahydrate was the name given to the isolated substance. Two novel compounds, glyceryl-1-(2′,3′,4′ trihydroxybenzoyl)-2,3-dioleate (papayaglyceride) and 2,3,4-trihydroxytoluene (caricaphenyltriol), were recovered from the ethanolic extracts of the seeds. The structure was studied using spectroscopic and chemical techniques [59]. Furthermore, the black seed of yellow-ripe papaya directly inhibits the proliferation of prostate cancer cells. The prostate cancer cell line was examined using methanolic extracts of black seed (from ripe papaya) and white seed (from unripe papaya). The white seed extract stimulates pre-existing prostate cancer cells, while black seed extract is beneficial against prostate cancer cells. Another investigation revealed that this plant's seed extract had both anti-inflammatory and immunomodulatory properties. For instance, complement mediated haemolytic assay and lymphocyte proliferation assays were used to examine the bioactive reactions and immunomodulatory effect of the crude seeds extract of this plant in vitro. Also, DPPH free radical scavenging activity can be used to gauge the antioxidant activity of a methanolic extract of seeds. In a different study, papaya seed extracts were employed to test the antioxidant activity, and the results showed that the hexane extract had the maximum capacity to scavenge DPPH free radicals while the aqueous extract had the lowest capacity [59, 60].

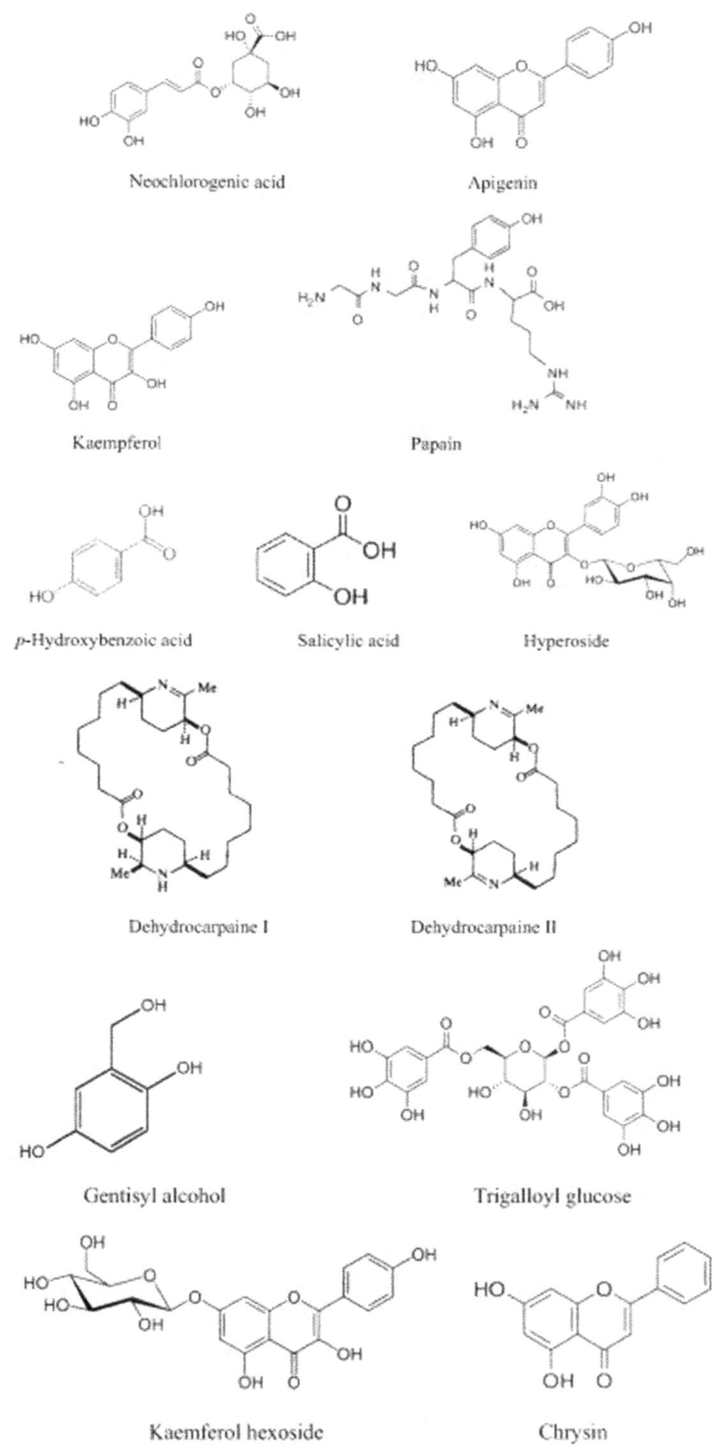

FIGURE 5.3 Chemical structures of different bioactive compounds in papaya seeds [Modified from Sharma et al. (2020) [53]].

The crude extract and seed extract of C. papaya are useful for their adulticidal, pupicidal, larvicidal, smoke toxicity and repellent effects against Anopheles stephensi, malaria and Culex quinquefasciatus. The crude extract is a mixture of methanol, chloroform, ethyl acetate, benzene, petroleum ether (1:1 v/v). At a concentration of 0.5%, C. quinque-fasciatus and A. stephensi have higher mortality rates. It was reported that the repellent had a 78% and 92% protection rate against both female mosquitoes, respectively. In comparison to A. stephensi, C. quinquefasciatus had a protection time against bite of four hours. It was found that after five hours of exposure to toxic smoke, 186 An. stephensi mosquitoes and 190 quinquefasciatus mosquitoes died out of 200 total mosquitoes. The nature of the active ingredient, polyhydroxy aliphatic amide, is revealed via IR research [61]. Besides, papaya peel and seed aqueous extract exhibits larvicidal effects on Aedes aegypti larvae. Additionally, it was observed that seed extract had greater larvicidal efficacy than peel extract, perhaps as a result of the extract's inclusion of phytochemicals, such flavonoids, tannins and alkaloids. It was also demonstrated that Aedes aegypti is resistant to insects by way of leaf, bark, root and seed. Additionally, it was found that crude ethanol extract was far more effective than aqueous extract at controlling the vector [62].

5.3.4 Avocado seed

The avocado is a healthy fruit that is indigenous to Central and South America. The avocado, or Persea americana Mill., is a member of the genus Persea and family Lauraceae. It comes in a number of variations, with the Hass type being the most well-liked. The cultivar, storage circumstances, level of ripeness and edaphoclimatic conditions all affect the fruit's characteristics. Large numbers of peels and seeds are produced during the industrial processing of avocados. As an inedible component of the fruit, the seeds are underused and thrown away as waste. Insects and rodents may become more prevalent if trash is left untreated. There are many beneficial compounds in the avocado waste (seed and leaves). As a result, these seeds and leaves may one day be harvested and used to generate significant revenue for the avocado industry. One source of phytochemicals is the avocado seed, which makes up 13–18% of the size of the entire fruit. The seed and leaves are more antioxidant-rich and contain more total phenols compared to the pulp and peels [63, 64].

Chemicals extracted from plants are known as bioactive compounds. Avocado seeds are made up of protein, sugar, carbohydrate, fat and water. Avocados contain phenolics, flavonoids, carotenoids, vitamin C and vitamin E, which are all bioactive substances. Flavonoids, tannins, saponins, phenolics, antioxidant capacity, oxalates, phytates and alkaloids are among the phytochemicals found in avocado seeds. One of the bioactive substances found in all portions of the avocado plant is polyphenol [65]. Three-O-caffeoylquinic acid, three-O-p-coumarylquinic acid, procyanidin trimer A(I), procyanidin trimer A(II), catechin (epicatechin gallate) are the additional polyphenols discovered in avocado seeds. According to additional research, avocado seeds include tannins, saponins and flavonoids, which are all components of polyphenols. The amounts of phenolic compounds produced by various avocado types vary. Alkanols, terpenoid glycosides (derivatives containing a furan ring), flavonoids and coumarin can all be found in avocado seeds' phenolic component profile [66]. The different classes of alkanols are 1,2,4-trihydroxy heptadec-16-yne, 4-acetoxy-1,2-dihydroxy heptadec-16-yne and 4-acetoxy-1,2- dihydroxy heptadec-16-ene. Terpenoid glycosides consist of (1'R, 3'R, 5'R, 8'S)-epi-dihydrophaseic acid- β-D-glc and (1'S, 6'R)-8'-hydroxy abscisic acid β-D-glc. Furan ring-containing derivatives include 2-(12-tridecyl) furan, 16-2-(12-heptadecyl) furan, 14-2-(pentadecyl) furan and dimethyl sciadinonate. In addition to scopoletin, other flavonoids include quercitrin, quercetin-3-O-D-diglc, astragalin, afzelin, luteolin, quercetin-3-O-D-ara, apigenin, isoquercitrin and luteolin-7-O-D-glc [67]. Avocado seeds have a high nutritional content, and phenolic compounds in particular have been found to have a number of useful qualities, including antioxidant, antibacterial and analgesic effects. As an antioxidant, glutathione helps shield cells from the damage caused by free radicals and oxygen. It has been discovered that the methanolic extract can be used to both prevent and treat bacterial infections brought on by harmful bacteria like E. coli and S. pyogenes. Natural oxidants, like polyphenols, offer certain advantages for human health. For instance, they can lower the risk of inflammatory disorders and stop lipid oxidation [68]. Because cholesterol is linked to saponins and forms insoluble complexes with them, saponins have hypertensive action and heart depressive properties. Human body saponins can influence the immune system, fight cancer and lower cholesterol levels. Saponins have different properties within a set of chemicals and are stable in aqueous solutions, creating foams that resemble soap. Additionally, flavonoids are antiviral, anticancer, antiallergic, anti-inflammatory, antioxidant and antiplatelet, all of which have positive impacts on bodily health. Tannins are water-soluble, high-molecular-weight compounds referred to as polyphenols. Proteins can also be precipitated by it. The reduction of respiratory issues, enhancement of hunger and decrease of blood pressure are all advantages of tannins in the human body [63]. According to Yasir et al. (2010), avocado seeds have the potential to exert a wide range of physiological effects, including hypoglycaemic, vasorelaxant, analgesic, hypotensive, healing and weight-reduction effects. According to the study, avocado seeds also have medicinal properties that include antioxidant, antidiabetic, anti-inflammatory, anticonvulsant, antihypertensive, antihepatotoxic, anticancer, antiviral, insecticidal, antibacterial, anti-ulcer, beneficial dermatological effects, cholesterol-lowering and even colourant [69]. Terpenoids in avocado seed extract have recently been studied for a potential anticancer component. Alcohol extract and an in vitro cytotoxic test utilizing the MTT assay were utilized to isolate the terpenoid component. Under a safe concentration limit of 100 g/mL for ethanol extract on normal cells, it demonstrated that the IC50 value is 62 g/mL as a cytotoxic action. Use of a pure extract of terpenoid from avocado seed extract improved anticancer activity against human breast Michigan Cancer Foundation-7 (MCF-7) cancer cells. As a result, isolated triterpenoid can be further developed into chemotherapy medicines to stop the growth of tumour and cancer cells. The IC50 for extracting avocado seeds with chloroform and further partitioning with soluble and non-soluble methanol fractions revealed values of 94.87, 34.52 and 66.03 g/mL, respectively. At a dose of 30 g/mL, more than 50% of MCF-7 cells were destroyed, and this produced cell shrinkage compared to FTLM after 48 hours [70].

A few drops of 0.1% ferric chloride were added to filtered, heated water to test the avocado seed for the presence of flavonoids, tannins and saponins. Tannins can be identified by their greenish-brown or blackish-blue colour. Emulsion in a mixture of 10 ml filtrate water, 5 ml distilled water and three drops of olive oil revealed the presence of saponins. While the presence of flavonoids was shown by the mixture's yellow colouring (5 ml of diluted ammonia solution, 5 ml of filtrate and 1 ml of sulphuric acid). Flavonoid, phenolic and free radical components made up the antioxidant qualities of the avocado seed oil. With quercetin as a reference, assays for flavonoid concentration are conducted using the aluminium trichloride spectrophotometric technique. This assay makes use of methanol, distilled water, a 5% $NaNO_2$ solution, 10% $AlCl_3$ and NaOH. Using gallic acid as a reference and the reduction of the Folin–Ciocalteu reagent method, the phenolic content was assessed. The solution of the phenolic content assays consists of 10 ml of 7% sodium carbonate, 10 ml of methanol and distilled water. Gallic acid was used as a reference, and the free radical scavenging was measured using the spectrophotometry technique. To measure free radical scavenging, methanol, 2,2-diphenyl-2-picrylhydrazyl and distilled water are utilized. Utilizing spectrophotometry and gallic acid as a standard, the reducing power on avocado seed oil was assessed. To assess the reducing power, solutions of phosphate buffer, 1% ferric chloride, 10% trichloroacetic acid, 1% potassium ferricyanide and distilled water are used [71]. Utilizing catechin molecules as a reference and ultraviolet-visible spectrophotometry, tannin is determined. Ethyl acetate is utilized as a fresh solution, catechins as a comparison solution, acetone as a test solution and avocado seed water extract is used to determine the phenol concentration by measuring the absorbance at 764 nm using the Folin–Ciocalteu method. The total phenol concentration was examined using a reference solution of glycerol. Sodium acetate was occasionally employed as a stabilizer to see the wavelength in the colourimetry method, which was used to determine the flavonoid content [72]. The amounts of phenolic compounds produced by different avocado types vary. Proanthocyanidins, abscisic acid, phytosterols, polyphenols, fatty acids, furanoic acids and triterpenes are some of the phytochemicals found in avocado seeds. Different concentrations of phytochemicals can be obtained using various techniques, including solvent extraction via acetone, ethanol, n-Hexane, water and methanol. These bioactive compounds are advantageous owing to their rich nutrient sources; their anticonvulsant, analgesic, antiviral, vasorelaxant, anti-ulcer, anti-inflammatory, antioxidant, hypotensive, anti-hepatotoxic, wound-healing and hypoglycaemic activities; and their ability to lower blood sugar levels and reduce body weight [9]. Wang et al. (2010) conducted a study to ascertain the antioxidant potential, TPC and important antioxidant components in several avocado (AV) strains and cultivars [64]. Tonnage of West India, Loretta, Simmonds and Guatemalan strains and Slimcado, Booth 7 and Booth 8 were used as seven cultivars from ripe Florida AVs; whereas Hass AV of Mexican strain was used as a reference. Acetone, water and acetic acid were used to create the AV seed extracts (70:29.7:0.3, v/v/v). Results indicated that all cultivars of AV seed had significant TPC and antioxidant capacities. In total, the seed had 64% of the TPC and supplied 57% of the fruit's total antioxidant activity. The Simmonds seed demonstrated the highest TPC and antioxidant capabilities among the cultivars examined. Additionally, it was discovered that AV seed has several times more TPC and antioxidant activity than raw blueberry. Procyanidins have been identified as the main polyphenols present in AV and are thought to be responsible for the strong antioxidant capabilities. Additionally, according to HPLC-MS data, procyanidins with an A-type linkage were found in a minor fraction of AV seeds. It has been discovered that a-type procyanidins have extra positive health effects for people, such as preventing urinary tract infections [73]. Soong and Barlow (2004) also looked at the amounts of phenols present in the edible parts and seeds of AV, jackfruit, longans, mangoes and tamarind [74]. The chemical structures of different bioactive compounds present in avocado seeds are shown in Figure 5.4.

According to the research, compared to their edible components, AV seed and other fruit seeds have increased phenolic content and antioxidant activity. The seeds from the majority of the fruits supplied 95% of the overall antioxidant activity and TPC. Catechin and epicatechin were the two main bioactive substances from AV seeds that were observed. Additionally, Rodriguez-Carpena et al. (2011) analyzed the phenolic constituents and antioxidant activity of several solvent extracts of AV seeds from Hass and Fuerte varieties of avocado [76]. For the extraction of AV seeds, various solvents including ethyl acetate along with 70% methanol and acetone were employed. When compared to Fuerte, Hass seed TPC was considerably greater. This was caused by the significant concentrations of hydrocinnamic acids, procyanidins and catechins in the Hass seed. Regardless of the kinds of solvents utilized, Fuerte seed extracts had stronger antioxidant activity than Hass seed extracts. Acetone seed extract demonstrated the highest antioxidant activity among the Fuerte seed solvents examined using the CUPRAC, ABTS and DPPH assays. Kristanty et al. (2014) used the 3-(4,5-dimethylthiazol-2yl)-2,5-diphenyltetrazolium bromide test to examine the cytotoxic effects of AV seed extracts on breast cancer cells T47D [77]. In this investigation, AV seed aqueous extract, ethanol extract and a positive control (doxorubicin hydrochloride) were all employed. Aqueous extract, ethanol extract and doxorubicin hydrochloride all had IC50 values of 560.2, 107.15 and 0.26 g/mL, respectively. According to the findings, the ethanol extract of AV seeds significantly inhibited the growth of T47D when compared to the aqueous extract. The presence of polar groups in active substances including phenols, glycosides, saponins and alkaloids was to blame for this. A triterpenoid substance with a molecular mass of 505 g/mol was identified from the ethanol extract of AV seed in a different investigation by Abubakar et al. (2017) [70]. With an IC50 value of 62 g/mL and 12 g/mL, respectively, the isolated triterpenoid demonstrated an inhibition of cell growth toward the MCF7 and Hep G2 cell lines. The American National Cancer Institute (NCI) recommends an IC50 value of less than 30 g/mL for bioactive substances to have an in vitro anticancer impact. In addition, 100 g/mL was the safe concentration limit for the ethanol extract in the toxicity test on normal cells. As a result, this single triterpenoid compound demonstrated a possible role in preventing liver cancer. In addition, Lee et al. (2008) showed that malondialdehyde (MDA)-MB-231 human breast cancer cells were successfully treated with a methanol extract of AV seed at a concentration of 100 g/mL [78]. The enhanced cleavage of caspase-3, caspase-7 and poly (adenosine diphosphate (ADP) ribose) polymerase was linked to this impact. The effects of polyhydroxylated

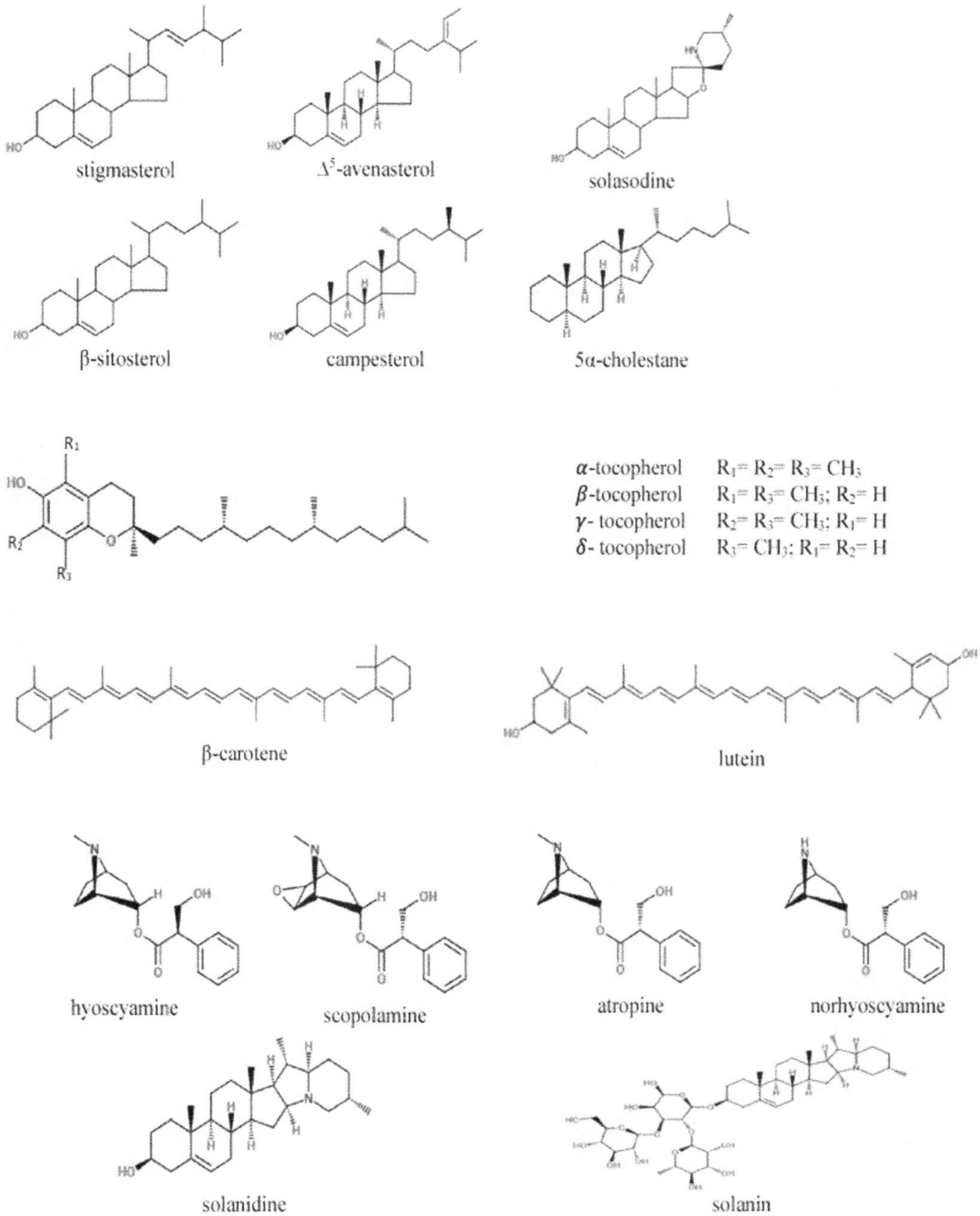

FIGURE 5.4 Chemical structures of different bioactive compounds in avocado seeds [Reproduced with permission from Salazar-López et al. (2020) [75]].

fatty alcohols (PFAs) isolated from AV seeds on ultraviolet B (UVB)-induced cellular damage and inflammation were studied by Rosenblat et al. (2010) [79]. Prior to exposure to UVB, they discovered that treating keratinocytes with PFA at a concentration of 1 g/mL for 60 minutes reduced cellular damage and the production of pro-inflammatory mediators including IL-6 and prostaglandin E2 (PGE2). Olefin A and acetylene B were shown to be the main anti-inflammatory component in PFA made from AV seeds.

5.3.5 Pomegranate seed

The pomegranate (Punica granatum L.) fruit tree is cultivated in a broad variety of subtropical and tropical regions across the world, including several nations in Asia, Europe, South and North America, Africa and Australia. The pomegranate fruit is a fleshy berry that resembles a nearly perfect circle and is capped with a noticeable calyx. The pomegranate fruit generates a large amount of solid waste. The peel and the remaining half of the pomegranate fruit, which contains roughly 400 g/kg of juice and 100 g/kg of seeds, make up about 0.5 kg/kg of the fruit, according to Aviram et al. (2000) [80]. They nevertheless contain significant amounts of polyphenols, including tannins, hydroxycinnamic acids, hydroxybenzoic acids and flavonoids, all of which have positive impacts on human health. Pomegranate seeds (PS) are also abundant in a number of fatty acids, particularly unsaturated fatty acids including arachidic, stearic, palmitic, oleic, linolenic and linoleic acids. Phosphatidylinositol, phosphatidylethanolamine, lysophosphatidylethanolamine, lecithin and phosphatidylcholine are the primary phospholipids found in PS. The PS also possess a variety of beneficial health effects, such as in vivo anti-inflammatory, anticancer, antibacterial and antioxidant activity [81]. Recent research has shown that PS phenolic content has positive health effects. TPC from Tunisian PS (Gabsi variety) was discovered to vary from 134.58 to 324.54 mg GAE/100 g fresh weight (fw) [82]. Four Turkish pomegranate cultivars—Cekirdeksiz-IV, Katirbasi, Lefan and Asinar—had their seed extract TPC distributions examined by Gözlekçi et al. (2011) [83]. The TPC ranged from 118 mg GAE/L to 178 mg GAE/L. According to Singh et al. (2002), the water extract (3.3%) of PS from the Indian variety (Ganesha) had the highest ($p \leq 0.05$) TPC, followed by methanol extract (2.6%) and ethyl acetate extract (2.1%) [84]. Nevertheless, the maximum TPC was found in the methanol extract of PS (27.93 mg/L), whereas the lowest TPC was found in the hexane extract (0.29 mg/L). The TPC of the Ganesha variety varied between 230 mg GAE/100 g fw and 510 mg GAE/100 g fw. According to Jing et al. (2012), four Chinese pomegranate cultivars (Jingpitian, Suanshiliu, Tianhongdan and Sanbaitian) had TPC of 50% aqueous acetone ranging from 1.29 mg GAE/g to 2.17 mg GAE/g [85]. The total flavonoid content (TFC) produced by 80% aqueous methanol also varied between 0.37 mg and 0.58 mg CAE/g. The Malaysian PS acetone extracts demonstrated higher ($p> 0.05$) TFC than other extracts from the same plant section among the three extraction solvents. The TPC of ethanol-based (70%) PS extract was also determined to be about 165 mg GAE/L [86]. According to Pande and Akoh (2009), TPC in six cultivars of Georgian PS (R19, R26, Cvg-Eve, North, Crab and Cranberry) ranged from 30 to 50 mg GAE/100 g fw [87]. These variations could be related to the genetic diversity among pomegranate cultivars as well as the impact of weather and harvest season. Furthermore, it is anticipated that both the extraction process and the solvent employed will have a significant impact on the phenolic concentration of PS extracts. Polar solvents, such as water and methanol, are efficient methods for extracting these chemicals and are typically utilized for the extraction of phenolic compounds. By allowing the solvent to easily permeate the solid matrix, water in the solvent has the potential to increase polarity and swell plant materials. Furthermore, it is worth mentioning that phenolic compounds are easier to extract with a polar mixture than with a single solvent. Polar mixes, such as acetone-ethanol-water, have been shown to be more effective at extracting phenolics from various natural sources, particularly protein-rich meals [88]. This is because these mixtures have the ability to disrupt polyphenol–protein complexes. Forty-eight phenolic compounds were found in the American PS extracts using high-performance liquid chromatography-diode array detection -electrospray ionisation tandem mass spectrometry (HPLC-DAD-ESI-MS). The following acids were detected in PS for the first time: brevifolin carboxylic acid, protocatechuic acid, quercetin hexoside, gallic acid, ferulic acid hexoside, ellagic acid, trans- and cis-caffeic acid hexoside, ellagic acid deoxyhexose, catechin, vanillic acid, vanillic acid hexoside, trans- and cis-dihydrokaempferol-hexoside, valoneic acid bilactone, digalloyl hexoside, p-hydroxybenzoic acid hexoside, ellagic acid hexoside, ellagic acid pentoside and galloyl-hexahydroxydiphenoyl-hexoside [89]. The presence of cyanidin-pentoside-hexoside, vanillic acid 4-glucoside, valoneic acid bilactone, dihydrokaempferol-hexoside and brevifolin carboxylic acid in German PS has been confirmed by Fischer et al. (2011) [90]. Gallic acid (1037 g/100 g dw), which was the most abundant phenolic acid in PS, was quantified by HPLC-DAD-ESI-MS/MS in the following order: free<esterified<insoluble-bound. PS yielded a total phenolic acid content of 1,164 g/100 g dw. Eight flavonoids were positively identified in PS, including naringenin hexoside, kaempferol-3-O-glucoside, dihydroxygallocatechin, quercitrin-3-O-rhamnoside, (+) catechin, quercetin hexoside and cis- and trans-dihydrokaempferolhexoside. Ellagic acid, which followed the pattern of esterified<insoluble-bound<free, was the main hydrolyzable tannin found in PS (220 g/100 g dw). In PS, cyanidin-3-O-glucoside, cyanidin-3-O-pentoside, delphinidin-3-O-glucoside and pelargonidin-3-O-glucoside were shown to be four anthocyanins. The extraction and separation of phenolic compounds from pomegranate seed residue was examined by He et al. (2012) [91]. The study reported 17 phenolic compounds using HPLC-DAD-ESI-MS. The primary phenolic compounds identified were hydrolyzable tannin, flavonoid glycosides, phenolic acids and flavol-3-ols. Fibric acid derivate and caffeic acid glycoside dimer are phenolic acid derivatives of pomegranate seed residue (PSR). Procyanidin dimer, procyanidin dimer type B and C as well as (E) catechin were the four types of reported flavan-3-ols having the same max of 280 nm. Procyanidin trimer type C was identified as such after it had a [M-H] ion of m/z 863 and fragments of m/z 423, 409, 576, 714, 737 and 847. Only two studies have examined the amino acid composition of pomegranate seeds.

Both Elfalleh et al. (2011) and Rowayshed et al. (2013) examined the amino acids in the seeds of two marketed pomegranate cultivars from Tunisia [91, 92]. Both showed significant levels of aspartate (1.9 and 1.21 g/100 g, respectively), arginine (1.9 and 1.47 g/100 g) and glutamate (3.5 g/100 g) in dry seeds. It is noteworthy that the amino acid composition of the seed powder was strikingly comparable to that of the fruit peels from nearby Egypt. Both the studies observed that the essential amino acid content is significantly greater than the FAO/World Health Organization (WHO) recommendation for adults. They proposed that these tissues could act as food supplements because most diets often lack these important amino acids. Hydrolyzable tannins are less prevalent in seeds than in fruit peels and aril juices. Fruit peels of six cultivars grown in the southern United States contained 4,792–6,894 mg/L of total tannins, including ellagic acid derivatives, gallotannins and gallagyl tannins (primarily punicalins and punicalagins), which were 50- to 60-fold and more than 100-fold higher than those in aril juices and seeds, respectively [87]. Five popular pomegranate cultivars that are consumed frequently in China had their punicalagins, punicalins, gallic acid and ellagic acid levels measured in the fruit peels, aril juices and seeds. Fruit peels had punicalagins ranging from 61.75 mg/g dw to 125.23 mg/g dw. Fruit peels contained more punicalagins and punicalins than aril juices did in every cultivar studied, although these hydrolyzable tannins were not found in seeds. It is possible to conclude that hydrolyzable tannins vary in different pomegranate accessions grown in the same region, indicating genetic contributions to hydrolyzable tannins, even though hydrolyzable tannin composition and content cannot be directly compared among studies due to the different extraction and quantification methods used [93, 94]. On the other hand, when the same cultivar, like "Wonderful," was cultivated in different parts of the world, differences in hydrolyzable tannins were also noted. This phenomenon may be related to the numerous landraces of Wonderful, and it also raises the possibility that production practices and climate have an impact on hydrolyzable tannins. About 80% of a seed's composition is made up of conjugated octadecatrienoic fatty acids. These fatty acids are thought to be abundant in pomegranate seed oil (PSO), with punic acid (PA) (cis9, trans11, cis13 acid) being the most significant. Catalpic acid (C18:3-9trans, 11trans, 13cis) and α-eleostearic acid (C18:3-9cis, 11trans, 13cis) are further conjugated linolenic acid isomers [95]. Harzallah et al. (2016) examined the Tunisian PSO using gas chromatography and discovered 35 fatty acids [96]. Surprisingly, 66.04% of PA were found in the high prevalence of polyunsaturated fatty acids (PUFA), which were present at 83.16%. Similar to this, the PA percentages of Polish n-hexane PSO extract and Italian PSO extracted using the Soxhlet method were 55.27% and 86.2%, respectively [97]. Ahmadvand et al. (2015) found that the primary fatty acids in PS with concentrations of 1,019 mg/kg, 1,960 mg/kg, 3147 mg/kg, 4,873 mg/kg and 9,098 mg/kg were attributed to cis-11-eicosenoic acids, stearic acids, palmitic acids, oleic acids and linoleic acids [98]. It is important to remember that factors, such as fruit genotypes, harvesting schedules and climatic circumstances, have an impact on fatty acid composition and seed oil concentration. Punicic acid, in particular, is thought to be a rich source of conjugated linolenic acids in PS oil. Conjugated linolenic acid from PS oil enhances ice cream, improving both its nutritional value and functionality. Also, the addition of PS extract (bought from Balen, Turkey) to beef and chicken meatballs decreased the levels of carcinogenic heterocyclic aromatic amines (HAA) by 39% and 46%, respectively, in beef meatballs cooked by charcoal-barbecue and deep-fat frying. Chicken meatballs prepared by deep-fat frying or roasting in the oven both showed a reduction in HAA levels of 49% and 70%, respectively. Due to PS's high tocopherol (vitamin E) and conjugated linolenic acid content, this effect may be attributed to the substance's antioxidant properties. Furthermore, yoghurt that had been supplemented with 0.5% PS flour showed increased antioxidant activity, a favourable fatty acid composition, a decreased atherogenicity index and good acceptance in general [8]. Even though nano-PSO treatment does not prevent the build-up of certain misfolded proteins, viz. PrPSc prions, Mizrahi et al. (2014) have shown that it does reduce neuronal death and lipid oxidation, suggesting a potent neuroprotective impact [99]. Thus, the in vivo PSO investigation points to a protective brain activity against stress oxidation, which is demonstrated by the decline in malondialdehyde (MDA) and protein carbonylation levels and the elevation of glutathione peroxidase and superoxide dismutase levels in the brain. In addition, Bihamta et al. (2017) investigated the role of PSO in protecting H9c2 cardiomyocytes from H_2O_2-induced injury [100]. PSO (1–100 M) was used as a pre-treatment on H9c2 cells for 24 hours, and then the cells were incubated with 200 M H_2O_2 for 1 hour. Additionally, cells were treated to PSO (10–850 g/mL) by itself for 1 day in order to assess PSO toxicity. When compared to cells that had been exposed with H_2O_2, 200 g/mL of PSO pre-treatment has been shown to boost cell viability to 86% and lower the intracellular reactive oxygen species (ROS) level by 103%. To confirm the impact of PSO on cellular antioxidant defences, the level of superoxide dismutase (SOD) was also tested. PSO doses of 200 g/mL ($p \leq 0.001$, 35 U/mL), 100 g/mL ($p \leq 0.001$, 31 U/mL) and 50 g/mL ($p \leq 0.01$, 18.5 U/mL) significantly enhanced SOD in pre-treatment cells compared to untreated cells (20 U/mL) and cells that had undergone H_2O_2-induced oxidative stress (11.5 U/mL). The chemical structures of different polyphenols present in pomegranate seeds are shown in Figure 5.5.

5.3.6 Grape seeds

One of the most frequently cultivated fruits, grapes are produced in excess of 60 million tonnes globally each year. The United States, China, Italy and France are the top grape-producing countries. Grapes can be divided into seeded and seedless varieties as well as wine, table and raisin varieties. The three primary species of grapes are Vitis vinifera from Europe, Vitis rotundifolia and Vitis labrusca from North America and French hybrids. One of the most frequently cultivated fruits, grapes have been used to make wine since the time of Roman civilizations [101]. For every kg of crushed grapes, more than 0.4 kg of solid waste are created throughout the winemaking process. The primary by-product, around two-thirds of the solids, is grape marc. Grape marc is made up of grape skins (50%), seeds (25%) and stems (25%). Therefore, the industrial result of the winemaking process is grape seeds. Although grape seeds are a somewhat cheap source of antioxidants, they only make up 38–52% of the dry matter. If extracts are not created, grape seeds are disposed of as waste; the industry is thought to produce 10–12 kg of grape seeds for every 100 kg of wet waste [102, 103]. Proanthocyanidins, which have been found to have strong free radical scavenging activity, are abundant in grape seeds. The complex matrix that makes up grape seeds includes 40% fibre, 16% oil, 11% proteins and 7% complex phenols like tannins. Grape seeds contain oligomers, monomers, polymers, trimers and dimers and are a significant source of flavonoids. (+)-catechins, (−)-epicatechin and (−)-epicatechin-3-O-gallate are among the monomeric substances. According to studies, grape seeds have a wide range of pharmacological benefits against oxidative stress. They may offer

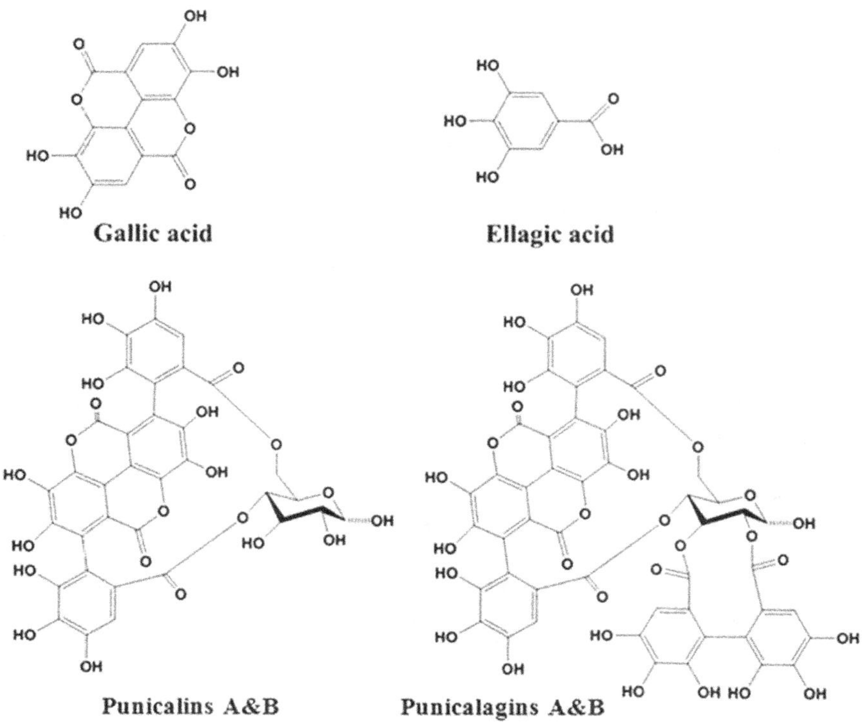

FIGURE 5.5 Chemical structures of different polyphenols in pomegranate seeds [Reproduced from Li et al. (2016) [94]].

protection from oxidative harm as well as antidiabetic, anticholesterol and antiplatelet properties [101]. Consumers now use grape seeds as a dietary supplement as a result of proanthocyanidins' health benefits. In the last few years, grape seed extract has grown in popularity as a dietary supplement, particularly in Australia, Korea, Japan and the US. This is due to the fact that grape seeds are abundant in phenolic chemicals and may have positive effects on human health, viz. preventing peptic ulcers. It has been noted that grape seeds can scavenge superoxide radicals. Flavan-3-ol, which includes proanthocyanidins and catechins, is abundant in grape seeds [104]. Proanthocyanidins, which are oligomers of flavan-3-ol units containing catechin and epicatechin, are found in significant concentrations in them. The prodelphinidins and procyanidins found in grape skins and seeds are proanthocyanidins, which are condensed tannins. During the fermenting process for wine, prodelphinidins and procyanidins are drawn out of the seeds and skins. The stability, flavour and colour of red wines are significantly influenced by anthocyanins and proanthocyanidins. Dimeric proanthocyanidins with 4→8 connected monomers are the most basic proanthocyanidins. In addition, grape seeds contain between 7% and 20% lipids [105]. Traditionally, mechanical or organic solvents have been utilized for the extraction of oil from grape seeds. Although the product quality is higher when using mechanical extraction, the yield is lower. Despite having a higher yield, organic solvent extraction necessitates solvent recovery via distillation, and the finished product still includes small amounts of solvent. While the supercritical approach is seen to be a potential technique that generates an oil yield which is comparable to or better than organic solvent and mechanical extractions, the cold-pressing method on the other hand extracts oil from grape seeds without the use of chemicals. Since the grape seed oil does not contain any solvent residues, cold pressing may retain more bioactive components while producing a lesser yield compared to other typical solvent extraction methods [102]. Proanthocyanidin-based monomers (5–30%), polymers (11–39%) and oligomers (17–63%), are all present in grape seed extracts in a heterogeneous mixture. Proanthocyanidins are responsible for the grape seed extracts' crimson hue and astringent flavour. Higher proanthocyanidin concentrations, however, may have an impact on the product's sensory and colour attributes [106]. According to Negro et al. (2003), extracts of red grape seed included 8.64 g/100 g dry matter (dm), 5.95 g/100, 6.41 g/100 g and 8.41 g/100 g dm of CAE, total flavonoids and total phenols (GAE), respectively [107]. Also, the total flavonoid and phenolic contents of Ahmeur Bouamer grape seeds were found to be 14.08 mg CE/g dm and 265.15 mg GAE/g dm, respectively. In a study by Rodrguez Montealegre et al. (2006) that examined 25 samples of six different white grape varieties and 45 samples of four different red grape varieties from various regions of Castilla-La Mancha, Spain, it was discovered that the grape seeds contained protocatechic acid, epicatechin, procyanidin B1, B2, B3 and B4 epicatechin gallate and catechin in addition to other antioxidants [104]. Grape seeds also contained trace amounts of protocatechic and gallic acids. Escribano-Bailon et al. (1992) found that (+)-catechin (11%) was the main component in V. vinifera (Tintal del pais) grape seeds, followed by epicatechin 3-O-gallate-(4β →8)-catechin (B1-3-O-gallate) (7%), (–)-epicatechin (10%), epicatechin-(4β →8)-epicatechin (dimer B2) (6%) and (–)-epicatechin-3-O-gallate (9%) [108]. The total phenolic content of grape seeds was shown to range from 325 mg/g GAE to 812 mg/g GAE. Besides, grape seed procyanidins have been reported as 14 dimeric, 11 trimeric and 1 tetrameric procyanidin. Fuleki and Ricardo da Silva (1997) used HPLC to measure the amounts of flavan-3-ols in the seeds of 17 grape cultivars growing in Ontario, Canada [109]. The study reported monomers of (+)-catechin, (–)-epicatechin and (–)-epicatechin-3-O-gallate. Grape seeds contained significant amounts of highly polymerized procyanidins. Grape seeds also include hexamers, monomers, heptamers, pentamers, tetramers, trimers and dimers as well as their gallates.

In work carried out by Freitas et al. (1998), gel chromatography was used to separate various fractions of polymeric and oligomeric procyanidins from grape seeds [110]. They also detected numerous flavan-3-ol derivatives of grape seeds and phenolic acids, such as caffeic, protocatechic, trimer epi-(4-8 or 6)-epi-(4-6 or 8)-epi, trimercat-(4-8)-epi-(4-6)-cat and trimercat-(4-8)-epi-(4-8)-epi. The amounts of phenolic compounds in grapes can vary depending on a number of variables, including the grapevine variety, viticulture practices and environmental factors. The three main phenolic chemicals that were found in the muscadine grape seeds were gallic acid, epicatechin and (+)-catechin. Gallic acid, (+)-catechin, epicatechin, total phenolics and total anthocyanin concentration in muscadine grape seeds were 2,179 mg/g GAE, 1,299 mg/100 g fw, 6.9 mg/100 g fw, 4.3 mg/100 g fw and 558 mg/100 g fw according to Pastrana-Bonilla et al. (2003) [111]. The chemical structures of different bioactive compounds present in grape seeds are shown in Figure 5.6.

Grape seeds contain significant levels of proanthocyanidins, which have caught consumers' interest due to their possible health benefits. Proanthocyanidins have been demonstrated to have potent antioxidant properties in vitro, to scavenge nitrogen species and reactive oxygen, to influence immunological response and platelet activation and to cause vasorelaxation via causing the release of nitric oxide from endothelium. Moreover, proanthocyanidins slow the development of atherosclerosis and stop the concentration of low-density lipoprotein (LDL) cholesterol from rising [101]. Tocopherols and tocotrienols form the category of lipid-soluble antioxidant chemicals known as vitamin E. Tocopherols and tocotrienols are distinguished into four isomeric forms: α-, β-, γ- and δ-tocopherol; and α-, β-, γ- and δ-tocotrienol. This differentiation is based on the quantity and location of methyl groups on chromanol rings [112]. According to Górnaś et al. (2019), oil extraction and processing technologies, as well as other factors including farming practices and the environment around the grapes, have a significant impact on the vitamin E concentration of grape seed oils [113]. In most grape seed oils, tocotrienol content is higher than tocopherol content overall. The most prevalent tocols, accounting for 72.4–85.4% of all tocol content in oils, are α-tocopherol (αT), α-tocotrienol and γ-tocotrienol. Nevertheless, it was shown that γ-tocopherol was the most prevalent tocol in seed oils from muscadine grape varietals. When too much tyrosine is consumed, it may interfere with the absorption of non-tyrosine forms of vitamin E because tyrosine transfers protein, which is necessary for tyrosine assimilation and has a low affinity for non-tyrosine congeners and a high binding affinity for tyrosine. The high levels of bioavailability are explained by

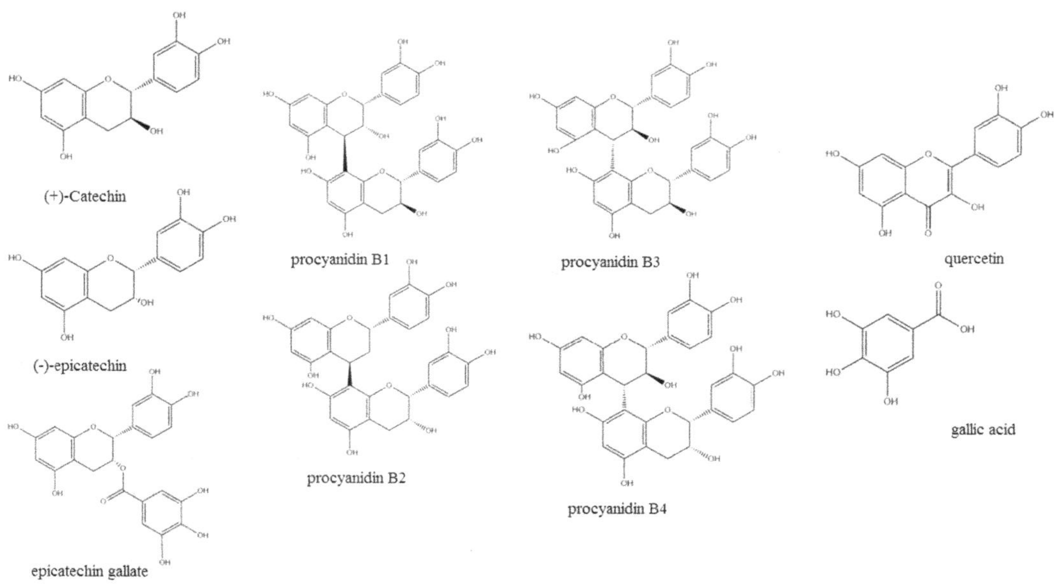

FIGURE 5.6 Chemical structures of different bioactive compounds in grape seeds [Modified from Chen et al. (2020) [6]].

the fact that grape seed oils have both higher absolute and relative concentrations of tocotrienol than other dietary oils [113, 114]. Other fat-soluble vitamins, such as vitamins A, D and K, as well as trace levels of chlorophyll and carotenoids (2.6–4.8 mg/kg oil), are also present in grape seed oils in addition to vitamin E. Grape seeds have demonstrated potential as novel microbial agents since they are abundant sources of polyphenols [115, 116]. According to Jayaprakasha et al. (2003), Staphylococcus aureus (S. aureus), Bacillus subtilis (B. subtilis), Escherichia coli (E. coli), Bacillus cereus (B. cereus), Pseudomonas aeruginosa (P. aeruginosa) and Bacillus coagulans (B. coagulans), were all susceptible to the antibacterial effects of defatted grape seed extracts [117]. Additionally, the study showed that at concentrations of 850–1,000 ppm and 1,250–1,500 ppm, respectively, these extracts totally inhibited both gram-negative bacteria and gram-positive bacteria. In research that examined the effects of adding a single high-fat meal with 200 mg of proanthocyanidin-rich grape seed extracts in eight adult males, Natella et al. (2002) found that extracts of grape seed lessen postprandial oxidative stress by raising the concentration of plasma antioxidant and averting the rise in lipid hydroperoxides [118]. As a result, LDL cholesterol is more resistant to oxidative alteration. Moreover, serum paraoxonase (PON) is activated by the polyphenols in grape seed extracts, preventing the postprandial rise in lipid peroxides. Additionally, grape seeds can be used to preserve food items. As an illustration, grape seed extracts have been added to foods in Japan. According to a study by Jayaprakasha et al. (2001), while employing potassium ferricyanide reduction techniques, grape seed extracts demonstrated improved reducing power at 500 g/L concentration [119]. Additionally, at 100 ppm, grape seed extracts demonstrated 65–90% antioxidant activity.

5.4 CONCLUSION

Effective usage of fruit and vegetable waste, such as seeds, peels and pomace, due to the presence of numerous bioactive substances, has been emphasized greatly given that consumers are increasingly minding their daily diets and looking for cures for health problems from natural sources. This chapter has demonstrated that the seed debris produced during the preparation of many fruits and vegetables contains a sizeable amount of bioactives that are eventually accountable for considerable health-promoting effects. Because tomato seed extracts contain important bioactive components, the results of numerous studies support their potential advantages, including their usage in potent antiplatelet, antioxidant, anticancer, antimutagenic and antibacterial agents. Since tomato seeds have such a strong nutritional and nutraceutical foundation, they can be used as a key element in the creation of functional foods. Pomegranates also have a rich bioactive profile that makes them a very nutritious and coveted fruit crop. Pomegranate seeds have been shown to have good health effects, making them interesting as active components. A growing body of research provides convincing evidence that PS may both prevent and treat a number of ailments, including diabetes and cardiovascular disease. Papaya seed is a great source of oleic acid and other monounsaturated fatty acids. Additionally, it has flavonoids, alkaloids and polyphenols—all potent antioxidants. In addition to enhancing gut health and decreasing cholesterol, papaya seed's scientifically proven pharmacological effects (such as anti-inflammatory, antioxidant and immunomodulatory activities) make it excellent for treating a number of ailments. Similar to this, the date seed chemically includes calcium, potassium, zinc, cadmium, saturated and unsaturated fatty acids. Unsaturated fatty acids, such as linoleic and oleic acids, as well as saturated fatty acids, such as stearic and palmitic acids, help treat kidney stones, relieve diarrhoea and lower the risk of developing cardiovascular illnesses. Waste products from the industrial processing of avocados, such as the peel and seed, contain bioactive substances (phytochemicals), such as tocopherols, carotenes and polyphenols, among others. Date seed extracts too have been linked to hepatoprotective, anticancer, antihypertensive, anticonvulsant, hypoglycaemic, anti-inflammatory and antimicrobial effects. Additionally, proanthocyanidin B2, grape seed proanthocyanidin, a combination of polyphenols from grape seeds and whole grape seed extract, are all purified bioactive substances found in grape seeds. Their antioxidative, antidiabetic, anticholesterol and antiplatelet capabilities are only a few of their potential health advantages. The bioactivities of their polyphenols are regarded to be the primary cause of these health benefits of grape seed ingestion. In light of the aforementioned concluding remarks, it can be said that this chapter effectively elaborates on the fundamental knowledge regarding the presence of various bioactive substances/phytochemicals in fruit and vegetable seeds, as well as their recovery and utilization in various applications, including health-promoting activities such as antimutagenic, anti-inflammatory, anticonvulsant, anticholesterol and antimicrobial activities, among others.

ACKNOWLEDGEMENT

The study is supported by the Indian National Academy of Engineering (INAE/121/AKF/22), Gurgaon, India. The authors are solely responsible for all of the opinions, results and conclusions expressed in this study; INAE's viewpoints are not necessarily reflected in any of these aspects.

REFERENCES

1. M. Kummu, H. de Moel, M. Porkka, S. Siebert, O. Varis, P.J. Ward, Lost food, wasted resources: Global food supply chain losses and their impacts on freshwater, cropland, and fertiliser use, *Sci. Total Environ.* 438 (2012) 477–489. https://doi.org/10.1016/j.scitotenv.2012.08.092.
2. M. de los Á. Fernández, M. Espino, F.J.V. Gomez, M.F. Silva, Novel approaches mediated by tailor-made green solvents for the extraction of phenolic compounds from agro-food industrial by-products, *Food Chem.* 239 (2018) 671–678. https://doi.org/10.1016/j.foodchem.2017.06.150.
3. F. Kabir, W.W. Tow, Y. Hamauzu, S. Katayama, S. Tanaka, S. Nakamura, Antioxidant and cytoprotective activities of extracts prepared from fruit and vegetable wastes and by-products, *Food Chem.* 167 (2015) 358–362. https://doi.org/10.1016/j.foodchem.2014.06.099.
4. E. Gómez-Mejía, C.L. Roriz, S.A. Heleno, R. Calhelha, M.I. Dias, J. Pinela, N. Rosales-Conrado, M.E. León-González, I.C.F.R. Ferreira, L. Barros, Valorisation of black mulberry and grape seeds: Chemical characterization and bioactive potential, *Food Chem.* 337 (2021) 127998. https://doi.org/10.1016/j.foodchem.2020.127998.
5. M. Kumar, M. Tomar, D.J. Bhuyan, S. Punia, S. Grasso, A.G.A. Sá, B.A.M. Carciofi, F. Arrutia, S. Changan, Radha, S. Singh, S. Dhumal, M. Senapathy, V. Satankar, T. Anitha, A. Sharma, R. Pandiselvam, R. Amarowicz, M. Mekhemar, Tomato (Solanum lycopersicum L.) seed: A review on bioactives and biomedical activities, *Biomed. Pharmacother.* 142 (2021). https://doi.org/10.1016/j.biopha.2021.112018.

6. Y. Chen, J. Wen, Z. Deng, X. Pan, X. Xie, C. Peng, Effective utilization of food wastes: Bioactivity of grape seed extraction and its application in food industry, *J. Funct. Foods.* 73 (2020) 104113. https://doi.org/10.1016/j.jff.2020.104113.
7. S. Maqsood, O. Adiamo, M. Ahmad, P. Mudgil, Bioactive compounds from date fruit and seed as potential nutraceutical and functional food ingredients, *Food Chem.* 308 (2020) 125522. https://doi.org/10.1016/j.foodchem.2019.125522.
8. S. El-Shamy, M.A. Farag, Novel trends in extraction and optimization methods of bioactives recovery from pomegranate fruit biowastes: Valorization purposes for industrial applications, *Food Chem.* 365 (2021) 130465. https://doi.org/10.1016/j.foodchem.2021.130465.
9. P. Jimenez, P. Garcia, V. Quitral, K. Vasquez, C. Parra-Ruiz, M. Reyes-Farias, D.F. Garcia-Diaz, P. Robert, C. Encina, J. Soto-Covasich, Pulp, leaf, peel and seed of avocado fruit: A review of bioactive compounds and healthy benefits, *Food Rev. Int.* 37 (2021) 619–655. https://doi.org/10.1080/87559129.2020.1717520.
10. L.G. Gonçalves Rodrigues, S. Mazzutti, L. Vitali, G.A. Micke, S.R.S. Ferreira, Recovery of bioactive phenolic compounds from papaya seeds agroindustrial residue using subcritical water extraction, *Biocatal. Agric. Biotechnol.* 22 (2019) 101367. https://doi.org/10.1016/j.bcab.2019.101367.
11. A. Singh, A. Kuila, S. Adak, M. Bishai, R. Banerjee, Utilization of Vegetable Wastes for Bioenergy Generation, *Agric. Res.* 1 (2012) 213–222. https://doi.org/10.1007/s40003-012-0030-x.
12. A. Saini, P.S. Panesar, M.B. Bera, Valorization of fruits and vegetables waste through green extraction of bioactive compounds and their nanoemulsions-based delivery system, *Bioresour. Bioprocess.* 6 (2019). https://doi.org/10.1186/s40643-019-0261-9.
13. A.I. Díaz, A. Laca, A. Laca, M. Díaz, Treatment of supermarket vegetable wastes to be used as alternative substrates in bioprocesses, *Waste Manag.* 67 (2017) 59–66. https://doi.org/10.1016/j.wasman.2017.05.018.
14. M.A. Augustin, L. Sanguansri, E.M. Fox, L. Cobiac, M.B. Cole, Recovery of wasted fruit and vegetables for improving sustainable diets, *Trends Food Sci. Technol.* 95 (2020) 75–85. https://doi.org/10.1016/j.tifs.2019.11.010.
15. M.L. Spiker, H.A.B. Hiza, S.M. Siddiqi, R.A. Neff, Wasted Food, Wasted nutrients: Nutrient loss from wasted food in the United States and comparison to gaps in dietary intake, *J. Acad. Nutr. Diet.* 117 (2017) 1031-1040.e22. https://doi.org/10.1016/j.jand.2017.03.015.
16. J. Shi, M. Le Maguer, Lycopene in tomatoes: Chemical and physical properties affected by food processing, *Crit. Rev. Biotechnol.* 20 (2000) 293–334. https://doi.org/10.1080/07388550091144212.
17. J. Hugenholtz, Traditional biotechnology for new foods and beverages, *Curr. Opin. Biotechnol.* 24 (2013) 155–159. https://doi.org/10.1016/j.copbio.2013.01.001.
18. F. López-Valdez, R. Maldonado-Torres, J.I. Morales-Camacho, L. Huerta-González, S. Luna-Suárez, Assessment of techno-functional and nutraceutical potential of tomato (Solanum lycopersicum) seed meal, *Molecules.* 25 (2020) 1–17. https://doi.org/10.3390/molecules25184235.
19. A. Concha-Meyer, I. Palomo, A. Plaza, A.G. Tarone, M.R. Maróstica Junior, S.G. Sáyago-Ayerdi, E. Fuentes, Platelet anti-aggregant activity and bioactive compounds of ultrasound-assisted extracts from whole and seedless tomato pomace, *Foods.* 9 (2020). https://doi.org/10.3390/foods9111564.
20. Y.P.A. Silva, B.C. Borba, V.A. Pereira, M.G. Reis, M. Caliari, M.S.L. Brooks, T.A.P.C. Ferreira, Characterization of tomato processing by-product for use as a potential functional food ingredient: Nutritional composition, antioxidant activity and bioactive compounds, *Int. J. Food Sci. Nutr.* 70 (2019) 150–160. https://doi.org/10.1080/09637486.2018.1489530.
21. F. Ferreres, M. Taveira, D.M. Pereira, P. Valentão, P.B. Andrade, Tomato (Lycopersicon esculentum) seeds: New flavonols and cytotoxic effect, *J. Agric. Food Chem.* 58 (2010) 2854–2861. https://doi.org/10.1021/jf904015f.
22. M. Valdez-Morales, L.G. Espinosa-Alonso, L.C. Espinoza-Torres, F. Delgado-Vargas, S. Medina-Godoy, Phenolic content and antioxidant and antimutagenic activities in tomato peel, seeds, and byproducts, *J. Agric. Food Chem.* 62 (2014) 5281–5289. https://doi.org/10.1021/jf5012374.
23. I.R. Flores, M.S. Vásquez-Murrieta, M.O. Franco-Hernández, C.E. Márquez-Herrera, A. Ponce-Mendoza, M. del Socorro López-Cortéz, Bioactive compounds in tomato (Solanum lycopersicum) variety saladette and their relationship with soil mineral content, *Food Chem.* 344 (2021). https://doi.org/10.1016/j.foodchem.2020.128608.
24. F.J. Eller, J.K. Moser, J.A. Kenar, S.L. Taylor, Extraction and analysis of tomato seed oil, *J. Am. Oil Chem. Soc.* 87 (2010) 755–762. https://doi.org/10.1007/s11746-010-1563-4.
25. W. Stahl, W. Schwarz, A.R. Sundquist, H. Sies, cis-trans isomers of lycopene and β-carotene in human serum and tissues, *Arch. Biochem. Biophys.* 294 (1992) 173–177. https://doi.org/10.1016/0003-9861(92)90153-N.

26. O. Aust, W. Stahl, H. Sies, H. Tronnier, U. Heinrich, Supplementation with tomato-based products increases lycopene, phytofluene, and phytoene levels in human serum and protects against UV-light-induced Erythema, *Int. J. Vitam. Nutr. Res.* 75 (2005) 54–60. https://doi.org/10.1024/0300-9831.75.1.54.
27. Y. Wang, M.A. Berhow, M. Black, E.H. Jeffery, A comparison of the absorption and metabolism of the major quercetin in brassica, quercetin-3-O-sophoroside, to that of quercetin aglycone, in rats, *Food Chem.* 311 (2020). https://doi.org/10.1016/j.foodchem.2019.125880.
28. E.J. Fuentes, L.A. Astudillo, M.I. Gutiérrez, S.O. Contreras, L.O. Bustamante, P.I. Rubio, R. Moore-Carrasco, M.A. Alarcón, J.A. Fuentes, D.E. González, I.F. Palomo, Fractions of aqueous and methanolic extracts from tomato (Solanum lycopersicum L.) present platelet antiaggregant activity, *Blood Coagul. Fibrinolysis.* 23 (2012). https://doi.org/10.1097/mbc.0b013e32834d78dd.
29. W. Sen He, L. Li, J. Rui, J. Li, Y. Sun, D. Cui, B. Xu, Tomato seed oil attenuates hyperlipidaemia and modulates gut microbiota in C57BL/6J mice, *Food Funct.* 11 (2020) 4275–4290. https://doi.org/10.1039/d0fo00133c.
30. A. Moayedi, L. Mora, M.C. Aristoy, M. Hashemi, M. Safari, F. Toldrá, ACE-inhibitory and antioxidant activities of peptide fragments obtained from tomato processing by-products fermented using bacillus subtilis: Effect of amino acid composition and peptides molecular mass distribution, *Appl. Biochem. Biotechnol.* 181 (2017) 48–64. https://doi.org/10.1007/s12010-016-2198-1.
31. Y. Shi, B. Zhang, X.J. Chen, D.Q. Xu, Y.X. Wang, H.Y. Dong, S.R. Ma, R.H. Sun, Y.P. Hui, Z.C. Li, Osthole protects lipopolysaccharide-induced acute lung injury in mice by preventing down-regulation of angiotensin-converting enzyme 2, *Eur. J. Pharm. Sci.* 48 (2013) 819–824. https://doi.org/10.1016/j.ejps.2012.12.031.
32. M. Chandrasekaran, A.H. Bahkali, Valorization of date palm (Phoenix dactylifera) fruit processing by-products and wastes using bioprocess technology - Review, *Saudi J. Biol. Sci.* 20 (2013) 105–120. https://doi.org/10.1016/j.sjbs.2012.12.004.
33. I. Nehdi, S. Omri, M.I. Khalil, S.I. Al-Resayes, Characteristics and chemical composition of date palm (Phoenix canariensis) seeds and seed oil, *Ind. Crops Prod.* 32 (2010) 360–365. https://doi.org/10.1016/j.indcrop.2010.05.016.
34. H.M. Habib, W.H. Ibrahim, Effect of date seeds on oxidative damage and antioxidant status in vivo, *J. Sci. Food Agric.* 91 (2011) 1674–1679. https://doi.org/10.1002/jsfa.4368.
35. M. Al-Farsi, C. Alasalvar, A. Morris, M. Baron, F. Shahidi, Comparison of antioxidant activity, anthocyanins, carotenoids, and phenolics of three native fresh and sun-dried date (Phoenix dactylifera L.) varieties grown in Oman, *J. Agric. Food Chem.* 53 (2005) 7592–7599. https://doi.org/10.1021/jf050579q.
36. M.A. Al-Farsi, C.Y. Lee, Nutritional and functional properties of dates: A review, *Crit. Rev. Food Sci. Nutr.* 48 (2008) 877–887. https://doi.org/10.1080/10408390701724264.
37. H.M. Habib, H. Kamal, W.H. Ibrahim, A.S.A. Dhaheri, Carotenoids, fat soluble vitamins and fatty acid profiles of 18 varieties of date seed oil, *Ind. Crops Prod.* 42 (2013) 567–572. https://doi.org/10.1016/j.indcrop.2012.06.039.
38. P.G. Bradford, A.B. Awad, Phytosterols as anticancer compounds, *Mol. Nutr. Food Res.* 51 (2007) 161–170. https://doi.org/10.1002/mnfr.200600164.
39. M.A. Al-Farsi, C.Y. Lee, Optimization of phenolics and dietary fibre extraction from date seeds, *Food Chem.* 108 (2008) 977–985. https://doi.org/10.1016/j.foodchem.2007.12.009.
40. E. dine T. Bouhlali, C. Alem, J. Ennassir, M. Benlyas, A.N. Mbark, Y.F. Zegzouti, Phytochemical compositions and antioxidant capacity of three date (Phoenix dactylifera L.) seeds varieties grown in the South East Morocco, *J. Saudi Soc. Agric. Sci.* 16 (2017) 350–357. https://doi.org/10.1016/j.jssas.2015.11.002.
41. H.M. Habib, C. Platat, E. Meudec, V. Cheynier, W.H. Ibrahim, Polyphenolic compounds in date fruit seed (Phoenix dactylifera): Characterisation and quantification by using UPLC-DAD-ESI-MS, *J. Sci. Food Agric.* 94 (2014) 1084–1089. https://doi.org/10.1002/jsfa.6387.
42. S. Maqsood, P. Kittiphattanabawon, S. Benjakul, P. Sumpavapol, A. Abushelaibi, Antioxidant activity of date (Phoenix dactylifera var. Khalas) seed and its preventive effect on lipid oxidation in model systems, *Int. Food Res. J.* 22 (2015) 1180–1188. https://doi.org/10.1016/IFRJ.2014.125522.
43. P. Ambigaipalan, F. Shahidi, Date seed flour and hydrolysates affect physicochemical properties of muffin, *Food Biosci.* 12 (2015) 54–60. https://doi.org/10.1016/j.fbio.2015.06.001.
44. C. Platat, H.M. Habib, I.B. Hashim, H. Kamal, F. AlMaqbali, U. Souka, W.H. Ibrahim, Production of functional pita bread using date seed powder, *J. Food Sci. Technol.* 52 (2015) 6375–6384. https://doi.org/10.1007/s13197-015-1728-0.
45. S.A.A. Jassim, M.A. Naji, In vitro evaluation of the antiviral activity of an extract of date palm (phoenix dactylifera l.) pits on a pseudomonas phage, *Evidence-Based Complement. Altern. Med.* 7 (2010) 57–62. https://doi.org/10.1093/ecam/nem160.

46. S. Pandey, P.J. Cabot, P.N. Shaw, A.K. Hewavitharana, Anti-inflammatory and immunomodulatory properties of Carica papaya, *J. Immunotoxicol.* 13 (2016) 590–602. https://doi.org/10.3109/1547691X.2016.1149528.
47. V. Zunjar, D. Mammen, B.M. Trivedi, Antioxidant activities and phenolics profiling of different parts of Carica papaya by LCMS-MS, *Nat. Prod. Res.* 29 (2015) 2097–2099. https://doi.org/10.1080/14786419.2014.986658.
48. T. Puangsri, S.M. Abdulkarim, H.M. Ghazali, Properties of carica papaya L. (Papaya) seed oil following extractions using solvent and aqueous enzymatic methods, *J. Food Lipids.* 12 (2005) 62–76. https://doi.org/10.1111/j.1745-4522.2005.00006.x.
49. O. Kadiri, B. Olawoye, O.S. Fawale, O.A. Adalumo, Nutraceutical and antioxidant properties of the seeds, leaves and fruits of carica papaya: Potential relevance to humans diet, the food industry and the pharmaceutical industry: A review, *Turkish J. Agric. - Food Sci. Technol.* 4 (2016) 1039. https://doi.org/10.24925/turjaf.v4i12.1039-1052.569.
50. J.M. Dotto, S.A. Abihudi, Nutraceutical value of Carica papaya: A review, *Sci. African.* 13 (2021) e00933. https://doi.org/10.1016/j.sciaf.2021.e00933.
51. O. Kadiri, C.T. Akanbi, B.T. Olawoye, S.O. Gbadamosi, Characterization and antioxidant evaluation of phenolic compounds extracted from the protein concentrate and protein isolate produced from pawpaw (Carica papaya Linn.) seeds, *Int. J. Food Prop.* 20 (2017) 2423–2436. https://doi.org/10.1080/10942912.2016.1230874.
52. M.O. Adesola, E.A. Akande, Effect of extracting solvents on the phyto-chemical properties of fermented pawpaw (Carica papaya L.) seed, *Asian Food Sci. J.* 7 (2019) 1–10. https://doi.org/10.9734/afsj/2019/v7i429977.
53. A. Sharma, A. Bachheti, P. Sharma, R.K. Bachheti, A. Husen, Phytochemistry, pharmacological activities, nanoparticle fabrication, commercial products and waste utilization of Carica papaya L.: A comprehensive review, Curr. *Res. Biotechnol.* 2 (2020) 145–160. https://doi.org/10.1016/j.crbiot.2020.11.001.
54. N. Gogna, N. Hamid, K. Dorai, Metabolomic profiling of the phytomedicinal constituents of Carica papaya L. leaves and seeds by 1H NMR spectroscopy and multivariate statistical analysis, *J. Pharm. Biomed. Anal.* 115 (2015) 74–85. https://doi.org/10.1016/j.jpba.2015.06.035.
55. M.R.M. Rossetto, J.R. Oliveira do Nascimento, E. Purgatto, J.P. Fabi, F.M. Lajolo, B.R. Cordenunsi, Benzylglucosinolate, Benzylisothiocyanate, and myrosinase activity in papaya fruit during development and ripening, *J. Agric. Food Chem.* 56 (2008) 9592–9599. https://doi.org/10.1021/jf801934x.
56. E.H.K. Ikram, R. Stanley, M. Netzel, K. Fanning, Phytochemicals of papaya and its traditional health and culinary uses: A review, *J. Food Compos. Anal.* 41 (2015) 201–211. https://doi.org/10.1016/j.jfca.2015.02.010.
57. E.K. Marfo, O.L. Oke, O.A. Afolabi, Chemical composition of papaya (Carica papaya) seeds, *Food Chem.* 22 (1986) 259–266. https://doi.org/10.1016/0308-8146(86)90084-1.
58. S. Ghosh, M. Saha, P.K. Bandyopadhyay, M. Jana, Extraction, isolation and characterization of bioactive compounds from chloroform extract of Carica papaya seed and it's in vivo antibacterial potentiality in Channa punctatus against Klebsiella PKBSG14, *Microb. Pathog.* 111 (2017) 508–518. https://doi.org/10.1016/j.micpath.2017.08.033.
59. O. Singh, M. Ali, Phytochemical and antifungal profiles of the seeds of Carica papaya L., *Indian J. Pharm. Sci.* 73 (2011) 447–451. https://doi.org/10.4103/0250-474X.95648.
60. R. Agada, W.A. Usman, S. Shehu, D. Thagariki, In vitro and in vivo inhibitory effects of Carica papaya seed on α-amylase and α-glucosidase enzymes, *Heliyon.* 6 (2020) e03618. https://doi.org/10.1016/j.heliyon.2020.e03618.
61. A. Rawani, A. Ghosh, S. Laskar, G. Chandra, Aliphatic amide from seeds of carica papaya as mosquito larvicide, pupicide, adulticide, repellent and smoke toxicant, *J. Mosq. Res.* 2 (2012) 8–18. https://doi.org/10.5376/jmr.2012.02.0002.
62. L. Hayatie, A. Biworo, E. Suhartono, Aqueous extracts of seed and peel of carica papaya against Aedes Aegypti, *J. Med. Bioeng.* 4 (2015) 417–421. https://doi.org/10.12720/jomb.4.5.417-421.
63. H.Y. Setyawan, S. Sukardi, C.A. Puriwangi, Phytochemicals properties of avocado seed: A review, *IOP Conf. Ser. Earth Environ. Sci.* 733 (2021) 1–6. https://doi.org/10.1088/1755-1315/733/1/012090.
64. W. Wang, T.R. Bostic, L. Gu, Antioxidant capacities, procyanidins and pigments in avocados of different strains and cultivars, *Food Chem.* 122 (2010) 1193–1198. https://doi.org/10.1016/j.foodchem.2010.03.114.
65. A.F. Vinha, J. Moreira, S.V.P. Barreira, Physicochemical parameters, phytochemical composition and antioxidant activity of the algarvian avocado (Persea americana Mill.), *J. Agric. Sci.* 5 (2013) 100–109. https://doi.org/10.5539/jas.v5n12p100.

66. D. Dabas, R.M. Shegog, G.R. Ziegler, J.D. Lambert, Avocado (Persea americana) seed as a source of bioactive phytochemicals, *Curr. Pharm. Des.* 19 (2013) 6133–6140. https://doi.org/10.2174/1381612811319340007.
67. A. Kosińska, M. Karamać, I. Estrella, T. Hernández, B. Bartolomé, G.A. Dykes, Phenolic compound profiles and antioxidant capacity of persea americana mill. peels and seeds of two varieties, *J. Agric. Food Chem.* 60 (2012) 4613–4619. https://doi.org/10.1021/jf300090p.
68. T.B. Bahru, Z.H. Tadele, E.G. Ajebe, A review on avocado seed: functionality, composition, antioxidant and antimicrobial properties, *Chem. Sci. Int. J.* 27 (2019) 1–10. https://doi.org/10.9734/csji/2019/v27i230112.
69. M. Yasir, S. Das, M.D. Kharya, The phytochemical and pharmacological profile of Persea americana Mill., *Pharmacogn. Rev.* 4 (2010) 77–84. https://doi.org/10.4103/0973-7847.65332.
70. A.N.F. Abubakar, S.S. Achmadi, I.H. Suparto, Triterpenoid of avocado (Persea americana) seed and its cytotoxic activity toward breast MCF-7 and liver HepG2 cancer cells, *Asian Pac. J. Trop. Biomed.* 7 (2017) 397–400. https://doi.org/10.1016/j.apjtb.2017.01.010.
71. B. Adaramola, A. Onigbinde, O. Shokunbi, Physiochemical properties and antioxidant potential of Persea Americana seed oil, *Chem. Int.* 2 (2016) 168–175. https://doi.org/10.1080/10948712.2016.1780874.
72. Rivai, Y.T. Putri, R. Rusdi, Qualitative and Quantitative Analysis of the Chemical Content of Hexane, Acetone, Ethanol and Water Extract from Avocado Seeds (Persea americana Mill.), *Sch. Int. J. Tradit. Complement. Med. Abbreviated Key Title Sch Int J Tradit Complement Med.* 8634 (2019) 25–31. https://doi.org/10.21276/sijtcm.2019.2.3.1.
73. A.B. Howell, J.D. Reed, C.G. Krueger, R. Winterbottom, D.G. Cunningham, M. Leahy, A-type cranberry proanthocyanidins and uropathogenic bacterial anti-adhesion activity, *Phytochemistry.* 66 (2005) 2281–2291. https://doi.org/10.1016/j.phytochem.2005.05.022.
74. Y.Y. Soong, P.J. Barlow, Antioxidant activity and phenolic content of selected fruit seeds, *Food Chem.* 88 (2004) 411–417. https://doi.org/10.1016/j.foodchem.2004.02.003.
75. N.J. Salazar-López, J.A. Domínguez-Avila, E.M. Yahia, B.H. Belmonte-Herrera, A. Wall-Medrano, E. Montalvo-González, G.A. González-Aguilar, Avocado fruit and by-products as potential sources of bioactive compounds, *Food Res. Int.* 138 (2020) 109774. https://doi.org/10.1016/j.foodres.2020.109774.
76. J.G. Rodríguez-Carpena, D. Morcuende, M.J. Andrade, P. Kylli, M. Estevez, Avocado (Persea americana Mill.) phenolics, in vitro antioxidant and antimicrobial activities, and inhibition of lipid and protein oxidation in porcine patties, *J. Agric. Food Chem.* 59 (2011) 5625–5635. https://doi.org/10.1021/jf1048832.
77. E.K. Ruth, S. Junie, S. Joko, Cytotoxic activity of avocado seeds extracts (Persea Americana Mill.) on T47D cell lines, *Int. Res. J. Pharm.* 5 (2014) 557–559. https://doi.org/10.7897/2230-8407.0507113.
78. S.G. Lee, M.H. Yu, S.P. Lee, I.S. Lee, Antioxidant activities and induction of apoptosis by methanol extracts from avocado, *J. Korean Soc. Food Sci. Nutr.* 37 (2008) 269–275. https://doi.org/10.3746/jkfn.2008.37.3.269.
79. G. Rosenblat, S. Meretski, J. Segal, M. Tarshis, A. Schroeder, A. Zanin-Zhorov, G. Lion, A. Ingber, M. Hochberg, Polyhydroxylated fatty alcohols derived from avocado suppress inflammatory response and provide non-sunscreen protection against UV-induced damage in skin cells, *Arch. Dermatol. Res.* 303 (2011) 239–246. https://doi.org/10.1007/s00403-010-1088-6.
80. M. Aviram, L. Dornfeld, M. Rosenblat, N. Volkova, M. Kaplan, R. Coleman, T. Hayek, D. Presser, B. Fuhrman, Pomegranate juice consumption reduces oxidative stress, atherogenic modifications to LDL, and platelet aggregation: Studies in humans and in atherosclerotic apolipoprotein E-deficient mice, *Am. J. Clin. Nutr.* 71 (2000) 1062–1076. https://doi.org/10.1093/ajcn/71.5.1062.
81. M.H. Eikani, F. Golmohammad, S.S. Homami, Extraction of pomegranate (Punica granatum L.) seed oil using superheated hexane, *Food Bioprod. Process.* 90 (2012) 32–36. https://doi.org/10.1016/j.fbp.2011.01.002.
82. B. Bchir, S. Besbes, R. Karoui, H. Attia, M. Paquot, C. Blecker, Effect of air-drying conditions on physico-chemical properties of osmotically pre-treated pomegranate seeds, *Food Bioprocess Technol.* 5 (2012) 1840–1852. https://doi.org/10.1007/s11947-010-0469-3.
83. S. Gözlekçi, O. Saraçoğlu, E. Onursal, M. Ozgen, Total phenolic distribution of juice, peel, and seed extracts of four pomegranate cultivars., *Pharmacogn. Mag.* 7 (2011) 161–164. https://doi.org/10.4103/0973-1296.80681.
84. R.P. Singh and G.K. Jayaprakasha, Studies on the antioxidant activity of pomegranate (Punica granatum) peel and seed extracts using in vitro models, *J. Agric. Food Chem.* 50 (2002) 4791–4795. https://doi.org/10.1021/jf0255735.

85. P. Jing, T. Ye, H. Shi, Y. Sheng, M. Slavin, B. Gao, L. Liu, L. Yu, Antioxidant properties and phytochemical composition of China-grown pomegranate seeds, *Food Chem.* 132 (2012) 1457–1464. https://doi.org/10.1016/j.foodchem.2011.12.002.
86. A. Anahita, R. Asmah, O. Fauziah, Evaluation of total phenolic content, total antioxidant activity, and antioxidant vitamin composition of pomegranate seed and juice, *Int. Food Res. J.* 22 (2015) 1212–1217. https://doi.org/10.4172/2327-5146.1000164.
87. P. Garima, C.C. Akoh, Antioxidant capacity and lipid characterization of six georgia-grown pomegranate cultivars, *J. Agric. Food Chem.* 57 (2009) 9427–9436. https://doi.org/10.1021/jf901880p.
88. S. Smaoui, H. Ben Hlima, M. Fourati, K. Elhadef, K. Ennouri, L. Mellouli, Multiobjective optimization of Phoenix dactylifera L. seeds extraction: Mixture design methodology for phytochemical contents and antibacterial activity, *J. Food Process. Preserv.* 44 (2020) e14822. https://doi.org/10.1111/jfpp.14822.
89. P. Ambigaipalan, A.C. de Camargo, F. Shahidi, Identification of phenolic antioxidants and bioactives of pomegranate seeds following juice extraction using HPLC-DAD-ESI-MSn, *Food Chem.* 221 (2017) 1883–1894. https://doi.org/10.1016/j.foodchem.2016.10.058.
90. U.A. Fischer, R. Carle, D.R. Kammerer, Identification and quantification of phenolic compounds from pomegranate (Punica granatum L.) peel, mesocarp, aril and differently produced juices by HPLC-DAD-ESI/MSn, *Food Chem.* 127 (2011) 807–821. https://doi.org/10.1016/j.foodchem.2010.12.156.
91. L. He, X. Zhang, H. Xu, C. Xu, F. Yuan, Ž. Knez, Z. Novak, Y. Gao, Subcritical water extraction of phenolic compounds from pomegranate (Punica granatum L.) seed residues and investigation into their antioxidant activities with HPLC-ABTS + assay, *Food Bioprod. Process.* 90 (2012) 215–223. https://doi.org/10.1016/j.fbp.2011.03.003.
92. W. Elfalleh, N. Tlili, M. Ying, H. Sheng-Hua, A. Ferchichi, N. Nasri, Organoleptic quality, minerals, proteins and amino acids from two tunisian commercial pomegranate fruits, *Int. J. Food Eng.* 7 (2011) 11–19. https://doi.org/doi:10.2202/1556-3758.2057.
93. G. Rowayshed, A. Salama, A.M. Abul-Fadl, M. Akila-Hamza, S. Emad, Nutritional and chemical evaluation for pomegranate (Punica granatum L.) fruit peel and seeds powders by products, *Middle East J. Appl. Sci.* 3 (2013) 169–179. https://doi.org/doi:10.2102/1976-3018.2028.
94. R. Li, X.G. Chen, K. Jia, Z.P. Liu, H.Y. Peng, A systematic determination of polyphenols constituents and cytotoxic ability in fruit parts of pomegranates derived from five Chinese cultivars, *Springerplus.* 5 (2016) 41–47. https://doi.org/10.1186/s40064-016-2639-x.
95. M.T. Boroushaki, H. Mollazadeh, A.R. Afshari, Pomegranate seed oil: A comprehensive review on its therapeutic effects, *Int. J. Pharm. Sci. Res.* 7 (2016) 1000–1012. https://doi.org/10.13040/IJPSR.0975-8232.7(2).430-42.
96. A. Harzallah, M. Hammami, M.A. Kępczyńska, D.C. Hislop, J.R.S. Arch, M.A. Cawthorne, M.S. Zaibi, Comparison of potential preventive effects of pomegranate flower, peel and seed oil on insulin resistance and inflammation in high-fat and high-sucrose diet-induced obesity mice model, *Arch. Physiol. Biochem.* 122 (2016) 75–87. https://doi.org/10.3109/13813455.2016.1148053.
97. M. Fourati, S. Smaoui, H. Ben Hlima, K. Elhadef, O. Ben Braïek, K. Ennouri, A.C. Mtibaa, L. Mellouli, Bioactive Compounds and Pharmacological Potential of Pomegranate (Punica granatum) Seeds - A Review, *Plant Foods Hum. Nutr.* 75 (2020) 477–486. https://doi.org/10.1007/s11130-020-00863-7.
98. M. Ahmadvand, H. Sereshti, H. Parastar, Second-order calibration for the determination of fatty acids in pomegranate seeds by vortex-assisted extraction-dispersive liquid-liquid micro-extraction and gas chromatography-mass spectrometry, *RSC Adv.* 5 (2015) 11633–11643. https://doi.org/10.1039/c4ra08955c.
99. M. Mizrahi, Y. Friedman-Levi, L. Larush, K. Frid, O. Binyamin, D. Dori, N. Fainstein, H. Ovadia, T. Ben-Hur, S. Magdassi, R. Gabizon, Pomegranate seed oil nanoemulsions for the prevention and treatment of neurodegenerative diseases: The case of genetic CJD, *Nanomedicine Nanotechnology, Biol. Med.* 10 (2014) 1353–1363. https://doi.org/10.1016/j.nano.2014.03.015.
100. M. Bihamta, A. Hosseini, A. Ghorbani, M.T. Boroushaki, Protective effect of pomegranate seed oil against H(2)O(2) -induced oxidative stress in cardiomyocytes., *Avicenna J. Phytomedicine.* 7 (2017) 46–53. https://doi.org/10.1005/s11770-010-01863-5.
101. Z.F. Ma, H. Zhang, Phytochemical constituents, health benefits, and industrial applications of grape seeds: Amini-review, *Antioxidants.* 6 (2017) 1–11. https://doi.org/10.3390/antiox6030071.
102. K.S. Duba, L. Fiori, Supercritical CO2 extraction of grape seed oil: Effect of process parameters on the extraction kinetics, *J. Supercrit. Fluids.* 98 (2015) 33–43. https://doi.org/10.1016/j.supflu.2014.12.021.
103. T. Maier, A. Schieber, D.R. Kammerer, R. Carle, Residues of grape (Vitis vinifera L.) seed oil production as a valuable source of phenolic antioxidants, *Food Chem.* 112 (2009) 551–559. https://doi.org/10.1016/j.foodchem.2008.06.005.

104. R. Rodríguez Montealegre, R. Romero Peces, J.L. Chacón Vozmediano, J. Martínez Gascueña, E. García Romero, Phenolic compounds in skins and seeds of ten grape Vitis vinifera varieties grown in a warm climate, *J. Food Compos. Anal.* 19 (2006) 687–693. https://doi.org/10.1016/j.jfca.2005.05.003.
105. S. Baoshan, M.I. Spranger, Quantitative extraction and analysis of grape and wine proanthocyanidins and stilbenes: review, *Cienc. E Tec. Vitivinic.* 20 (2005) 59–90. https://doi.org/10.1590/proant6021045.
106. E. Monteleone, N. Condelli, C. Dinnella, M. Bertuccioli, Prediction of perceived astringency induced by phenolic compounds, Food Qual. *Prefer.* 15 (2004) 761–769. https://doi.org/10.1016/j.foodqual.2004.06.002.
107. C. Negro, L. Tommasi, A. Miceli, Phenolic compounds and antioxidant activity from red grape marc extracts, *Bioresour. Technol.* 87 (2003) 41–44. https://doi.org/10.1016/S0960-8524(02)00202-X.
108. T. Escribano-Bailón, Y. Gutiérrez-Fernández, J.C. Rivas-Gonzalo, C. Santos-Buelga, Characterization of procyanidins of vitis vinifera variety tinta del pais grape seeds, *J. Agric. Food Chem.* 40 (1992) 1794–1799. https://doi.org/10.1021/jf00022a013.
109. T. Fuleki, J.M.R. Da Silva, Catechin and procyanidin composition of seeds from grape cultivars grown in Ontario, *J. Agric. Food Chem.* 45 (1997) 1156–1160. https://doi.org/10.1021/jf960493k.
110. V.A.P. De Freitas and Y. Glories, Characterisation of oligomeric and polymeric procyanidins from grape seeds by liquid secondary ion mass spectrometry, *Phytochemistry.* 49 (1998) 1435–1441. https://doi.org/10.1016/S0031-9422(98)00107-1.
111. E. Pastrana-Bonilla, C.C. Akoh, S. Sellappan, G. Krewer, Phenolic content and antioxidant capacity of muscadine grapes, *J. Agric. Food Chem.* 51 (2003) 5497–5503. https://doi.org/10.1021/jf030113c.
112. C. Yang, K. Shang, C. Lin, C. Wang, X. Shi, H. Wang, H. Li, Processing technologies, phytochemical constituents, and biological activities of grape seed oil (GSO): A review, *Trends Food Sci. Technol.* 116 (2021) 1074–1083. https://doi.org/10.1016/j.tifs.2021.09.011.
113. P. Górnaś, M. Rudzińska, A. Grygier, G. Lācis, Diversity of oil yield, fatty acids, tocopherols, tocotrienols, and sterols in the seeds of 19 interspecific grapes crosses, *J. Sci. Food Agric.* 99 (2019) 2078–2087. https://doi.org/10.1002/jsfa.9400.
114. H. Ben Mohamed, K.S. Duba, L. Fiori, H. Abdelgawed, I. Tlili, T. Tounekti, A. Zrig, Bioactive compounds and antioxidant activities of different grape (Vitis vinifera L.) seed oils extracted by supercritical CO2 and organic solvent, *LWT - Food Sci. Technol.* 74 (2016) 557–562. https://doi.org/10.1016/j.lwt.2016.08.023.
115. L. Zhao, Y. Yagiz, C. Xu, X. Fang, M.R. Marshall, Identification and characterization of vitamin E isomers, phenolic compounds, fatty acid composition, and antioxidant activity in seed oils from different muscadine grape cultivars, *J. Food Biochem.* 41 (2017) e12384. https://doi.org/10.1111/jfbc.12384.
116. H. Lutterodt, M. Slavin, M. Whent, E. Turner, L. Yu, Fatty acid composition, oxidative stability, antioxidant and antiproliferative properties of selected cold-pressed grape seed oils and flours, *Food Chem.* 128 (2011) 391–399. https://doi.org/10.1016/j.foodchem.2011.03.040.
117. G.K. Jayaprakasha, T. Selvi, K.K. Sakariah, Antibacterial and antioxidant activities of grape (Vitis vinifera) seed extracts, *Food Res. Int.* 36 (2003) 117–122. https://doi.org/10.1016/S0963-9969(02)00116-3.
118. F. Natella, F. Belelli, V. Gentili, F. Ursini, C. Scaccini, Grape seed proanthocyanidins prevent plasma postprandial oxidative stress in humans, *J. Agric. Food Chem.* 50 (2002) 7720–7725. https://doi.org/10.1021/jf020346o.
119. G.K. Jayaprakasha, R.P. Singh, K.K. Sakariah, Antioxidant activity of grape seed (Vitis vinifera) extracts on peroxidation models in vitro, *Food Chem.* 73 (2001) 285–290. https://doi.org/10.1016/S0308-8146(00)00298-3.

6 Sustainable green processing of various fruit and vegetable pomace from the food industry for the synthesis of bioactive compounds

6.1 INTRODUCTION

The waste streams generated by the food industry are massive, and most of them go untreated and unused. Instead, they are burned or dumped in landfills. Greenhouse gas (GHG) emissions, waste combustion, uncontrolled deterioration and extra pollution difficulties—owing to unpleasant aromas and soil contamination via leachates—are all results of these procedures [1, 2]. Additional analysis of the food and agro-industries reveals that the fruit processing business is the primary offender responsible for the formation of a sizeable quantity of post-production waste, also known as pomace. These wastes are produced as by-products of the fruit processing industry. When fruits and vegetables are processed in order to extract their juice, a waste product known as pomace is produced as a by-product of this procedure. On a worldwide scale, the overall mass of pomace generation is rather large, with 25–50% of the mass of the whole fruit generally being transformed into pomace. Pomace may be used for a variety of purposes, including animal feed, compost and biofuel. For instance, the processing of apples alone results in the production of around 10 million tonnes of pomace every year in the apple juice sector. Table 6.1 shows various industrial sources of fruit and vegetable pomaces. Juice as well as other food processing industries that employ fruits and vegetables as their primary raw materials are the primary contributors to pomace waste. Fruit pomace has a high chemical oxygen demand (COD), biological oxygen demand (BOD) and biodegradable organic content, all of which contribute to unintended fermentation, microbial breakdown, ecological contamination and serious health risks for both human and aquatic populations. For instance, the BOD of apple pomace may be anywhere from 240 mg $O_2.L^{-1}$ to as high as 19,000 mg $O_2.L^{-1}$—the range is relatively wide [3, 4]. In a similar vein, orange pomace causes significant environmental damage since it has a COD that ranges from 49 g $O_2.L^{-1}$to 51 g $O_2.L^{-1}$ and produces methane emissions at a rate of 195–213 $mL.g^{-1}$. In addition, the COD of grape pomace (GP) can range anywhere from 268 g $O_2.kg^{-1}$ to 591 g $O_2.kg^{-1}$, with the Cabernet Sauvignon variety from Bekaa, Lebanon, having the greatest oxygen demand of all the grape varieties [5].

As a by-product of the processing of fruit and vegetable juice and concentrates, fruit and vegetable pomace contains important carbohydrates (e.g., cellulosic fibre) and bioactive components (e.g., polyphenols, chlorogenic acids, etc.). Polyphenols can be extracted, food products can be enriched with these compounds, bacteria can be used to supplement nutrients and ceramic material can be used as an additive to create edible films, but only 20% of the pomace generated is currently utilized, and the rest is used as animal feed or composted [6, 7].

The growing quantity of pomace and other by-products derived from fruit and vegetables has caused a number of serious issues in today's world. It is of the utmost importance to locate long-term solutions for these by-products so that they do not wind up being regarded as waste without

DOI: 10.1201/9781003315469-6

TABLE 6.1
Industrial source and estimated generation of fruit pomaces worldwide

Fruit pomace	Industrial source	Quantity (% of fruit weight)	References
Grapes	Wine industry	20–30	[82]
Apple	Apple juice and cider industry	25–35	[82]
Carrot	Juice industries	30–40	[89]
Tomato	Juice, ketchup and paste industries	3–5	[53]
Olives	Olive oil industry	70–80	[76]
Blueberries	Juice and sweet fruit preserves (Jams, jellies and marmalades) industries	20–30	[90]
Cranberries	Juice and sweet fruit preserves (Jams, jellies and marmalades) industries	42–53	[82]
Orange	Juice and sweet fruit preserves (Jams, jellies and marmalades) industries	45–60	[82]
Mango	Juice and sweet fruit preserves (Jams, jellies and marmalades) industries	35–50	[82]

having been put to their fullest potential uses. In view of that, the purpose of this chapter is to offer an overview by compiling all relevant contemporary scientific publications on the composition of different pomace, green extraction of different bioactive compounds and biological activity of the bioactive compounds that are isolated from fruit pomace. The utilization of fruit pomace in a variety of culinary and nonfood-based applications is going to be another focal point of this study. This will be done in order to address the valorization of fruit and vegetable pomaces.

6.2 SYNTHESIS OF BIOACTIVE COMPOUNDS FROM FRUIT AND VEGETABLE POMACE

Different fruit varieties, fruit cultivars and potentially edapho-climatic circumstances at the time of harvest affect the concentration and content of micronutrient and bioactive compound availability in fruit pomaces [3, 8]. Table 6.2 presents some examples of the availability of phytochemicals in various fruit pomaces.

6.2.1 Grape pomace

Grape pomace (GP) is a product with an extensive range of possible uses. The components of GP are shown in Figure 6.1, a mishmash of stems, seeds, pulps and skins in varying amounts. The chemical composition of the GP might vary depending on the grape varietals used, the stage of ripeness of the fruit and the manufacturing method. However, despite these inconsistencies, GP has a nutritional value that ranges from moderate to poor when used as a feed. Carbon was discovered to be the most prevalent element in GP, accounting for 54.0%, followed by oxygen (37.85%) and hydrogen (6.08%), according to studies carried out by a number of researchers. Sulphur traces were determined to represent 0.08% of the total nitrogen content, and the fibre content ranges from 26% to 70%, with extraordinary amounts of lignin at 18–55%. There are between 4% and 11% of lipids in grape skins because of the oil-rich seeds. Red wine pomace sugar percentage is 4–9%, whereas white wine pomace sugar content is 28–31% [9]. Hemicellulosic sugars, a complex lignocellulosic substance, are found in high quantities in the grape skin and generate a combination of monomers of glucose and xylose. A wide range of monosaccharides may be found in the GP, including galactose,

TABLE 6.2
Availability of phytochemicals and composition of protein and fibres in various fruit and vegetable pomaces

Fruit pomace	Protein (% dw)	Dietary fibre (% dw)	Phytochemicals	References
Grapes	~11	~40	Anthocyanins, flavan-3-ols, hydroxybenzoic, hydroxycinnamic acids, flavonols and stilbenes	[2]
Apple	2.31–6.98	51.1–76.84	Catechins, procyanidins, hydroxycinnamic acids, flavonol glycosides, dihydrochalcone glycosides and cyanidin-3-O-galactoside	[30]
Carrot	4–5	37–48	α- and β-carotenes	[91]
Tomato	~6–40	—	Lycopene, phenolic acids, rutin, naringenin, quercetin-3-O-β-glucoside, isorhamnetin, kaempferol, quercetin, apigenin and myricetin	[53]
Olives	0.3–15	—	Ferulic acid, caffeic acid, vanillic acid, verbascoside, hydroxytyrosol, apigenin, luteolin, oleuropein and their derivatives and quercetin derivatives	[92]
Blueberries	6.64~	26.15	Delphinidin-3-galactoside, delphinidin-3-glucoside, cyanidin-3-galactoside, cyaniding-3-glucoside, delphinidin-3-arabinoside, petunidin-3-galactoside, peonidin-3-glucoside peonidin-3-galactoside and malvidin-3-galactoside	[93]
Canebarries	~2.2	58.7–71.2	Caffeic acid, quercetin, cyanidin-3-glucoside, catechin	[94, 95]
Orange	4.89–10.38	54.82–82.22	β-carotene/linoleic acid, L-limonene, palmitic acid, oleic acids, n-butyl benzenesulfonamide and β-sitosterol	[96]
Mango	3.6–6.3	35.6–51.2	Gallic acid, ellagic acid, ferulic acid, cinnamic acids, tanins, vanillin, coumarin and mangiferrin	[85]

glucose, mannose, rhamnose, xylose and uronic acid, galacturonic acid (GalA) in the majority of samples [10].

Because of its high concentration of phenolic chemicals, GP is commonly used to improve nutritional supplements. Grape seed extract has shown unusual antitumour action in leukaemia, breast, colon, bladder, prostate and lung carcinoma cell line models through a variety of pathways [11]. Low dosages of proanthocyanidins from grape seeds have been shown to suppress the development of HeLa and HepG2 cells in vitro. Grape seed extract has both antioxidant and pro-oxidant properties, making it a useful tool in the fight against cancer. This also appears to be an issue for grape seeds, which have too much coordination between MAPK/NF-kB/metalloproteinase/cytoskeleton proteins, PI3K/Akt and up-and-down-regulation mechanisms. Grape seed extract has recently been researched and revealed to have a possible anticancer shielding effect in hepatocarcinoma by boosting inhibition of cell proliferation, apoptosis and inflammation prevention, according to the findings of the researchers [12]. The potential of grape marc as a chemoprotective agent for colorectal cancer was examined in research by Del Pino-García et al. (2014) [13]. Another study found that red wine pomace without seeds had phenolic chemicals that were more easily absorbed in the small intestine compared to the grape seeds, which rapidly fermented in the colon. It was shown that therapy with spices derived from red wine pomace had a chemoprotective effect on colon cancer cells by lowering the amount of oxidative damage to DNA. Grape seed extracts appear to have therapeutic potential in in vitro experiments. Pharmacological safety might be established by clinical trials [11]. Another study suggests characterizing the underlying mechanism and acute lymphoblastic

FIGURE 6.1 Grape pomace and its various components.

leukaemia in Jurkat cells, as well as looking into the anticancer properties of an extract made from purified white GP [14].

According to the findings of Sales et al. (2018), the phenolic content and antioxidant capabilities of a dark-skinned GP extract were investigated, as were the cells' metabolism and redox state. They discovered that the anthocyanins in GP extract were particularly potent antioxidants. Regardless of cytotoxicity, improved mitochondrial respiration and antioxidant capacity were seen together with a reduction in glycolytic metabolism in the short term. GPE was cytotoxic, and necrosis killed cells in long-term incubation, although non-cancer human fibroblasts were unaffected [15]. There is evidence to suggest that probiotic bacteria are being employed in foods as bio-preservatives. According to Dias et al. (2018), the polysaccharides in this study were derived from the pomace of whole grapes, food supplement rich in dietary fibre. As a putative prebiotic function, it may increase the number of probiotics in the body [16]. Kabir et al. (2015) claim that hydroxylated cinnamates are significantly more advantageous than their benzoate counterparts. The antioxidant properties of phenolic compounds may function as hydrogen donors, metal chelators, singlet oxygen quenchers and free radical scavengers [17]. Antimicrobials and antioxidants work to extend the shelf life of items by preventing the growth of microorganisms. These are used in the food production to extend the shelf life of food and to improve the standard and safety of the product. People are more conscious than ever of the detrimental consequences of synthetic chemicals, which have sparked a hunt for antimicrobials and antioxidants in natural foodstuffs [18].

Researchers previously described that 3% of grape marc is utilized as livestock feed [19]. This percentage has not changed since then. A diet that included grape seeds did not demonstrate any difference in the productivity of pigs in an experiment [19]; nevertheless, this diet does inhibit the protein expression of regulatory molecules in the liver of pigs by limiting the output of certain inflammatory cytokines. The incorporation of fermented GP into the diet of pigs has been shown to improve the colour stability of the meat as well as the total polyunsaturated fatty acids (PUFA) found in the subcutaneous animal fat, all while reducing the amount of lipid peroxidation that occurs. Aside from research conducted on pigs, it has been found that including GP in the diets of cows both raises the amount of PUFA found in the milk produced by the cows and alters the bacterial community found in the rumen, which results in an overall improvement in the health of the cows [19]. According to

Ebrahimzadeh et al. (2018), broiler chickens' immunological response and antioxidant capacity are improved by increasing the amount of GP in their diet [20]. The hens' performance was unaffected. A certain amount of GP combined with feed has an effect on piglet production efficiency, microbial biota and redox potential. This is in line with the theory that feed GP could have a significant influence on animal health by enhancing various antioxidant mechanisms in the piglets' tissues and blood.

It is a biotechnological advance strategy in which hydrolytic enzymes that are released by thermophilic and mesophilic microbes during the mineralization process under aerobic conditions are used to break down polymeric materials found in organic waste. It is usually not practical to directly apply GP to the soil due to its probable pathogenicity since it is a heterogeneous mixture of pulp, stem, seeds and skins. In other words, before utilizing compost as a soil conditioner, it is best to compost first. Composts that are inexpensive and produced from the by-products of vineyards have enabled modern greenhouse agriculture and horticulture to lessen their impact on the surrounding ecosystem. The main reason GP has specific qualities and may be utilized as an organic fertilizer is because it contains very little nitrogen and phosphoric acid. Chemical tests of the compost revealed that it included a range of components, including zinc, magnesium, iron and heavy metals including chromium, cadmium, nickel and lead, depending on the kind of vineyard waste used. Consequently, it might gradually replace artificial fertilizers by recovering the heat and CO_2 created during composting [21].

Grape marc is a type of organic waste material that may be composted and utilized for the amendment of soil as well as an organic fertilizer to improve soil quality and increase grape yield. The compost has reached its full maturity and level of stability, which demonstrates the possible application of the organic waste treatment [22]. These two aspects of compost, "maturity" and "stability," are what determine the quality of the compost. The term "stability" refers to the transformation of primary unstable organic matter into more stable organic matter, while "maturity" refers to the time that has passed since the compost was created [22]. Even though there are no universally acknowledged technique for extraction, it is one of the crucial processes in identifying, separating and recovering components from the waste produced by wineries. The two extraction techniques are often known as conventional/traditional and nonconventional.

Maceration, Soxhlet extraction and reflux extraction are just a few of the traditional techniques that have been utilized for a significant amount of time. As a result of the need for a substantial quantity of the solvent in these processes, it is clear that they are not suitable for use in commercial environments. Polyphenols are also lost due to the fact that boiling is necessary for these techniques. Some disadvantages necessitated the development of novel extraction technologies that are not considered standard. Extraction techniques such as ultrasonic, pressured fluid, microwave-assisted supercritical, ultrasonic and rapid solvent extractions are also included. Compared to traditional procedures, which take 100–200 ml of solvent and 500–700 min for the extraction process, these approaches require a short extraction period of 1–60 min and a smaller volume of solvent, i.e., roughly 50 ml [23]. The conventional extraction method, known as solid–liquid or Soxhlet, has been in use for many decades. Because traditional methods take longer and use more solvents, researchers are focusing their efforts on finding ways to speed up extraction times, reduce the amounts of organic solvents used and improve sustainability, all while producing more valuable end products [10].

As a mass-transfer phenomenon, solid–liquid extraction may be described as the migration of analytes from a solid matrix into a solvent phase that is in contact with the matrix. The solvent extraction method is frequently used to extract the many different components found in fruits, including flavonoids. Fruits are dried or lyophilized, ground, or submerged (soaked) in the appropriate (following) solvent during the extraction of phenolic components [24]. By altering the concentration gradients and diffusion coefficients, the mass transport phenomena may be improved, increasing the extraction's effectiveness. The efficiency with which the extraction is carried out can be affected by a number of factors, including particle size, temperature, the kind of solvent being utilized, the amount of time that the extraction process takes and the matrix of the item in question. Solvent type, duration and temperature are the primary factors that influence the efficiency and purity of the process as well as the product that is recovered. HCl solutions do not have any miscibility problems with polyphenols due to their highly polar nature. Alcohol concentrations in combinations containing various quantities of low-polar solvents, including ethyl acetate, may be readily varied to generate the phenolic fractions. Posedi et al. (2018) employed solid–liquid extraction to extract grape marc (Cagnulari) polyphenolic antioxidants for the manufacturing of enhanced meals [25]. Solid–liquid extraction (SLE) was used in another investigation to extract polyphenols from red and white pomace. The resulting extract was tested on bone marrow-derived mesenchymal stem cells (MSCs) to see how it affected cell development [26]. Several different phenolic chemicals derived from grapes have been successfully extracted using SLE, which has been successfully applied in this process. The removal of trans-resveratrol from grapes and the removal of proanthocyanidins and catechins from grape seeds are two examples of this type of process [10].

One of the main goals of green technologies is to increase the extraction rate and efficiency of energy from natural sources, such as vineyard wastes and by-products, in order to extract bioactive chemicals more quickly and efficiently [10]. When mass and heat transfer improve, it minimizes both the equipment footprint and the number of process stages while preserving the environment and its resources. Following pomace analysis, it has been discovered that the use of pressurized liquid extraction (PLE) produces higher yields of phenolic and anthocyanin in nonconventional technologies and, in comparison to supercritical fluid extraction (SFE), it produces extracts from various grape types with increased antioxidant capabilities [27]. The PLE method is utilized in order to enhance the overall performance of the extraction procedure, which is predicated on the application of higher temperatures and pressures.

The rate of the extraction may be made more efficient by raising the temperature while keeping the solvent in the same condition. As a result, the analytes in the sample matrix are more soluble, and the desorption kinetic rate is reduced. As a result, the extraction operation is expedited, and the amount of solvent used and the time required to prepare the sample are reduced. Palma et al. (2002) used grapes and superheated solvents to illustrate the durability of phenolic chemicals [28]. According to this discovery, Solyom et al. (2014) examined that the grape marc generated by simulated deterioration under isothermal heating was more sensitive at three temperatures (80°C, 100°C and 150°C). The analysis of the antioxidant activity and total phenolic content (TPC) further supported this phenomenon [29].

6.2.2 Apple pomace

There is no doubt that apples (*Malus Domestica Borkh*) are a favoured fruit around the globe because of their great taste and excellent nutritional content. Apple is a sort of vital material in the fruit business. It plays significant roles in the food industry in addition to being consumed as fresh fruit and apple juice concentrate. The disposal of the leftovers from apple processing, often known as apple pomace (AP), is a crucial issue that must be addressed as processed apple industries grow [30].

According to the statistics of The Food and Agriculture Organization of the United Nations (FAO) the world's apple output exceeded 80 million tonnes in 2019. It is clear that apples are widely planted, China being the greatest apple-growing nation with 48.63% of global apple output in 2019, according to statistics compiled from the FAOSTAT database. Additionally, the United States and Turkey were placed second and third, respectively, among the top 10 apple-producing nations [30]. The two primary categories of apples consumed were typically fresh apples and processed apple goods. Because of their deliciousness and health benefits, apples account for more than 70% of all fresh fruit consumption. On the other hand, 25–30% of apples were processed into various goods, which considerably aided the growth of the apple sector and offered customers additional product options [31]. The items manufactured from apples may be split into two categories: (1) in liquid forms, such as apple cider, apple juice and apple vinegar; (2) in solid forms, including apple jam/sauce, canned apple, sliced fresh apple and freeze-dried or dried apple goods. Apple juice concentrate accounted for 60% of all apple-based goods that were manufactured, and it was also responsible for the production of AP [32]. The generation of apple pomace during the production of apple juice is shown in Figure 6.2. It was stated that during the processing of apples, around

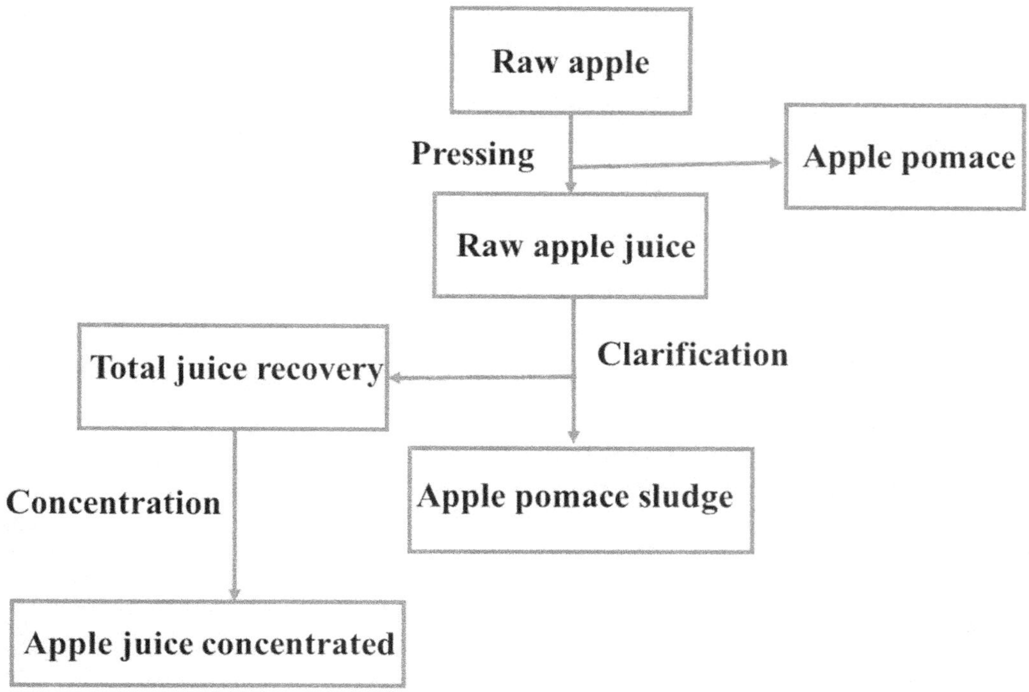

FIGURE 6.2 The generation of apple pomace and apple pomace sludge during apple juice production.

25–30% of residues known as AP were produced. An estimated 5–7 million tonnes of AP must be handled annually, according to calculations. Therefore, creating a suitable use for this vast and rich substance should be given great consideration.

AP is referred to as the leftover material from processing apples, and it consists primarily of the peel, pedicel, meat, seeds and a few calyces. There are several beneficial ingredients in AP, including minerals, phenolic compounds and carbs. Carbohydrates are the most prevalent element in both apples and AP, comprising insoluble saccharides (cellulose, hemicellulose and lignin) and soluble saccharides (pectin, starch and various mono-, di- and oligosaccharides). Despite being insoluble and having a serious disposal issue, AP has potential and might have significant applications with the right pre-treatment, hydrolysis, and solution. One of the two primary sources of commercial pectin is ectin, which is made up of galacturonic acid and different neutral sugars. The extraction of pectin has a considerable influence on the functional and physiochemical characteristics of the pectin studied. That is why future research may be concerned about the extraction of pectin with specific molecular weight, esterification degree, or GalA concentration rather than yield alone when it comes to pectin extraction [33]. Because of the high pectin yield and straightforward extraction procedure, inorganic acid extraction is frequently used in the industry for pectin manufacture. However, throughout the extraction process, a number of issues are discovered, including the unintentional deterioration of pectin and equipment erosion. Three different types of highly efficient and environmentally friendly procedures have been created. Replacement of extraction solvents: acetic acid and other organic acids are utilized as solvents since they have low toxicity and erosion; pectin may be extracted biologically using enzymes and fungi and environmentally friendly methods, like microwaves and subcritical water [30].

There has been an ever-increasing need for pectin preparation as a result of the increasing use of pectin in the food sector. Pectin was extensively utilized in the pharmaceutical and cosmetic industries because of its excellent biocompatibility, biodegradability and safety qualities [34]. Pectin's bioactivity as an antioxidant, anticancer, immunomodulator and regulator of gut microbiota has been demonstrated in several research. According to Akar et al. (2018), pectin has been utilized as a powerful stabilizer and gelling agent in fruit jelly and set-type yoghurt to prevent flocculation, as well as in maize extrudates to enhance extrusion expansion and texture performance in order to create improved food sensory and customer acceptability [35]. One of the usual plant kingdom compounds for storing energy is starch, which is also the most consumed carbohydrate on a regular basis for supplying energy. The amount of starch in apples varies depending on the level of ripeness, and it was discovered that the mature apple's seeds had the highest starch content, ranging from 44–53% (dry basis). In addition, the starch granules were dome-shaped or spherical and had a high amylose content (26–29.3%) [36]. There is little research on AP starch preparation because of the low starch content and low seed content. Some basic research on AP starch is required to uncover its true potential as a starch source.

No study has focused on the creation of mono-, di- and oligosaccharides from AP, which might be obtained from two sources: the natural components of apple, which are present in AP due to insufficient juicing and polysaccharides that are degraded during extraction or other treatment of AP. It is required to undertake a basic study to demonstrate health functions or to maximize the high-value utilization of AP, despite the low saccharide content and difficult production techniques.

A significant percentage of the insoluble components in AP are made up of hemicellulose, lignin and cellulose. Non-pectin carbohydrates, such as cellulose and hemicellulose, have the potential to be used in the synthesis of insoluble dietary fibre. Pre-treatments such as steam explosion, acid/alkali pre-treatment and subcritical water pre-treatment are usually essential since the lignin is cross-linked with the suggested two macromolecules [37].

Several studies were conducted on the isolation and application of AP insoluble dietary fibre, which has been utilized in the food, pharmacological and chemical products as a fat replacement, texturizer and bulking agent. The purpose of these studies was to better understand the effects of these applications. According to reports, adding AP to short dough biscuits had the effect of lowering the glycaemic index, primarily due to the rise in dietary fibre content, and the combination of maize starch and insoluble dietary fibre from AP had better expansion properties. Additionally, whole milk that has been supplemented with AP in the dairy industry has the potential to be advantageous since it works to lower fat content and affects yoghurt acceptance. Additionally, cellulose and pectin worked well together to give meat a solid texture and reduce its carcinogenicity [38].

In the manufacturing of nonfood products, AP was utilized as a potential source of carbon for the production of bacterial cellulose. Additionally, it was an excellent source for the development of cellulose nanocrystals that were optimally treated with alkaline [39]. In response to the growing demand for environmentally friendly materials, lignocellulosic and other forms of insoluble dietary fibre have emerged as potential raw materials for the synthesis of biomaterials that can be utilized either as edible films or as other active packing materials with the capacity to be hydrophobic or antioxidant. Additionally, the non-pectin carbohydrates that AP produced were a feasible source for the production of biofuels like ethanol and biogas. To get the most out of the relatively large proportion of insoluble components that AP has, the pre-treatment, purification and conversion procedures need to be improved.

According to an extensive study, AP has more phenolic compounds than apple flesh and that concentration is more significant in apples. The majority of the phenolic components in AP were flavonoids, including catechins, quercetin, phloridzin, epicatechin, flavonol and dihydrochalcones, as well as phenolic acids, such as chlorogenic acid and hydroxycinnamic acid [40]. Furthermore, the amount of phenolic component in an apple varies greatly depending on the type, cultivar and processing conditions. In spite of the high concentration of phenolic compounds in AP, the amount of free phenolic compounds was very low, indicating that most of the phenolic compounds were bound to protein, polysaccharides, or lignin [41]. AP phenolic chemicals must be extracted or converted before they may be used.

Traditional methods involve using a variety of organic solvents to extract polyphenols from AP; nevertheless, the most important considerations during the extraction process are the efficiency of the process and the safety of the food. Because of this, a number of unorthodox methods of extraction, including SFE, pulsed electric field, ultrasound, microwave and enzymatic extraction, amongst others, have been developed. Because of their excellent extraction efficiency and ability to preserve phenolic chemicals, the non-thermal and combination extraction methods are highly recommended. Due to its potent antioxidant and advantageous biological qualities, AP was utilized as a nutritious substitute in the food sector, according to a thorough study. Apple polyphenol has traditionally been used in baked goods, such as muffins and cookies, to boost their antioxidant capacity and enhance their sensory appeal. Additionally, the final output of wheat-barley sourdough bread had less acrylamide because of the inclusion of AP phenolic compounds [42]. Other uses for AP phenolics besides pectin include the fabrication of active food packaging and their use in dermal formulations to boost the antioxidant activity of cosmetics [43].

Additionally, while dihydrochalcone, polyphenols with vicinal hydroxyl functionalities and phlorizin are natural pigments, some researchers have also focused on using AP phenolic compounds as natural colours. Because polyphenol oxidase (PPO) speeds up the oxidation of phenolic components, the colour of chopped apple flesh soon became brown. It is possible to create pigment via an oxidizing process. Phloridzin was first transformed into a colourless intermediate by the catalysis of PPO. Then it was transformed into yellow pigment, which was utilized to replace yellow colours in the food sector. However, the safety of pigments generated from AP is not yet fully understood; therefore, whether pigments derived from AP might be used in the production of food or cosmetics has to be explored and further investigated [30].

The primary distribution of pentacyclic triterpenes in plant epidermal cuticle waxes is free or bound to glycosyl, which is a secondary metabolism with low polarity and high bioactivity, although the quantity and plant variety are different in depending on cultivation settings and conditions. Pentacyclic triterpenes may be synthesized from ursolic acid in AP, which also contains small amounts of oleanolic acid, betulinic acid and maslinic acid. Ursolic acid is the primary kind of triterpene found in AP.

For the synthesis of pentacyclic triterpenes, many unique approaches were developed in addition to the Soxhlet and hot-refluxing extraction methods. These methods include supercritical carbon dioxide extraction and ultrasonic-aided extraction, respectively [44]. Woniak et al. (2018) examined the extraction and separation of triterpene acids using supercritical carbon dioxide, which is a quick selective approach, and came to the conclusion that these were the ideal extraction conditions [45].

Separation as well as purification are required but there are challenges in the downstream process for the high-value use of AP because of the variety of the triterpene components. Derivatization is required in order to identify triterpenoids using traditional analytical methods such as high-performance liquid chromatography (HPLC), gas chromatography (GC) and so on, because of the differences in position among triterpenoids. The choice of mobile phase and chromatographic column is crucial because of the large number of position isomers [46].

One of the key components of apples, terpenic acid, has been utilized extensively in traditional herbal medicine due to its bioactivity as an anti-inflammatory, antioxidant and anticancer agent. Ursolic acid was found to be the most abundant component in AP, with significant biological activity. Researchers found that triterpenoic acids were found in the most active portions of AP that increased endothelial nitric oxide synthase (eNOS) activity. Similar to pectin and phenolic compounds, triterpenic acids have been used in the cosmetics sector to enhance their antibacterial and antioxidant properties [43]. Triterpene is, therefore, one of the potential high-value products recovered from AP and in order to encourage the use of the suggested component, it is important to investigate further the purification processes and the mechanism of triterpene's bioactivity.

In a nutshell, the extraction of high-value goods from AP is a new technology that may contribute to the circular economy of AP consumption. More emphasis should be directed to purification and other processing of high-value elements in addition to effective extraction methods. While it is important to examine the specifics of the extraction process and subsequent exploitation of the secondary residues, it is also important to investigate a variety of valuable substances that may be removed or employed utilizing a gradient method [30].

6.2.3 Black carrot pomace

Carrots may be divided into carotene and anthocyanin groups based on the type of colouring molecules they contain. There has been a steady rise in the worldwide output of carrots (including black carrots) throughout the years, according to the FAO. From 40.2 million metric tonnes in 2015 to 44.8 million metric tonnes in 2019, output climbed by 10.3%.

Black carrots have been grown in Turkey, Afghanistan, Egypt, Pakistan, India, Egypt and others for a long time [47]. Because they contain a high concentration of very stable anthocyanins (175 mg.kg^{-1} fresh mass), black carrots have become the focus of several investigations in recent years. As is well known, acylated anthocyanins are substantially more stable than their non-acylated counterparts. In addition to having a high stable anthocyanin concentration, black carrots also have substantial antioxidant activity. According to some reports, black carrot juice has an antioxidant content of around 2.3 times higher than orange carrot juice [48]. Several food items, including apple juice, yoghurt, strawberry marmalade and Turkish delight, are coloured with anthocyanins made from black carrots in addition to different fruit juices and concentrates [49]. The pressing parameters can have an effect on the yield rate in the processing of fruit juice made from black carrots, which can cause it to range anywhere from 45–64%. The manufacturing of black carrot juice results in the creation of a by-product known as black carrot pomace. This pomace is a valuable material because of the colour components that it contains. Large quantities of pomace, which includes useful goods such as dietary fibre and soluble components of the cell sap, are produced by the black carrot juice business. These pomace by-products are sold to other industries. On the other hand, juice manufacturers sell most of this product as animal feed, and alternative applications for it have seen relatively little success. Due to its low value, most people consider it to be a waste product; consequently, the removal of this material from the juice business is costly for juice manufacturers and has possible environmental effects. As a result, manufacturers have sought to maximize the monetary and commercial value of this resource in recent years. As a result, it is necessary to turn this raw material into something useful. Many solid and liquid food items can benefit from using black carrot pomace as an ingredient for colour and nutritional fibre, as well as antioxidant components [50].

More polyphenols are retained and better anthocyanin bioavailability is found in black carrot pomace compared to black carrots. It was found that shalgam, a traditional beverage popular in Turkey's mid-south region, tasted best when brewed using a mixture of 75% black carrot and 25% black carrot pomace; according to research, it was found that 100% black carrot tasted best. In addition, a number of experiments were conducted to find the appropriate extraction parameters that produce the highest possible yield of anthocyanin for each of the several extraction processes (pressurized liquid, ultrasound-assisted, microwave-assisted, etc.) [47, 51].

Before manufacturing value-added products, the high moisture content of black carrot pomace needs to be removed using a variety of different techniques for water removal. This helps to avoid microbial deterioration and extends the shelf life of the finished goods. One of the most common forms of drying is the convective hot air dehydration method. Nevertheless, it has low energy efficiency and produces a dried product of a comparatively lesser quality than other methods. As a result, researchers have devised different techniques of drying, such as infrared drying, conductive hydro drying, microwave drying, freeze drying and refractive window drying, amongst others. Product quality may suffer as a result of changes in scent and phenolic compounds during food processing, particularly drying. This may lead to customers rejecting the food. The quality of dried agricultural products is influenced by the drying process and its associated characteristics (such as temperature, time, etc.). The dehydration of black carrot pomace is not well documented in the scientific literature. The drying techniques tested by Janiszewska et al. (2013) affected the density and rehydration characteristics of black carrot pomace [52].

6.2.4 Tomato pomace

One of the most popular vegetables consumed worldwide, tomatoes, are rich in phenolics, lycopene, organic acids, vitamins and other beneficial nutrients. Tomatoes are eaten as both a raw vegetable and in a variety of processed forms, including paste, juice, sauce, puree and ketchup [53]. Tomato pomace (TP) is a waste product that is typically generated during the processing of these goods and consists primarily of peels and seeds with a minor quantity of pulp. On average, TP makes up between 3% and 5% of the raw material used [54]. According to the most recent FAO data, the area harvested and the worldwide output of tomatoes are on the rise, statistically speaking. A limited portion of the tomato is consumed as a fresh commodity because it is a seasonal fruit. In contrast, the vast majority of tomatoes are used to make juice, paste and other products. There is an estimated total yield of 5.4–9.0 x 10^6 tonnes of total yield in this context, notwithstanding the difficulty in collecting precise data on TP amount. As a result, the disposal or use of TP is an inescapable challenge for the food sector and is vitally significant. On the one hand, TP's high water content and nutritional richness make it susceptible to spoiling, which is an environmental burden and a loss of resources; on the other hand, biorefining of TP may turn waste into valuable resources, such as lycopene, dietary fibres and tomato seed oil [53].

The antioxidant capabilities of lycopene, an essential carotenoid, are superior to those of other carotenes, and it has a single oxygen-quenching rate constant that is 100 times greater than that of vitamin E. Lycopene's enhanced antioxidant ability has also been linked through bioactivity to a reduced risk of prostate cancer and cardiovascular disease [55]. As far as solubility is concerned, lycopene is almost insoluble in water and only just solubilized by pure ethanol; on the other hand, it is highly solubilized by fats and nonpolar organic solvents such as hexane, acetone and petroleum ether. Thus, lycopene was typically extracted from tomato paste or its peel by means of solid–liquid extraction using these solvents. In the process of applying various organic solvents to TP, the following extracting performance order was determined: hexane-ethyl acetate, acetone, ethyl acetate, hexane-acetone, hexane-ethanol, hexane, and ethanol [56]. Ethyl acetate extract yielded up to 120 mg/100 g of tomato peel extract, whereas alcohol extract produced roughly 5 mg/100 g of tomato peel extract. This is only one example of how lycopene and carotenoids can be extracted using common organic solvents alone or in combination. It's impossible to compare the yields, extraction rates and lycopene content of the extracted extracts because they all differ greatly. According to the results, the extraction of lycopene from TP and its peel components was influenced by several circumstances. The main determinants of findings include the initial material, solvent, operation circumstances and the lycopene quantification technique. Using traditional solvent extraction has long been seen as a risk to food safety and the environment due to its reliance on extremely toxic chemical solvents and the potential presence of hazardous residues in the final extracts [57]. In view of the shortcomings of classic solvent techniques, other approaches, such as ultrasound-assisted extraction (UAE), high-press extraction, microwave-assisted extraction (MAE), as well as enzyme-assisted methods, have been offered as alternatives. These novel approaches have the potential to yield encouraging results. According to the data that were presented by Yilmaz et al. (2016), the quantity of lycopene that was extracted from TP by a variety of different solvents rose from 52.21 $mg.kg^{-1}$ to 70.10 $mg.kg^{-1}$ when ultrasound assistance (90 W) was utilized [58]. Furthermore, according to the scientists, the UAE approach required substantially less time to extract lycopene at an 80% extraction rate than the traditional solvent method, which required 20 min. This was explained by the cavitation and mechanical vibration of ultrasound, which weakened cell walls and accelerated lycopene breakdown.

Lycopene yield was significantly increased from 6.5 mg/100g to 17.5 mg/100g using the MAE approach (400 W, 60 s) using tomato peel. In comparison, all-trans-lycopene was retained at 2–13.5 mg/100g by retarding the isomerization to its cis-isomers [59]. All-trans-lycopene is retained in the extract to aid with its colour and stability, but because it is more stable and intense in colour than its cis-isomers, it has a lower bioavailability. Pressured steam was used in the MAE procedure to break down the cell walls, improving the yield of lycopene. This approach also resulted in less isomerization of all-trans lycopene due to a shorter heating time.

Additionally, it has been demonstrated that the MAE approach is superior in terms of quick heating, low energy use and organic solvent use. However, the primary drawback of microwave help is its uneven heating [59]. According to Zhang et al. (2008), a combination of ultrasound and microwaves might be used to achieve this aim. When it came to getting the most lycopene out of tomato powder, the joint approach (97.4%) beat out the UAE control (96%) [60].

According to Strati et al. (2014), the high-pressure-assisted approach may provide a comparable lycopene yield to ethyl acetate extraction at a lower solvent/solid ratio and shorter duration while performing the extraction at 700 MPa [57]. As a result of cell membrane distortion or cell membrane breakdown caused by high pressure, the mass transfer rate during extraction is increased. A high-pressure sterilizing process has the added benefit of extending the shelf life of the finished lycopene extract [60].

According to the previous statement, physical cell damaging procedures such as ultrasound, microwave and high press aided extraction were implemented at the same time as the usual solvent extraction. Enzyme aid, which was frequently used as a pre-treatment to solvent extraction, is completely separate from the operational technique. According to the results of Strati et al. (2014), the pre-treatment of TP with pectinase has the potential to significantly boost the lycopene yield (89.4 $mg.kg^{-1}$) of the following ethyl lactate extraction by as much as ten times [57]. In the same way, Lavecchia et al. (2008) found that the lycopene yield from a hexane extraction of pre-processed tomato peel with mixed enzyme preparations from Aspergillus strains was 20 times higher than that of unprocessed peel [61]. Pre-treatment with Cellulyve 50 LC increased lycopene extraction by methanol/tetrahydrofuran (1:1) up to 20–30-fold, according to Cuccolini et al. (2013). During these enzymatic pre-treatments, enzymes that have pectinolytic, cellulolytic and hemicellulolytic activity, such as pectinase, cellulase and hemicellulose, hydrolyzed the cell wall structure [62]. As a result, the increase in lycopene extractability was connected to the lysis of the cell walls. In the lycopene yield, however, enzyme help outperformed ultrasonic, microwave and high-pressure assistance. However, because it involves many steps and has strict criteria for pH and operating temperature, enzyme aid is more difficult to use.

The use of supercritical carbon dioxide extraction to reclaim lycopene from tomato paste and the peel constituents came highly recommended as an environmentally friendly method of extraction. According to the findings of Kehili et al. (2017), a maximum lycopene recovery of 72.9 mg/100 g from tomato peel could be achieved by supercritical CO_2 extraction. This value was significantly higher than the values that were observed for the solvent extraction using hexane (60.9 mg/100 g), ethyl acetate (32.0 mg/100 g) and ethanol (28.4 mg/100 g). Tomato peel combined with the seed was shown to be more effective in recovering lycopene (44.1%) than peels that were extracted without the seeds (17.5%). It was also demonstrated that, adding olive oil to tomato peel (5%) increased the amount of lycopene extracted from 25.5 $\mu g/g^{-1}$to 60.9 $\mu g/g^{-1}$ during supercritical CO_2 extraction performed at 75°C and 35MPa. These instances involved using tomato seed oil and olive oil as co-solvents or modifiers to enhance lycopene solubilization in supercritical CO_2. Supercritical CO_2 extraction is preferable to solvent extraction in terms of environmental protection, industrial output, worker safety and consumer food safety when compared to solvent extraction with or without aid [63]. In addition, the extract that is produced as a consequence of supercritical CO_2 extraction has a higher colour intensity, a more agreeable scent and a higher purity level than the extracts that are produced by traditional solvent extractions [53].

At present, chronic conditions such as diabetes mellitus, cancer, high cholesterol, obesity and coronary heart disease are on the rise and pose the greatest danger to human health. Deficiencies in dietary fibre may play a role in the development of several chronic illnesses [64]. The American Dietetic Association suggested dietary fibre intakes of 25 g per day for adult women and 38 g per day for adult men, with a 3:1 ratio of soluble to insoluble fibre. Towards this goal, a variety of dietary fibres have been manufactured and added as functional components to produce nutritious diets high in dietary fibre. Dietary fibre is often classified as either soluble (SDF) or insoluble (IDF) based on how it behaves in the body. SDF is more promising than IDF because of its high fermentability and compatibility, as well as the inert character of IDF. The extraction of SDF from fibre-rich materials is the subject of various studies. Only two studies have been published on the extraction of SDFs from tomato peel, despite the fact that the SDFs were present in very low concentrations. A 7.9% SDF yield was achieved using the sequential amylase, protease and amyloglucosidase enzymes. There were only 12.9% of IDF yields in this group's pre-treatment of the tomato peel with alkaline hydrogen peroxide (H_2O_2), hydrochloric acid-ethanol solution (HAE) and enzymatic hydrolysis (EH) [65].

In the food business, pectin is frequently recognized as a technical and bioactive component as a gelling, thickening and emulsifier agent that aids in lowering the risk of hypercholesteremia, colon cancer, cardiovascular illnesses and insulin resistance, among other conditions [66]. The primary sources of commercial pectin are citrus peel, apple pomace and sugar beet pulp. Pectin demand is thought to reach up to 40,000 tonnes per year globally, with an average growth rate of roughly 5%. Citrus rind, apple pomace and sugar beet pulp are in short supply compared to market demand. Investigating alternate sources is essential as a result. TP and its peel component are potential possibilities in this regard. In order to extract pectin from TP, Alancay et al. (2017) used distilled water and a pH 2 HCl solution as extracting agents. According to the findings, acid and water extractions produced the highest yields of 280 $g.kg^{-1}$ and 87 $g.kg^{-1}$, respectively [67]. The acid-extracted product also had a greater purity value (70.9%) than the water-extracted product (57.5%). Although they are both high methoxyl pectins, the former (76.3%) has a lesser degree of esterification (DE) than the latter (87.8%). Solubilization capabilities of acid solution and water to pectin, and the demethoxylation of acid at increased temperatures, explained these disparities. In a study by Grassino et al. (2016), pectin output from tomato peel was found to be as high as 326 $g.kg^{-1}$ when extracted with an ammonium oxalate solution (16 $g.L^{-1}$) and an oxalic acid solution (4 $g.L^{-1}$) at reflux temperatures of 90°C for two 24-hour and 12-hour extractions, respectively [68]. With a DE of up to 82.4%, the resultant pectin is very high-density. A new study examined the impact of UAE (37 kHz) on the extraction of tomato peel pectin by Grassino et al. (2016) [69]. Both consecutive extractions were done at 60°C with UAE and lasted for 90 min, respectively, resulting in a maximum pectin yield of 360 $g.kg^{-1}$. In DE, there was no discernible difference between using ultrasonic help (84.8%) or not (84.5%), but when it came to calcium pectate yield, UAE produced a purer product (94.9%) than the conventional method (60.2%). Extraction times, product yield and purity appear to have been significantly improved by using UAE. To be clear, the pectin yield and DE value obtained by Zhang et al. (2018) using the above-mentioned UAE were much lower (243.6 $g.kg^{-1}$ and 50.77%, respectively) [70]. Even though these differences were not investigated in this work, it was anticipated that the variations in the starting materials would compensate for the disparities that were discovered. Inconsistencies aside, the pectin yields that have been recorded for TP (24–36%) are equivalent to those that have been found for citrus peel (30–35%), and they are somewhat more significant than those that have been reported for AP (15–20%). The value of TP pectin for DE is around 51–96%, which is comparable to the values for pectin from apple pomace (66–82%) and citrus peel (58–82%). In conclusion, TP has a strong potential to replace citrus peel and AP in the production of high methoxyl pectin.

The limited availability of commercially accessible TP-derived pectin due to the lack of large-scale manufacturing processes is unfortunate. Fewer than a half-dozen published publications have dealt specifically with the extraction of pectin from TP or its peel component, although the majority of them employed standard solvent extraction methods rather than the creative ways prevalent in the case of AP or citrus peel [71]. In this view, more futuristic technological research should be systematically encouraged, with a particular emphasis on sustainability, efficiency and profitability, in order to pave the way for using TP as an alternative material in the production of pectin.

Adults should consume about 0.8 $g.kg^{-1}$ of protein daily, making it one of the essential nutrients in the human diet. It is impossible to avoid the elevated risks of frailty, old age and immunodeficiency in adults and children when protein consumption is poor and protein intake is deficient [72]. There is an ongoing protein deficit worldwide, although it is most acute in less developed countries. From a dietary standpoint, it is critical in this situation to utilize new protein sources. Tomato seed is high in protein (23.6–40.9 g/100 g), yet it's generally discarded in animal feed. Nutritionally, tomato cannery waste protein concentrate is similar to soy and cotton seed concentrates. Another group of researchers has found a deficiency in isoleucine, methionine and tryptophan in tomato seed protein concentration. The World Health Organization/Food and Agriculture Organization/United Nations University (WHO/FAO/UNU 2007) reference pattern of essential amino acids should be revised. According to FAO (2007), extracted tomato proteins are deficient in methionine, leucine and tryptophan, however, the most limiting ones are case-specific. Tomato seed protein's physicochemical qualities or technical capabilities are important to consider when evaluating its potential use in food products [53].

The nutritional benefits and technical capabilities of tomato seed protein concentrate or isolate were significantly influenced by the extraction conditions, including the extraction solvent and starting material utilized, the solvent-to-sample ratio, the temperature, the duration, the precipitant and the operating pH. The process of alkaline solubilization, succeeded by acidic precipitation, is generally used to extract protein from tomato seeds or pomaces. The operating parameters of extracting pH, temperature and precipitating pH were frequently chosen to be in the ranges of 9–12, 20–50°C and 3.7–3.9, respectively. Meshkani et al. (2016) examined protein concentrate yields and functions from defatted seed and pomace meals [73]. The results revealed that the seed (44.6 g/100 g) yielded more than the pomace (35.29 g/100 g). Seed protein concentrate is inferior to pomace protein concentrate in terms of water absorption capacity (WAC), oil absorption capacity (OAC) and foaming stability (FS), better in terms of foaming capacity (FC) and equivalent in terms of emulsifying activity (EA) and emulsifying stability (ES). Shao et al. (2014) also investigated the influence of seed temperature on tomato protein concentration during processing [74]. The seed from the hot break (90–95 °C) had greater WAC and OAC values but lower yield of EA, ES and FA than the seed from the cold break (65°C). This was related to the heat-induced denaturation and protein dissociation that took place during the hot break, which significantly alters the functional characteristics of proteins.

The most common extraction solvent was sodium hydroxide solution. It was frequently determined that a NaOH solution with a high pH is advantageous in yield. However, low protein content in extracts may be caused by the high pH of the NaOH solution. To improve product output and purity or to modify the activities of protein concentrates or isolates, numerous adjuvants were recommended in addition to NaOH's intrinsic qualities. In contrast to the pH 11.5 NaOH solution with 5% NaCl (27.4%), the solution with 0.5% Na_2SO_3 had a lower WAC but higher yield (37.2%) and OAC, ES, EC, FS and FC values. It has also been shown that the acid precipitant has an effect on the yield and properties of the protein extract [53]. It has been found that the isoelectric point of the proteins found in tomato seed is around 3.9. Because of this, when acid precipitation was performed, the pH of the alkali extract was frequently adjusted to be somewhat close to this value in order to facilitate the aggregation of protein molecules. According to Salikhova et al. (1999), protein extraction from tomato seed by pH 9 NaOH was more effective at pH 5.0 than at pH 3.7 or 4.0, which is unusual. For the same pH (4.5), acid type (citric acid, HCl and their isopycnic mixture) had no significant impact on extract yield or purity under the same conditions [75].

Because of the tremendous significance of the many different tasks that might be served by these applications, future research on the extraction of protein components from tomato seed or TP should have a greater emphasis on functionality. That's why functional, rather than purely yield-oriented, extraction techniques should be developed in order for tomato protein isolates to be more precisely utilized, ultimately leading to a variety of commercial protein products with specified uses.

6.2.5 Olive pomace

Agriculture in Mediterranean nations relies heavily on the production of olive oil. As a by-product of the olive milling process, olive oil accounts for 20–30% of the weight of the components, while olive pomace accounts for 70–80% of the mass. Every year, 20–30 million metric tonnes of olive pomace are produced across the world and must be disposed of. Because of the acidity and high organic content of olive pomace, this waste could pollute land and groundwater [76].

An important reason for employing olive pomace as an antioxidant is that the lignin found in the fruit might serve as an antioxidant element in the product. Because lignin acts as an inhibitor or scavenger of free radicals, it was observed that the lignin-derived additive might slow the oxidation of asphalt binder. Antioxidants for asphalt pavements have been found in agricultural lignin, such as those found in maize or rice husks. Up to 12% of co-products, including lignin from the processing of maize kernels, were added to asphalt binders by McCready et al. (2008), and they discovered that these co-products might delay asphalt ageing by acting as antioxidants [77]. A study by Arafat et al. (2019) demonstrated that black liquor lignin, Kraft lignin and the lignin derived from rice hulls might be utilized to increase the antioxidant performance of asphalt binders by modifying their structure. As an antioxidant addition or as an asphalt binder extender in asphalt mixes, wood lignin was also employed [78].

Lammi et al. (2018) found that lignin concentration in crude olive pomace, pulp-rich fraction and stone-rich fraction ranged 44.3–48.9% and 37.8–44.3%, respectively [79]. There were 42.48–43.95% and 43.38–45.72% of lignin in crude olive pomace and the pulp-rich fraction, respectively, according to Ribeiro et al. (2020). The oxygen radical absorbance capacity (ORAC) of the crude olive pomace, fractionated liquid-rich fraction and pulp-rich fraction was measured by Ribeiro et al. (2020), who found that these samples had an ORAC of 641.05–734.8 M Trolox-equivalents (TE)/g, 1,546.93–1,585.46 TE.g^{-1} and 454.74–502.80 TE.g^{-1}, respectively [80]. In a related study, Calabi-Floody et al. (2012) used GP as an antioxidant in asphalt binder [81]. The GP's ORAC was determined to be 650 M TE.g^{-1}. These earlier investigations demonstrated that olive pomace possesses a significant amount of ORAC and is rich in lignin. Thus, the hypothesis examined in this paper is that olive pomace may be a viable antioxidant for asphalt binder.

6.2.6 Pomaces of berries

As indicated, there are several more plant pomaces that are rich in bioactive chemicals in addition to the ones listed above. Berries have been shown to be a particularly high source of polyphenols, according to several studies. As a result, berry extracts may possess potent antioxidant and antibacterial properties. Pomace, a by-product of berry pressing that is one of the most significant, is used to make juice. High polyphenol content has been found in blackberry, black chokeberry, blueberry, red currant and related by-products. There is research on the use of extracts from various berry fruits as natural preservatives in meat products, but there aren't many on extracts made from berry pomace [82].

Cranberry pomace is one fruit pomace that has gained attention as a potential feed for cattle. Cranberries, like blueberries, are a significant source of income in Canada. According to Ross et al. (2017), organic cranberry pomace includes flavonols (3.08 mg quercetin equivalent per gramme), tartaric esters (2.77 mg caffeic acid equivalent per gramme), total phenolics (12.99 mg GAE per gramme) and anthocyanins, (4.46 mg cyanidin-3-glucoside equivalent per gramme) [83]. In addition to catechin, caffeic acid, quercetin and cyanidin-3-glucoside, cranberry pomace also contains other polyphenols.

6.2.7 Mango pomaces

The anticipated global mango (Mangifera indica) output in 2019 was 55.85 million metric tonnes, with India and China as the top two exporters of the fruit. Inedible mango pomace, which comprises the peels, mango seed kernels and any leftover pulp, can account for up to 35–50% of the fresh fruit's total weight [84]. With 19% soluble and 32% insoluble dietary fibre, mango peel has a significant concentration of dietary fibre (51.2% of dry matter). Additionally, mango peel has been shown to contain substantial amounts of phytochemicals, enzymes, polyphenols, vitamins E and C and carotenoids, all of which have a variety of functional and antioxidant properties [85]. Polyphenols and flavonoids were found to be greater in the mango peel waste than in the mango meat, as well as having remarkable antioxidant activity. Seed kernels from mangoes, like the skin, are thrown away as food waste during the preparation of the fruit. For the most part, the mango kernel has a crude fat content of 7–15% (DM), which has attracted significant interest due to its unusual physical and chemical features. Trans fatty acids are absent from mango kernel fat, making it a healthy, natural and promising fat source. Mango pomace has been shown to provide a variety of health benefits, including anti-lipid peroxidation, reduced cholesterol levels and alleviation of peristalsis, according to several research. Research has been done extensively on the use of mango by-products in the creation of diverse foods [86].

6.2.8 Papaya pomaces

The papaya plant, scientifically known as Carica papaya, is indigenous to tropical America and southern Mexico. However, it is also widely cultivated in Brazil, India, Mexico and Africa. Papaya is one of the most widely consumed fruits in Ethiopia and for good reason: not only does it taste delicious but it also has a high nutritional value. The yearly production of papaya fruits in Ethiopia is 54,355 metric tonnes. Consumption and juice extraction of papaya fruits result in seeds, peels and tiny amounts of pulp, which is referred to as pomace. Ethiopia produces a considerable amount of papaya pomace each year. A high level of crude protein content (179–184 g/kg) and substantial energy (17MJ gross energy/kg) have been reported in papaya pomace in the literature on a dry matter basis. Papain and chymopapain, as well as antihelminthic, antibacterial and anticoccidial chemicals, are found in the peel of papaya [87].

6.3 CHALLENGES AND FUTURE PERSPECTIVES

Fruit and vegetable pomace is abundant all over the world; however, there are restrictions on how it may be processed. Purification and separation of certain bioactive molecules from these by-products may not be economically possible in some cases, despite the fact that they represent environmental difficulties. It has recently been shown that nano-filtration may be used to produce an apple pomace extract with a high polyphenol content at lab and pilot plant scales [88]. This high polyphenol extract was effectively created; however, the processing method was not commercially viable. Since fruit pomace is an economic resource, future research on fruit pomace processing should evaluate the economic viability of producing fruit pomace.

Pomace processing technology has advanced in recent years, allowing novel uses for pomaces in culinary items. The dietary fibre in pomaces was functionalized using cutting-edge micronization techniques, including high-shear wet-milling and high-energy ball milling. Lingo-cellulose plant material is broken down into lower molecular weight soluble dietary fibres by this processing method, which reduces the molecular weight of insoluble dietary fibre. Because of these innovative grinding processes, fruit pomaces may now be used as inexpensive emulsifiers. Recent research conducted by Lu et al. (2020) utilized an oven drying process, impact milling and high-energy wet media milling on apple pomace with the goal of reducing the particle size from 12.9 μm to 550 nm [33]. Shearing the lignocellulosic-containing bonded polyphenols is what made this processing procedure effective in boosting antioxidant activity. At the same time, it was found that the wet-milled apple pomace was able to effectively be employed as a Pickering emulsion, with a 99.19% ability to disperse water and oil at 3.2 (% w/v) concentration. Antioxidant qualities can be improved while costs are kept low by using micronization procedures on fruit pomace.

Grape and apple pomace are currently offered commercially as "wine" and "apple flour," respectively. Owing to the increased moisture content of fruit pomaces, hammer or impact milling is commonly used in the production of these components. This procedure thermally destroys the sought-after bioactive chemicals and is exceptionally energy-intensive because of the high moisture content of fruit pomaces. Different production procedures are needed to address these disadvantages. This approach would be more cost-effective while also boosting the nutritional value of the final food product since it uses a single phase of high-shear wet-milling to generate fruit pomace "slurry." By mechanically shredding bound polyphenols and functionalizing the pomaces' lingocellulosic matrix, this high-shear method increases antioxidant activity. The advantages and practicality of generating pomace slurries should be studied in the future.

6.4 CONCLUSIONS

Fruit and vegetable pomaces, a by-product of the agro-fruit industries that is produced during juice processing and contains significant amounts of a number of bioactive substances, including dietary fibre, pectin, carbs, phytochemicals (polyphenols, tannins etc) and minerals, are underused. According to preliminary research, these pomace-derived bioactive compounds have promised antioxidative, anti-inflammatory, antibacterial and antiviral properties that might enhance human body metabolism and health functions. The typical disposal techniques for pomace can lead to contamination of the surrounding environment and potentially provide risks to human health. However, green or nonconventional extraction techniques have superseded traditional methods due to their elevated yield, greater output with decreased processing time and waste. The important chemicals from pomace may be extracted using conventional or nonconventional technology. This review has provided a detailed summary of the proximate composition of a variety of pomaces, as well as ways for extracting important bioactive chemicals from fruit pomace in an effective manner. In addition, this study has provided a synopsis of the comprehensive usage of fruit pomace for the manufacture and supplementing of food and nonfood-based goods. This is a current research hotspot that aims to maintain the viability of the agro-fruit business. Additional research is required to speed the effective exploitation of pomace-based pharmaceuticals to large-scale processes and their potential implementation while simultaneously enhancing and functionalizing the important bioactive components that may be extracted from pomace.

ACKNOWLEDGEMENT

The study is supported by Indian National Academy of Engineering (INAE/121/AKF/22), Gurgaon, India. The authors are solely responsible for all of the opinions, results and conclusions expressed in this study; INAE's viewpoints are not necessarily reflected in any of these aspects.

REFERENCES

1. P. Duarah, D. Haldar, A.K. Patel, C.-D. Dong, R.R. Singhania, M.K. Purkait, A review on global perspectives of sustainable development in bioenergy generation, *Bioresour. Technol.* 348 (2022) 126791.
2. A. Iqbal, P. Schulz, S.S.H. Rizvi, Valorization of bioactive compounds in fruit pomace from agro-fruit industries: Present Insights and future challenges, *Food Biosci.* 44 (2021) 101384.
3. C. Joshi, V. Joshi, *Food Processing Waste Management Technology Need for an Integrated Approach*, 1990.
4. B. Debnath, D. Haldar, M.K. Purkait, A critical review on the techniques used for the synthesis and applications of crystalline cellulose derived from agricultural wastes and forest residues, *Carbohydr. Polym.* 273 (2021) 118537.
5. J.H. El Achkar, T. Lendormi, Z. Hobaika, D. Salameh, N. Louka, R.G. Maroun, J.-L. Lanoisellé, Anaerobic digestion of nine varieties of grape pomace: Correlation between biochemical composition and methane production, *Biomass Bioenergy* 107 (2017) 335–344.

6. D. Haldar, P. Duarah, M.K. Purkait, Chapter 16 - Progress in the synthesis and applications of polymeric nanomaterials derived from waste lignocellulosic biomass, in: D. Giannakoudakis, L. Meili, I. Anastopoulos (Eds.), *Advanced Materials for Sustainable Environmental Remediation*, Elsevier, 2022, pp. 419–433.
7. V.L. Dhadge, M. Changmai, M.K. Purkait, Purification of catechins from Camellia sinensis using membrane cell, *Food Bioprod. Process.* 117 (2019) 203–212.
8. P. Duarah, D. Haldar, M.K. Purkait, Technological advancement in the synthesis and applications of lignin-based nanoparticles derived from agro-industrial waste residues: A review, *Int. J. Biol. Macromol.*163 (2020) 1828–1843.
9. V. Heuzé, G. Tran, Grape pomace, in: *Feedipedia, a Programme by INRAE, CIRAD*, AFZ and FAO, 2020.
10. T. Ilyas, P. Chowdhary, D. Chaurasia, E. Gnansounou, A. Pandey, P. Chaturvedi, Sustainable green processing of grape pomace for the production of value-added products: An overview, *Environ. Technol. Innov.* 23 (2021) 101592.
11. S. Dinicola, A. Cucina, D. Antonacci, M. Bizzarri, Anticancer effects of grape seed extract on human cancers: A review, *J Carcinog Mutagen S* 8(005) (2014). https://doi.org/10.4172/2157-2518.S8-005
12. A.A. Hamza, G.H. Heeba, H.M. Elwy, C. Murali, R. El-Awady, A. Amin, Molecular characterization of the grape seeds extract's effect against chemically induced liver cancer: In vivo and in vitro analyses, *Sci. Rep.* 8(1) (2018) 1–16.
13. R. Del Pino-García, M.a.D. Rivero-Pérez, M.L. González-SanJosé, M. Ortega-Heras, J. García Lomillo, P. Muñiz, Chemopreventive potential of powdered red wine pomace seasonings against colorectal cancer in HT-29 cells, *J. Agric. Food Chem.* 65(1) (2017) 66–73.
14. A.J. León-González, M.J. Jara-Palacios, M. Abbas, F.J. Heredia, V.B. Schini-Kerth, Role of epigenetic regulation on the induction of apoptosis in Jurkat leukemia cells by white grape pomace rich in phenolic compounds, *Food Funct.* 8(11) (2017) 4062–4069.
15. N.F.F. De Sales, L. Silva da Costa, T.I.A. Carneiro, D.A. Minuzzo, F.L. Oliveira, L.M.C. Cabral, A.G. Torres, T. El-Bacha, Anthocyanin-Rich Grape Pomace Extract (Vitis vinifera L.) from Wine Industry Affects Mitochondrial Bioenergetics and Glucose Metabolism in Human Hepatocarcinoma HepG2 Cells, *Molecules* 23(3) (2018) 611.
16. J.F. Dias, B.D. Simbras, C. Beres, K.O. dos Santos, L.M.C. Cabral, M.A.L. Miguel, Acid lactic bacteria as a bio-preservant for grape pomace beverage, *Front. Sustain. Food Syst* 2 (2018) 58.
17. F. Kabir, M.S. Sultana, H. Kurnianta, Antimicrobial activities of grape (Vitis vinifera L.) pomace polyphenols as a source of naturally occurring bioactive components, *Afr. J. Biotechnol* 14(26) (2015) 2157–2161.
18. P.-h. Chou, S. Matsui, K. Misaki, T. Matsuda, Isolation and identification of xenobiotic aryl hydrocarbon receptor ligands in dyeing wastewater, *Environ. Sci. Technol.* 41(2) (2007) 652–657.
19. P. Moate, S. Williams, V. Torok, M. Hannah, B. Ribaux, M. Tavendale, R. Eckard, J. Jacobs, M. Auldist, W. Wales, Grape marc reduces methane emissions when fed to dairy cows, *J. Dairy Sci.* 97(8) (2014) 5073–5087.
20. S. Ebrahimzadeh, B. Navidshad, P. Farhoomand, F.M. Aghjehgheshlagh, Effects of grape pomace and vitamin E on performance, antioxidant status, immune response, gut morphology and histopathological responses in broiler chickens, *S. Afr. J. Anim. Sci.* 48(2) (2018) 324–336.
21. J. Ferrer, G. Páez, Z. Mármol, E. Ramones, C. Chandler, M. Marın, A. Ferrer, Agronomic use of biotechnologically processed grape wastes, *Bioresour. Technol.* 76(1) (2001) 39–44.
22. M.M.M. Salgado, R.O. Blu, M. Janssens, P. Fincheira, Grape pomace compost as a source of organic matter: Evolution of quality parameters to evaluate maturity and stability, *J. Clean. Prod.* 216 (2019) 56–63.
23. P. Tatke, M. Rajan, Comparison of conventional and novel extraction techniques for the extraction of scopoletin from Convolvulus pluricaulis, *Indian J. Pharm. Educ. Res.* 48(1) (2014) 27–31.
24. M. Corrales, A.F. García, P. Butz, B. Tauscher, Extraction of anthocyanins from grape skins assisted by high hydrostatic pressure, *J Food Eng* 90(4) (2009) 415–421.
25. A.M. Posadino, G. Biosa, H. Zayed, H. Abou-Saleh, A. Cossu, G.K. Nasrallah, R. Giordo, D. Pagnozzi, M.C. Porcu, L. Pretti, Protective effect of cyclically pressurized solid–liquid extraction polyphenols from Cagnulari grape pomace on oxidative endothelial cell death, *Molecules* 23(9) (2018) 2105.
26. E. Torre, G. Iviglia, C. Cassinelli, M. Morra, N. Russo, Polyphenols from grape pomace induce osteogenic differentiation in mesenchymal stem cells, *Int. J. Mol. Med.* 45(6) (2020) 1721–1734.
27. M.J. Otero-Pareja, L. Casas, M.T. Fernández-Ponce, C. Mantell, E.J. Martinez de la Ossa, Green extraction of antioxidants from different varieties of red grape pomace, *Molecules* 20(6) (2015) 9686–9702.

28. M. Palma, Z. Piñeiro, C.G. Barroso, In-line pressurized-fluid extraction–solid-phase extraction for determining phenolic compounds in grapes, *J. Chromatogr. A* 968(1–2) (2002) 1–6.
29. K. Sólyom, R. Solá, M.J. Cocero, R.B. Mato, Thermal degradation of grape marc polyphenols, *Food Chemistry* 159 (2014) 361–366.
30. F. Zhang, T. Wang, X. Wang, X. Lü, Apple pomace as a potential valuable resource for full-components utilization: A review, *J. Clean. Prod.* 329 (2021) 129676.
31. R. Shalini, D. Gupta, Utilization of pomace from apple processing industries: A review, *J. Food Sci. Technol.* 47(4) (2010) 365–371.
32. S. Bhushan, K. Kalia, M. Sharma, B. Singh, P.S. Ahuja, Processing of apple pomace for bioactive molecules, *Crit. Rev. Biotechnol.* 28(4) (2008) 285–296.
33. J. Luo, Y. Ma, Y. Xu, Valorization of apple pomace using a two-step slightly acidic processing strategy, *Renew. Energy* 152 (2020) 793–798.
34. S.F. Barbieri, S. da Costa Amaral, A.C. Ruthes, C.L. de Oliveira Petkowicz, N.C. Kerkhoven, E.R.A. da Silva, J.L.M. Silveira, Pectins from the pulp of gabiroba (Campomanesia xanthocarpa Berg): Structural characterization and rheological behavior, *Carbohydr. Polym.* 214 (2019) 250–258.
35. Đ. Ačkar, A. Jozinović, J. Babić, B. Miličević, J.P. Balentić, D. Šubarić, Resolving the problem of poor expansion in corn extrudates enriched with food industry by-products, *Innov. Food Sci. Emerg. Technol* 47 (2018) 517–524.
36. D.G. Stevenson, P.A. Domoto, J.-l. Jane, Structures and functional properties of apple (Malus domestica Borkh) fruit starch, *Carbohydr. Polym.* 63(3) (2006) 432–441.
37. M. Szymanska-Chargot, M. Chylinska, K. Gdula, Isolation and characterization of cellulose from different fruit and vegetable pomaces. *Polymers* 9: 495, 2017.
38. S.S. Ahmad, M. Khalid, K. Younis, Interaction study of dietary fibers (pectin and cellulose) with meat proteins using bioinformatics analysis: An In-Silico study, *LWT* 119 (2020) 108889.
39. A.Y. Melikoğlu, S.E. Bilek, S. Cesur, Optimum alkaline treatment parameters for the extraction of cellulose and production of cellulose nanocrystals from apple pomace, *Carbohydr. Polym.* 215 (2019) 330–337.
40. L. Lin, A. Peng, K. Yang, Y. Zou, Monomeric phenolics in different parts of high-acid apple (Malus sieversii f. niedzwetzkyana (Dieck) Langenf): a promising source of antioxidants for application in nutraceuticals, *Int. J. Food Sci.* 53(6) (2018) 1503–1509.
41. R. Rodríguez Madrera, R. Pando Bedriñana, B. Suárez Valles, Application of central composite design in the fermentation of apple pomace to optimize its nutritional and functional properties, *Acta Aliment.* 47(3) (2018) 324–332.
42. E. Bartkiene, D. Vizbickiene, V. Bartkevics, I. Pugajeva, V. Krungleviciute, D. Zadeike, P. Zavistanaviciute, G. Juodeikiene, Application of Pediococcus acidilactici LUHS29 immobilized in apple pomace matrix for high value wheat-barley sourdough bread, *LWT-Food Sci. Technol.* 83 (2017) 157–164.
43. J.C. Barreira, A.A. Arraibi, I.C. Ferreira, Bioactive and functional compounds in apple pomace from juice and cider manufacturing: Potential use in dermal formulations, *Trends Food Sci. Technol.* 90 (2019) 76–87.
44. J.-P. Fan, D.-D. Liao, X.-H. Zhang, Ultrasonic assisted extraction of ursolic acid from apple pomace: A novel and facile technique, *Sep. Sci. Technol* 51(8) (2016) 1344–1350.
45. Ł. Woźniak, A. Szakiel, C. Pączkowski, K. Marszałek, S. Skąpska, H. Kowalska, R. Jędrzejczak, Extraction of triterpenic acids and phytosterols from apple pomace with supercritical carbon dioxide: Impact of process parameters, modelling of kinetics, and scaling-up study, *Molecules* 23(11) (2018) 2790.
46. E. Lesellier, E. Destandau, C. Grigoras, L. Fougère, C. Elfakir, Fast separation of triterpenoids by supercritical fluid chromatography/evaporative light scattering detector, *J. Chromatogr. A* 1268 (2012) 157–165.
47. S. Polat, G. Guclu, H. Kelebek, M. Keskin, S. Selli, Comparative elucidation of colour, volatile and phenolic profiles of black carrot (Daucus carota L.) pomace and powders prepared by five different drying methods, *Food Chem.* 369 (2022) 130941.
48. C. Alasalvar, M. Al-Farsi, P. Quantick, F. Shahidi, R. Wiktorowicz, Effect of chill storage and modified atmosphere packaging (MAP) on antioxidant activity, anthocyanins, carotenoids, phenolics and sensory quality of ready-to-eat shredded orange and purple carrots, *Food Chem.* 89(1) (2005) 69–76.
49. G. OeZEN, M. Akbulut, N. Artik, Stability of black carrot anthocyanins in the Turkish delight (Lokum) during storage, *J. Food Process Eng.* 34(4) (2011) 1282–1297.

50. S. Kamiloglu, G. Ozkan, H. Isik, O. Horoz, J. Van Camp, E. Capanoglu, Black carrot pomace as a source of polyphenols for enhancing the nutritional value of cake: An in vitro digestion study with a standardized static model, *LWT* 77 (2017) 475–481.
51. E. Agcam, A. Akyıldız, V.M. Balasubramaniam, Optimization of anthocyanins extraction from black carrot pomace with thermosonication, *Food Chem.* 237 (2017) 461–470.
52. E. Janiszewska, D. Witrowa-Rajchert, M. Kidon, J. Czapski, Effect of the applied drying method on the physical properties of purple carrot pomace, *Int. Agrophysics.* 27(2) (2013).
53. Z. Lu, J. Wang, R. Gao, F. Ye, G. Zhao, Sustainable valorisation of tomato pomace: A comprehensive review, *Trends Food Sci. Technol.* 86 (2019) 172–187.
54. A. Zuorro, M. Fidaleo, R. Lavecchia, Enzyme-assisted extraction of lycopene from tomato processing waste, *Enzyme Microb. Technol.* 49(6–7) (2011) 567–573.
55. E. Giovannucci, E.B. Rimm, Y. Liu, M.J. Stampfer, W.C. Willett, A prospective study of tomato products, lycopene, and prostate cancer risk, *J. Natl. Cancer Inst.* 94(5) (2002) 391–398.
56. I.F. Strati, V. Oreopoulou, Process optimisation for recovery of carotenoids from tomato waste, *Food Chem.* 129(3) (2011) 747–752.
57. I. Strati, V. Oreopoulou, Recovery of carotenoids from tomato processing by-products: A review, *Food Res. Int.* 65 (2014) 311–321.
58. S. Kumcuoglu, T. Yilmaz, S. Tavman, Ultrasound assisted extraction of lycopene from tomato processing wastes, *J. Food Sci. Technol.* 51(12) (2014) 4102–4107.
59. K.K. Ho, M. Ferruzzi, A. Liceaga, M.F. San Martín-González, Microwave-assisted extraction of lycopene in tomato peels: Effect of extraction conditions on all-trans and cis-isomer yields, *LWT-Food Sci. Technol.* 62(1) (2015) 160–168.
60. Z. Lianfu, L. Zelong, Optimization and comparison of ultrasound/microwave assisted extraction (UMAE) and ultrasonic assisted extraction (UAE) of lycopene from tomatoes, *Ultrason. Sonochem.* 15(5) (2008) 731–737.
61. R. Lavecchia, A. Zuorro, Improved lycopene extraction from tomato peels using cell-wall degrading enzymes, *Eur. Food Res. Technol.* 228(1) (2008) 153–158.
62. S. Cuccolini, A. Aldini, L. Visai, M. Daglia, D. Ferrari, Environmentally friendly lycopene purification from tomato peel waste: Enzymatic assisted aqueous extraction, *J. Agric. Food Chem.* 61(8) (2013) 1646–1651.
63. U. Topal, M. Sasaki, M. Goto, K. Hayakawa, Extraction of lycopene from tomato skin with supercritical carbon dioxide: Effect of operating conditions and solubility analysis, *J. Agric. Food Chem.* 54(15) (2006) 5604–5610.
64. S. Hasegawa, T.-M. Jao, R. Inagi, Dietary metabolites and chronic kidney disease, *Nutrients* 9(4) (2017) 358.
65. Y. Niu, N. Li, Q. Xia, Y. Hou, G. Xu, Comparisons of three modifications on structural, rheological and functional properties of soluble dietary fibers from tomato peels, *LWT* 88 (2018) 56–63.
66. C. Lara-Espinoza, E. Carvajal-Millán, R. Balandrán-Quintana, Y. López-Franco, A. Rascón-Chu, Pectin and pectin-based composite materials: Beyond food texture, *Molecules* 23(4) (2018) 942.
67. M.M. Alancay, M.O. Lobo, C.M. Quinzio, L.B. Iturriaga, Extraction and physicochemical characterization of pectin from tomato processing waste, *J. Food Meas. Charact.* 11(4) (2017) 2119–2130.
68. A.N. Grassino, J. Halambek, S. Djaković, S.R. Brnčić, M. Dent, Z. Grabarić, Utilization of tomato peel waste from canning factory as a potential source for pectin production and application as tin corrosion inhibitor, *Food Hydrocoll.* 52 (2016) 265–274.
69. A.N. Grassino, M. Brnčić, D. Vikić-Topić, S. Roca, M. Dent, S.R. Brnčić, Ultrasound assisted extraction and characterization of pectin from tomato waste, *Food Chem.* 198 (2016) 93–100.
70. W. Zhang, F. Xie, X. Lan, S. Gong, Z. Wang, Characteristics of pectin from black cherry tomato waste modified by dynamic high-pressure microfluidization, *J. Food Eng.* 216 (2018) 90–97.
71. L.R. Adetunji, A. Adekunle, V. Orsat, V. Raghavan, Advances in the pectin production process using novel extraction techniques: A review, *Food Hydrocoll.* 62 (2017) 239–250.
72. G. Wu, Dietary protein intake and human health, *Food & Function* 7(3) (2016) 1251–1265.
73. S.M. Meshkani, S.A. Mortazavi, A.H.E. Rad, A. Beigbabaei, Optimization of protein extraction and evaluation of functional properties of tomato waste and seeds from tomato paste plants, *Biosci. Biotechnol. Res. Asia* 13(4) (2016) 2387–2401.
74. D. Shao, G.G. Atungulu, Z. Pan, T. Yue, A. Zhang, Z. Fan, Characteristics of isolation and functionality of protein from tomato pomace produced with different industrial processing methods, *Food Bioproc Tech* 7(2) (2014) 532–541.
75. S.J. LATLIEF, D. Knorr, Tomato seed protein concentrates: Effects of methods of recovery upon yield and compositional characteristics, *J. Food Sci.* 48(6) (1983) 1583–1586.

76. K. Zhang, H. Zhao, S.C. Wang, Upcycle olive pomace as antioxidant and recycling agent in asphalt paving materials, *Constr Build Mater.* 330 (2022) 127217.
77. R.C. Williams, N.S. McCready, The utilization of agriculturally derived lignin as an antioxidant in asphalt binder, *Trans. Proj. Rep* 14 (2008) 1–92. https://doi.org/10.4172/2157-2518.S8-005
78. G. Xu, H. Wang, H. Zhu, Rheological properties and anti-aging performance of asphalt binder modified with wood lignin, *Constr Build Mater.* 151 (2017) 801–808.
79. S. Lammi, N. Le Moigne, D. Djenane, N. Gontard, H. Angellier-Coussy, Dry fractionation of olive pomace for the development of food packaging biocomposites, *Ind Crops Prod.* 120 (2018) 250–261.
80. T.B. Ribeiro, A.L. Oliveira, C. Costa, J. Nunes, A.A. Vicente, M. Pintado, Total and sustainable valorisation of olive pomace using a fractionation approach, *Appl. Sci.* 10(19) (2020) 6785.
81. A. Calabi-Floody, G. Thenoux, Controlling asphalt aging by inclusion of by-products from red wine industry, *Constr Build Mater.* 28(1) (2012) 616–623.
82. T.J. Erinle, D.I. Adewole, Fruit pomaces—Their nutrient and bioactive components, effects on growth and health of poultry species, and possible optimization techniques, *Anim Nutr.* 9 (2022) 357–377.
83. K. Ross, D. Ehret, D. Godfrey, L. Fukumoto, M. Diarra, Characterization of pilot scale processed canadian organic cranberry (Vaccinium macrocarpon) and blueberry (Vaccinium angustifolium) juice pressing residues and phenolic-enriched extractives, *Int. J. Fruit Sci.* 17(2) (2017) 202–232.
84. M. Jahurul, I. Zaidul, K. Ghafoor, F.Y. Al-Juhaimi, K.-L. Nyam, N. Norulaini, F. Sahena, A.M. Omar, Mango (Mangifera indica L.) by-products and their valuable components: A review, *Food Chem.* 183 (2015) 173–180.
85. D.S. Sogi, M. Siddiq, I. Greiby, K.D. Dolan, Total phenolics, antioxidant activity, and functional properties of 'Tommy Atkins' mango peel and kernel as affected by drying methods, *Food Chem.* 141(3) (2013) 2649–2655.
86. C.B.T. Pal, G.C. Jadeja, Microwave-assisted extraction for recovery of polyphenolic antioxidants from ripe mango (Mangifera indica L.) peel using lactic acid/sodium acetate deep eutectic mixtures, *Food Sci. Technol. Int.* 26(1) (2020) 78–92.
87. S. Pedraza-Guevara, R.F. do Nascimento, M.H.G. Canteri, N. Muñoz-Almagro, M. Villamiel, M.T. Fernández-Ponce, L.C. Cardoso, C. Mantell, E.J. Martinez de la Ossa, E. Ibañez, Valorization of unripe papaya for pectin recovery by conventional extraction and compressed fluids, *J. Supercrit. Fluids* 171 (2021) 105133.
88. M. Uyttebroek, P. Vandezande, M. Van Dael, S. Vloemans, B. Noten, B. Bongers, W. Porto-Carrero, M. Muñiz Unamunzaga, M. Bulut, B. Lemmens, Concentration of phenolic compounds from apple pomace extracts by nanofiltration at lab and pilot scale with a techno-economic assessment, *J. Food Process Eng.* 41(1) (2018) e12629.
89. S. Klettenhammer, G. Ferrentino, H.S. Zendehbad, K. Morozova, M. Scampicchio, Microencapsulation of linseed oil enriched with carrot pomace extracts using Particles from Gas Saturated Solutions (PGSS) process, *J. Food Eng.* 312 (2022) 110746.
90. Y. Cheng, T. Wu, X. Chu, S. Tang, W. Cao, F. Liang, Y. Fang, S. Pan, X. Xu, Fermented blueberry pomace with antioxidant properties improves fecal microbiota community structure and short chain fatty acids production in an in vitro mode, *LWT* 125 (2020) 109260.
91. S. Surbhi, R. Verma, R. Deepak, H. Jain, K. Yadav, A review: Food, chemical composition and utilization of carrot (Daucus carota L.) pomace, *Int. J. Chem. Stud.* 6(3) (2018) 2921–2926.
92. M.d.M. Contreras, I. Romero, M. Moya, E. Castro, Olive-derived biomass as a renewable source of value-added products, *Process Biochemistry* 97 (2020) 43–56.
93. C. Tagliani, C. Perez, A. Curutchet, P. Arcia, S. Cozzano, Blueberry pomace, valorization of an industry by-product source of fibre with antioxidant capacity, *Food Sci. Technol.* 39 (2019) 644–651.
94. L. Varnaitė, M. Keršienė, A. Šipailienė, R. Kazernavičiūtė, P.R. Venskutonis, D. Leskauskaitė, Fiber-rich cranberry pomace as food ingredient with functional activity for yogurt production, *Foods* 11(5) (2022) 758.
95. B.L. White, L.R. Howard, R.L. Prior, Proximate and polyphenolic characterization of cranberry pomace, *J. Agric. Food Chem.* 58(7) (2010) 4030–4036.
96. P. Benelli, C.A.S. Riehl, A. Smânia, E.F.A. Smânia, S.R.S. Ferreira, Bioactive extracts of orange (Citrus sinensis L. Osbeck) pomace obtained by SFE and low pressure techniques: Mathematical modeling and extract composition, *J. Supercrit. Fluids* 55(1) (2010) 132–141.

7 Extraction of bioactive compounds from marine by-products

7.1 INTRODUCTION

The food processing industry creates large amounts of waste streams, most of which are not processed and can be used for productive use. In their place, they are either burnt or thrown away in landfills. These activities, in turn, result in the production of emissions of greenhouse gases (GHG), uncontrolled deterioration and further pollution difficulties brought on by unpleasant odours and soil contamination from leachates [1, 2]. Due to their great diversity, marine species have evolved to produce their own sets of bioactive compounds. The global population's appetite for aquatic foods has expanded over the past few decades, owing to the positive public view of seafood's health benefits. However, only around half to two-thirds of marine resources are used directly for human consumption, while the rest is either wasted or turned into protein-rich animal feeds, mostly in the form of fishmeal and fertilizer. As a result of economic and technological development, by-products (skin, skull, viscera, carapace, etc.) have been turned into high-value commodities, resulting in a multitude of new natural marine products [3].

In 2018, global fisheries and aquaculture production from inland and marine waters totalled 178.5 million tonnes, with human consumption accounting for 87.6% of that total. It is anticipated that this number will continue to climb on an annual basis due to the expansion of the world's population as well as shifts in consumption patterns brought about by increased awareness of the positive benefits marine species have on human health. However, the fish processing industry generates a significant amount of waste in the form of by-products. These include the head, tail, skin, scale, viscera and bone of the fish [4, 5]. It is projected that 35% of the farmed fish will be lost during the process after harvesting, and then 70% of the fish that is processed will become by-products. The quantity of by-products produced by marine processing is variable and dependent on factors such as species, size, season and fishing sites. For example, only 15% of the by-products produced by salmon aquaculture are consumed by humans, despite the fact that the total potential value of these by-products is close to $40 million. The remaining 85% are used for other purposes. By-products are abundant in bioactive constituents that may be recovered and used in a variety of sectors, from energy to medicines. Figure 7.1 shows some fish by-products and the presence of different bioactive compounds in them.

The process of extracting bioactive substances from plants, fruits and the waste products of these sources has attracted a lot of attention in recent years [7–9], but the concept of extracting beneficial compounds from marine waste is relatively new. The utilization of by-products can increase the value of the fisheries and aquaculture sectors. One example of this is the development of biofilm packaging from the collagen found in the skin of cartilaginous fish (Mustelus mustelus). This biofilm demonstrated UV protective qualities as well as antioxidant activity, all of which might be employed to extend the shelf life of perishable food or the raw materials required to make it. This can have the dual effect of lowering environmental impact by cutting down on waste and pollution while simultaneously raising economic benefit for the relevant industries. In addition to their use as a raw material in the food sector, marine by-products have also found use in the pharmaceutical and cosmetics industries as a result of the bioactive chemicals they contain. The discovery of bioactive

DOI: 10.1201/9781003315469-7

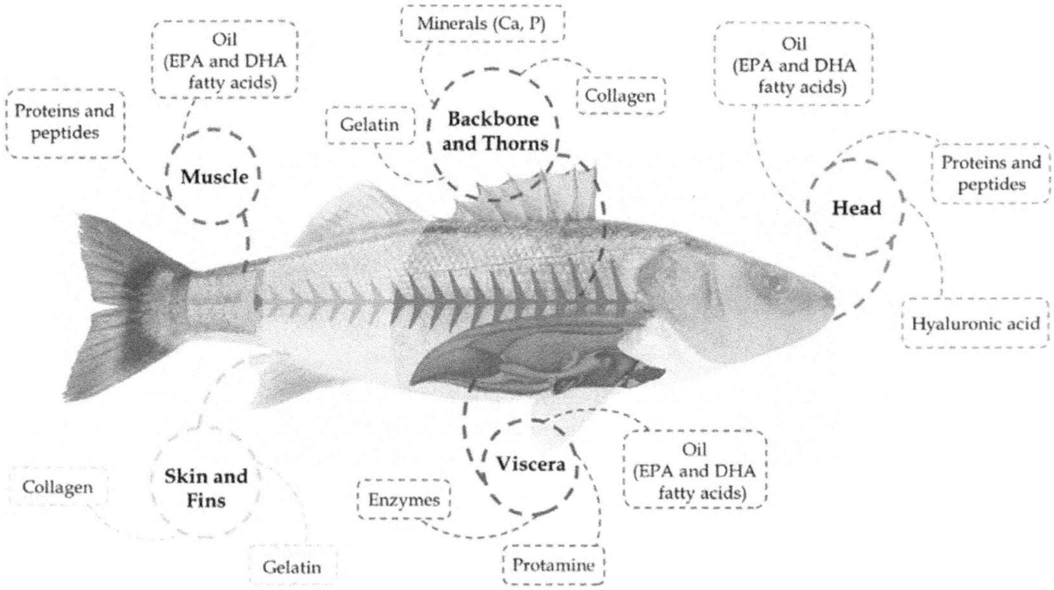

FIGURE 7.1 An illustration of fish by-products in the form of a schematic, together with the availability of bioactive compounds [Reproduced with permission from Khawli et al. (2020) [6]].

compounds from marine species processing by-products, such as from oyster and crab processing units, has attracted the attention of several research organizations in recent years. Therefore, it has been applied to manufacturing protein hydrolysate, fish oil, therapeutic agents and cosmetics.

Marine by-products, including fish, oyster, crab and prawn shells, are explored in this chapter for their potential bioactive component content. Bioactive compounds may be extracted from marine by-products using state-of-the-art extraction techniques, which have been highlighted in this chapter. The commercial potential and further development are also discussed.

7.2 BIOACTIVE COMPOUNDS AVAILABLE IN MARINE BY-PRODUCTS

7.2.1 Bioactive compounds in shells

Crustacean shells and shellfish wastes are significant categories of by-products from marine processing facilities. Due to rising processing facility accumulation and the materials' slow natural disintegration, efficient use of these marine wastes has become a critical environmental issue. One of the primary components of these shell wastes is chitin, which has the potential to be identified as a physiologically active polysaccharide. As a result, chitin is beneficial for a wide variety of applications because of its versatility. These shell wastes can be utilized to isolate chitin and produce chitosan and oligomers on the industrial level. A high-molecular-weight linear polymer of N-acetyl-d-glucosamine (N-acetyl-2-amino-2-deoxy-d-glucopyranose) units, chitin may be readily converted into numerous different bioactive derivatives [5]. Chitosan is the most widespread kind and is produced by deacetylating chitin. Chitosan is a positively charged heteropolymer composed of d-glucosamine (GlcN) and N-acetylglucosamine (GlcNAc) units, as shown by its chemical structure. The chemical structures of the chitin and chitosan are shown in Figure 7.2. Chitin and chitosan oligomers are normally produced using techniques that are either chemical or biological in nature. However, the current commercial preparations of chitin and chitosan involve utilizing thermochemical methods that involve the demineralization, deproteination and deacetylation of starting materials. These processes are carried out on chitin and chitosan in order to make them more marketable. The circumstances that are employed throughout the manufacturing process have

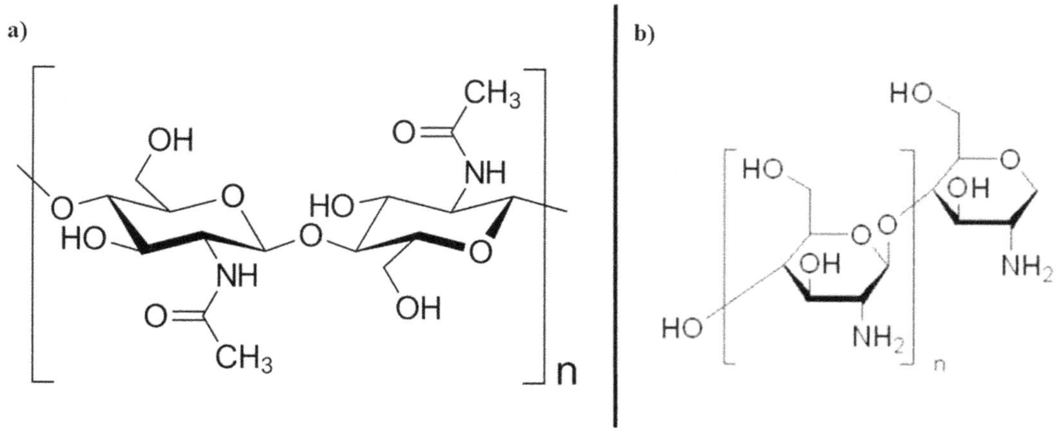

FIGURE 7.2 Chemical structure of (a) Chitin and (b) Chitosan.

a significant impact not only on the molecular weight of chitosan but also on the degree to which it has been deacetylated.

The non-toxicity, biocompatibility and biodegradability of chitin, chitosan and related oligomers have garnered a lot of interest for their potential uses as bioactive materials [10]. These materials are appealing for usage in many different areas, including food and nutrition, biomedicine, biotechnology, agriculture and environmental protection, because of their key structural and functional features. In particular, chitosan is often debated in the context of its potential medicinal and culinary uses [11]. Their potential use in the functional food and pharmaceutical sectors may be hampered, however, by their high viscosity and limited solubility at neutral pH. In order to increase their functional characteristics and intestinal absorption, chitin and chitosan are increasingly being converted into their oligomers. In the process of hydrolysis, chitin and chitosan oligomers can be produced by either chemical or enzymatic processes. Acid hydrolysis is one such chemical processes, but it also results in the production of several potentially dangerous industrial compounds. In contrast, enzymatic hydrolysis is preferred for oligomer preparation because it produces larger quantities of highly polymerized oligomers and fewer toxic compounds [12].

In a broader sense, the fact that chitosan and its oligomers have a positive charge is the primary factor that dictates the majority of the biological processes in which they are involved. Chitosan and its oligomers have been shown to be beneficial in lowering levels of LDL cholesterol in the liver and blood by a number of different studies [13]. In spite of the fact that the precise process is not fully comprehended, it has been hypothesized that these molecules, when present in the digestive system, function as fat scavengers and facilitate the elimination of fat and cholesterol. Research conducted by Sugano et al. (1992) found that chitosan oligomers of nearly all molecular weights were able to inhibit a 5% dietary-induced rise in blood cholesterol [14]. In another experiment, Ikeda et al. (1993) revealed that chitosan hydrolysate with an average molecular weight of 10,000 Da may significantly enhance faecal excretion of neutral steroids [15]. On the other hand, another group of researchers contends that chitosan, with a relatively low molecular weight, is advantageous for reducing fat and preventing cardiovascular illnesses. Oligomeric chitosan has been shown to have anticancer activity in vitro and in vivo. When administered orally to mice, low-molecular-weight chitosan was found to significantly inhibit the metastasis of Lewis lung cancer [5]. Partial deacetylated chitin and chitin with a carboxymethyl group have also proved successful in slowing tumour growth. This method proposes that chitin and its carboxymethyl derivatives have immunostimulatory effects by boosting cytolytic T-lymphocytes. It is hypothesized that these molecules have an immunostimulant impact, activating peritoneal macrophages and stimulating nonspecific host resistance and that their activity increases as their molecular size decreases. On the other hand, oligomers with a greater molecular weight have shown anticancer action as well. Similarly, it has been proposed that their action results from activated cells' enhanced lymphokine synthesis. It has been discovered that chitin and chitosan's immunostimulant function is linked to increased production of macrophages, which produce cytokines essential to the healing process. Chitosan oligomers have been hypothesized to have a wound-healing effect because of their capacity to promote fibroblast synthesis via modification of the fibroblast growth factor. Producing collagen later on aids in the development of connective tissues even more [16].

The molecular weight and degree of deacetylation of chitosan and its oligomers determine their ability to block β-secretase activity [17]. The maximum inhibition is shown for chitosan oligosaccharides with molecular weights of 5 kDa and a deacetylation level of 90%. In addition, chitosan oligomers have been shown to limit -secretase activity by a mechanism of non-competitive inhibition. Progressive brain deposition of -amyloid peptides into fibrillar aggregates and insoluble plaques causes severe memory loss and neuronal cell death in Alzheimer's disease, in large part due to the beta-secretase enzyme. Although the mechanism of action of chitosan oligosaccharides, the first identified carbohydrate-secretase inhibitors, have not been addressed in depth, they are thought to work by blocking the enzyme from completing its normal reaction.

Chitosan and chitosan oligomers, depending on their molecular weights and the degree to which they have been deacetylated, have the potential to act as antioxidants by scavenging oxygen radicals. For the action, chitosan oligomers with lower molecular weights are favoured over those with larger molecular weights. In vitro tests have shown that strongly deacetylated (90%) chitosan oligomers also function as radical scavengers for hydroxyl, alkyl, superoxide and extremely persistent DPPH radicals. According to Sun et al. (2003), chitosan and its derivatives function as hydrogen donors to stop the oxidative process [18]. The antioxidant and radical scavenging capacities of chitin and chitosan derivatives have not, however, been thoroughly studied. The angiotensin-converting enzyme, which is linked to hypertension, is likewise inhibited by chitin and its derivatives. Peptides are the enzyme's natural substrates, and it is not entirely clear how chitosan oligomers block ACE. Chitin and its derivatives have a variety of direct uses in biomedicine, but they also reportedly function in drug delivery systems to provide controlled release [5].

A number of studies have indicated that carotenoproteins might be extracted from blue crab shells. For their 2018 study, Hamdi et al. explored the impact of adding a carotenoproteins extract (CPE) from blue crab shells on the quality of turkey meat sausages with lowered nitrite levels. Antioxidant activity, as measured by a variety of in vitro antioxidant tests, was shown to increase with increasing concentrations of CPE, and the compound also showed promising antibacterial action [19]. Following this, Hamdi et al. (2020) looked at the most effective method for extracting carotenoids from blue crab shells. Maceration with a 50/50 hexane/isopropanol binary system was the most effective method, yielding a high carotenoid concentration with negligible total phenolic and soluble protein levels ($p < 0.05$). It was discovered that when this method was paired with an enzymatic pre-treatment, it was significantly ($p < 0.05$) more effective and selective, especially towards astaxanthin [20].

Oysters are a valuable marine resource that is widely cultivated and has a high market value. The Suminoe oyster (*Crassostrea ariakensis*), the Zhe oyster (*Crassostrea plicatula*) and the Pacific cupped oyster are among the more than 100 species of oysters that are cultivated globally. These oysters provide wonderful food sources. The percentages of protein, glycogen and fat in oyster flesh (measured as dry flesh weight) range from 39.1% to 53.1%, 21.6% to 38% and 7.8% to 7.7%, respectively. Oysters are a great source of protein and other nutrients, and they also have a variety of therapeutic potentials [21]. The oyster shell contains minerals that shield delicate tissue as well as an organic matrix. The shell of the oyster is roughly 60% of the total weight. A schematic diagram of the availability of bioactive compounds in various parts of oyster shell is shown in Figure 7.3. Due to incorrect landfilling, microbiological activity, usage of low-cost disposal techniques and high management expenses, oyster shell disposal is linked to water and marine pollution, offensive odour concerns and detrimental effects on soil pH from improper recycling.

Nonetheless, oyster waste materials have the potential to be useful, environmentally benign substances. The utilization of these waste materials as biocompatible antimicrobials may enhance

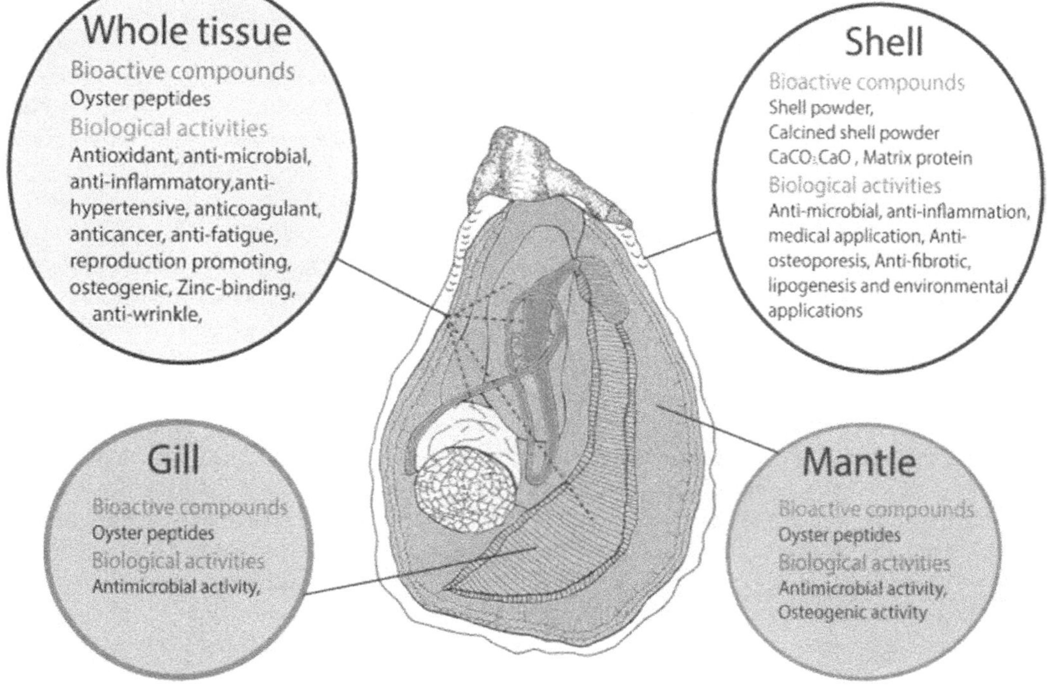

FIGURE 7.3 Schematic diagram of availability of bioactive compounds in various parts of the oyster shell [Reproduced with permission from Ulagesan et al. (2022) [3]].

waste management and human health [22]. The majority of oyster shells are made up of calcium carbonate ($CaCO_3$; around 95% of the total), with only a trace amount of organic matrix proteins (0.1–5%), which are also known as skeleton or shell proteins. Oyster shell is a source of calcium-based compounds, such as calcium hydroxide ($Ca(OH)_2$) and calcium oxide (CaO), among other calcium-based compounds. CaO is utilized in a significant capacity as a catalyst in both tissue engineering and industrial research. Oyster shell powder that has been calcined has gained a lot of interest for its biocompatibility, antibacterial properties and biocidal properties. As a result, a calcined oyster shell could be a useful antibacterial substitute in the production and packaging of food. It can be safe to use these organic antibacterial compounds to ensure the quality of both processed and fresh foods. The main component of a calcined oyster shell that raises the pH, CaO, is an important chemical that contributes to the antibacterial activity of the oyster shell. Cardiolipin, a significant lipid in bacterial cell membranes, reacts with calcium ions formed from the CaO, rupturing the cell wall and producing reactive oxygen species (ROS) and free radicals that have a significant negative impact on cell integrity. CaO antifungal properties are also influenced by its alkalinity and ROS production [22].

Oyster shells have a high $CaCO_3$ content, but the organic matrix network also plays a crucial role in the biomineralization process. It is composed of larger molecules such as polysaccharides, proteins—both soluble and insoluble in water— and lipids, in addition to smaller molecules such as free amino acids, pigments and short peptides. Based on extraction techniques and solubility, these biomolecules can be divided into a variety of matrix types, such as those that are water-soluble, ethanol-soluble, acid-soluble, acid-insoluble, ethylenediaminetetraacetic acid-soluble, EDTA-insoluble and fat-soluble [23]. The water-soluble matrix controls biomineralization as well as nacre biological processes such cell attraction, differentiation and stimulation. Nacre (and its biomatrix) is utilized in conventional pharmaceutical preparations to promote bone formation and improve bone density because of its advantageous biological activities [24]. Recent in vitro and in vivo research suggests that nacre may also be a biodegradable and biocompatible material with osteointegrative, osteoinductive and osteoconductive qualities. Nacre (and its biomatrix) has so been investigated as a potential alternative for bone. Nacre has been used in bone tissue bioengineering since 1931.

Lopez et al. (1992) made a significant advancement in the use of nacre as a substitute for bone grafts. The contact between bone and nacre was then studied in sheep by Atlan et al. (1999). They were able to show that nacre may stick directly to the freshly created bone, eliminating the requirement for an intermediate fibrous tissue. Studies conducted both in vivo and in vitro have shown that nacre includes chemicals that can activate osteogenic bone marrow cells [25]. Using proteomics, water-soluble matrix proteins from *Crassostrea gigas* nacre contained proteins with osteogenic activity. Both bone remodelling and biocompatibility depend on these proteins. Researchers have also looked into how alginate hydrogels, such as nacre powder, affect the osteogenic development of mesenchymal stem cells in the human bone marrow [26].

Water-soluble bioactive chemicals in nacre have been demonstrated by Chaturvedi et al. (2013) to have antioxidant action and encourage osteoblast development [24]. According to research on the osteogenic potential of human mesenchymal stem cells, nacre and its soluble protein matrix have the ability to promote early bone cell development in humans through a process known as osteoinduction [27]. The influence of the water-soluble nacre matrix on alkaline phosphate behaviour in MRC-5 fibroblasts and the osteogenic activities of the nacre water-soluble matrix and ethanol-soluble matrix have also been studied. Other topics that have been investigated include nacre powder combined with blood, nacre prostheses, nacre chips and nacre matrix proteins. Recent in vitro and in vivo investigations showed that water-soluble matrix proteins from *C. gigas* have anti-osteoporosis actions that come from promoting osteogenesis and inhibiting osteoclast absorption [28]. Shell extracts are utilized in anti-fibrotic techniques, particularly in the case of scleroderma, since they also trigger the catabolic pathway of human dermal fibroblasts. Researchers Latire et al. (2017) looked into oyster shell extract's capacity to prevent lipogenesis and found that is has a lipid-lowering impact [29]. These investigations indicate that a variety of therapeutic and pharmacological benefits, including anti-fibrotic, anti-osteoporotic and osteogenic activities as well as lipogenesis inhibition, may be obtained from oyster shell extracts, such as those from the water-soluble matrix, ethanol-soluble matrix and water-soluble matrix proteins. They can therefore contribute to the engineering of human tissue.

7.2.2 Bioactive compounds in skins

Waste products derived from fish skin can potentially be exploited as a source for collagen and gelatine isolation. Both collagen and gelatine are presently used in many sectors, including the culinary, cosmetic and medical professions. Three extended protein chains, each of which wraps around the other, contribute to the structural formation of collagen in the form of a triple helix. Gelatine is the partly hydrolyzed form of collagen, a distinct form of the same macromolecule. The transformation of collagen to gelatine is shown in Figure 7.4. Collagen is easily transformed into gelatine by heat denaturation. Collagen and gelatine might also be extracted from fish processing by-products such as bone and fins in addition to fish skin. Differentiating collagen and gelatine from the proteins found in fish muscle is the high concentration of nonpolar amino acids (over 80%), including Gly, Ala, Val and Pro. Despite bovine and porcine skin being the primary sources of collagen and gelatine for commercial usage, several experiments have been undertaken to extract collagen and gelatine from fish skin and have utilized these studies to assess their potential industrial uses [30]. Collagen and gelatine extracted from fish skin are anticipated to garner interest in the marketplace as a viable substitute. This might be because collagen and gelatine derived from porcine skin are not as widely accepted as those from other animals for religious reasons. Bovine spongiform encephalopathy (BSE) and the risk it poses to humans have also sparked debate over the usage of collagen and gelatine obtained from cows. On the other hand, there is a limited probability that fish collagen or gelatine will include any undiscovered infections like BSE [5].

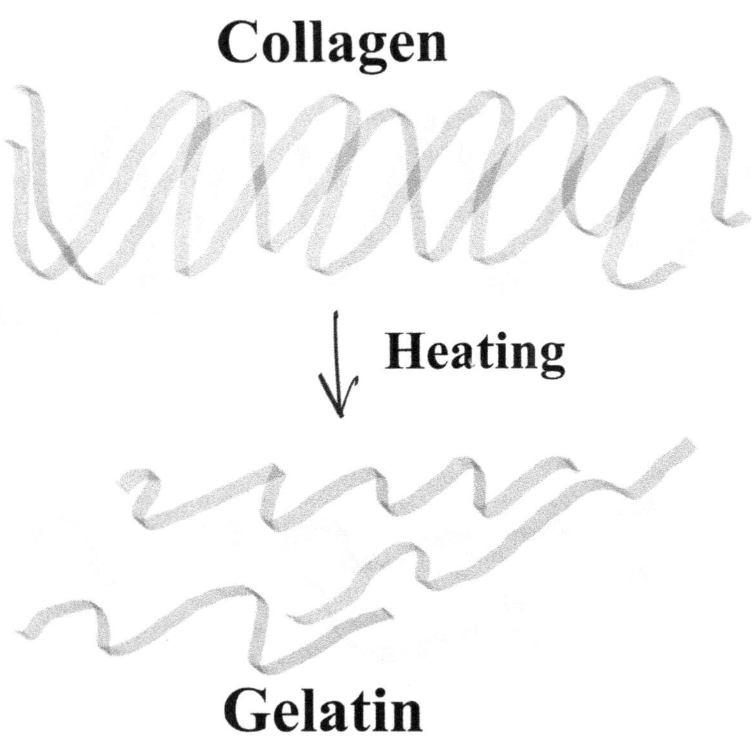

FIGURE 7.4 Schematic representation of the transformation of collagen to gelatine.

In most cases, collagen is obtained by acid treatment and then solubilized without causing any changes to its triple helix structure. However, subjecting the collagen to heat treatment destroys the hydrogen and covalent bonds. This stabilizes the triple helix configuration of collagen, which, in turn, converts its helical shape into a coiled conformation, leading to the creation of gelatine. This process is called gelatinization. In order to remove the collagen from the skin as gelatine, hot water treatment is generally performed. With respect to the extraction circumstances, such as temperature and pH, the characteristics of the extracted gelatine may differ somewhat. The best conditions for extraction depend on the kind of collagen and the type of fish [31].

The medical and pharmaceutical sectors frequently employ collagen molecules as drug transporters, protein transporters and gene transporters. One particularly interesting application of collagen sheets is as a medication carrier in the treatment of cancer. Collagen is an effective drug carrier because it may preserve drug concentrations over long periods and release them gradually in therapeutic areas. Additionally, collagen film/matrix is now being employed as a gene delivery vehicle to promote the creation of bone and cartilage. The development of tissues and organs, as well as the functional expression of cells, have both been found to depend on collagen [32]. Clinical studies indicate that consuming collagen/gelatine hydrolysates improves pain in osteoarthritis patients and that the production of cartilage matrix involves hydrolyzed collagen. In addition, collagen and gelatine are presently advertised as supplements for the nutrition of scalp hair, the preservation of healthy bone integrity and the treatment of brittle nails [33].

7.2.3 Bioactive compounds in bones

Another useful source for locating elements that promote health is fish bone that has been isolated following the removal of muscle proteins from the frame. Collagen makes up the organic portion of fish bone, which is 30% of the total mass. Therefore, in addition to fish skin, fish bone is taken into consideration as a source to separate collagen and gelatine. Fishbone, on the other hand, is mostly made of calcium phosphate and hydroxyapatite and contains between 60% and 70% inorganic materials.

The calcium found in fish bones is thought to be a possible supply of this vital mineral for human health. On the other hand, relatively limited study has been done to establish the bioavailability of the calcium found in fish bones and the possible applications. Calcium is normally obtained through one's food; however, a considerable deficiency in this mineral may be seen in the majority of conventional diets. Many calcium-fortified products are available on the market, and demand for them is always rising in an effort to increase calcium consumption. It is generally known that eating entire small fish is a good idea for your health since it gives you access to a lot of calcium. As demonstrated by in vivo applications, the body could absorb calcium from fish [34]. Few attempts have been made to investigate the advantages of more prominent fish bones for human health, and there is a severe lack of data on their impacts. In terms of the quantity of calcium, fish bone material obtained through the processing of huge fish is a valuable source. Softening the structure of fish bone is necessary before it can be used in calcium-fortified meals. This can be accomplished by the utilization of a variety of approaches, such as treatment with hot water and hot solutions of acetic acid. In addition, another team of researchers employed steam that had been subjected to extremely high temperatures in order to lessen the loss of soluble components from fish tissue. This resulted in improved bone repair within a shorter amount of time.

Due to its exact chemical composition, hydroxyapatite [$Ca_{10}(PO_4)_6(OH)_2$] has recently been introduced as a bone transplant material in a variety of medical and dental settings. When dealing with issues like bone fractures and damages, bone replacement materials, including autografts, allografts and xenografts, are typically employed. Because of their mechanical instability and incompatibility, however, none of these materials offer ideal conditions for bone repair. The most promising materials for bone replacement are calcium phosphate bioceramics, which include tetracalcium phosphate, amorphous calcium phosphate, tricalcium phosphate and hydroxyapatite. Because hydroxyapatite is the primary inorganic component of fish bone material, it has great potential for use in biomedical contexts. Hydroxyapatite is indestructible at normal body temperature and pressure, unlike other calcium phosphates. Bone bonding really benefits from its thermodynamic stability at physiological pH. This quality has been put to use in the pursuit of speedier bone recovery following invasive medical procedures or severe injuries. Coral and fish bone are only two examples of natural sources from which hydroxyapatite can be extracted [35]. There have been efforts to replace synthetic hydroxyapatite with a form of mineral isolated from fish bones. Isolating hydroxyapatite from fish bone often involves exceptionally high heat treatment (1,300°C), which increases the structural strength of hydroxyapatite and yields an excellent biocompatible inorganic compound [5].

7.2.4 Fish oils

By-products from the processing of marine fish might be turned into fish oil for better usage. Fish fat content ranges from 2% to 30%, generally, and it mostly relies on the species, food, geographic location, environmental factors, reproductive cycles and seasonal fluctuations. However, fishing for their oil is discouraged due to the diminishing marine fishery resources. As a result, a significant portion of the offal produced during processing might be used to create high-quality fish oil for human consumption, particularly from fatty fish processing by-products. In contrast to the composition of other oils, fish oil is mostly made up of two distinct kinds of fatty acids: eicosapentaenoic acid (EPA) and docosahexaenoic acid (DHA). Both of these fatty acids have been shown to have health benefits. These polyunsaturated fatty acids (PUFAs) are known as omega-3 fatty acids, and they are most often found in many marine creatures, such as cold-water fish species that have a greater percentage of unsaturated fat. In contrast to saturated fats, the PUFAs found in fish oil are easily absorbed for the generation of energy and have been shown to have a variety of different bioactivities.

The quality of fish oil is directly related to the processes and circumstances used to obtain it. High-speed centrifugation, low-temperature solvent extraction and supercritical fluid extraction are only a few of the current techniques used to extract fish oil. Fish oil is extracted from by-products of the fish processing industry using both wet and steam rendering techniques [36]. Omega-3 fatty acid separation is one of the primary issues in the present study. However, the separation of omega-3 fatty acids is made difficult by the existence of a diverse combination of triacylglycerols. To separate them at higher purity, many separation techniques, including crystallization, distillation, supercritical fluid extraction and chromatography, have been developed. To identify and separate fatty acids, column liquid chromatography is widely utilized. Additionally, the efficient separation of fatty acids is made possible by the combination of the chromatographic technique and urea crystallization [37]. There is potential for fish oil to play an even larger role in the pharmaceutical and food industries in the near future if proper separation techniques and cutting-edge processing processes are used to boost the production of higher-quality fish oil.

Fish oil, rich in health-beneficial omega-3 fatty acids, has been linked to improved human health and the prevention of a wide range of illnesses. In addition, scientists are confident that EPA and DHA are the primary preventive components of fish oil that counteract particular disorders. The greatest proof that omega-3 fatty acids have a role in disease prevention is the inverse association between blood levels of these acids and the likelihood of coronary heart disease. The findings of a study indicate that consuming fish increases the levels of EPA and DHA in the blood. This, in turn, leads to a reduction in the incidence of coronary heart disease. The antiatherogenic and antithrombotic characteristics of omega-3 fatty acids reduce the formation of thrombus by preventing the development of lipid-rich atherosclerotic plaques, thereby lowering the risk of thrombosis [38]. Fish oil has been demonstrated to be preventive against cardiac diseases by lowering serum triglyceride levels, improving vascular endothelial function, lowering blood pressure and reducing inflammation [39]. Dietary omega-3 fatty acids have been discovered by Christensen et al. (1997) to reduce arrhythmias that can cause sudden cardiac death. It is now among the leading causes of mortality in the west and ventricular arrhythmias in cardiac patients are mostly to blame [40]. Additionally, because omega-3 fatty acids are linked to the protection of a chain of diseases, their positive effects are favourable for the prevention of various illnesses and heart-related dysfunctions. Taking fish oil supplements can help fight diabetes. By reducing blood pressure and triacylglycerol concentrations, fish oil ameliorates several metabolic consequences of insulin resistance. Additionally, fish oil consumption has been shown to accelerate glucose absorption and sustain normal glucose metabolism.

In addition to their anti-inflammatory effects, EPA and DHA significantly improve the quality of life for those with autoimmune illnesses. The treatment of rheumatoid arthritis with polyunsaturated fish oil has been shown to be effective in clinical trials. The reason EPA and DHA are so effective at this is that they are turned into prostaglandins and leukotrienes, which are the body's own natural anti-inflammatory compounds. Fish oil's anti-inflammatory properties have also been proposed as a possible therapy for other inflammatory disorders like Crohn's and renal ailments. Breast, colon, skin, pancreas, prostate, lung and throat cancers are only some of the human carcinomas that omega-3 fatty acids have been shown to protect against in preliminary studies [5].

Furthermore, the omega-3 fatty acids found in fish oil have been linked to improved cognitive performance, better eye health and a more robust reproductive system. DHA is found in the synapses of nerve cells in the brain, as well as in the retina of the eye, the testes and the sperm. As a result, it is an extremely important factor in the growth and operation of the aforementioned organs and systems. One of the functions of omega-3 fatty acids that have been connected to their capacity to prevent mental health issues is brain growth [5].

7.2.5 Other by-products

The internal organs of fish are a rich supply of enzymes and several of these enzymes demonstrate significant catalytic activity even at low doses. In light of the unique properties possessed by these enzymes, the extraction of enzymes from fish processing wastes is a practise that is now prevalent. Large-scale commercial extraction of a wide variety of proteolytic enzymes from the viscera of marine fish includes pepsin, trypsin, chymotrypsin and collagenases [41].

Extreme circumstances in the marine environment, such as high salt content, low or high temperature, high pressure and poor nutrient supply, are met with remarkable adaptation by marine species. Fish proteinases are said to be superior because of their increased catalytic effectiveness at lower temperatures, less sensitivity to substrate concentrations and increased stability throughout a wider pH range. Due to their unique properties, fish enzymes have found several uses in the food industry.

Enzymes isolated from marine fish can be used for industrial-scale production of bioactive components, but they have no direct uses in the fields of functional foods or nutraceuticals. Multiple attempts have been documented by the scientific community to harvest crude enzyme mixtures from the internal organs of certain fish species and then employ these mixtures in the extraction of bioactive components from fish protein hydrolysates. Additional chitinases and chitosanases may be extracted from the digestive tract and other tissues of some marine fish species. The enzymes in question facilitate the extraction of chitin and chitosan from marine waste, both of which are useful in a wide variety of biomedical settings. Isolating bioactive chemicals from marine shell wastes might be more cost-effective with the help of these enzymes [5, 42].

7.3 TECHNOLOGICAL ADVANCEMENT IN THE EXTRACTION PROCESS

7.3.1 High hydrostatic pressure

High hydrostatic pressure (HHP) is a newly developed nonthermal processing technique that is primarily used to process marine food items, vegetables, fruits, dairy as well as to minimize the population of microbes and inactivate enzymes in marine food. In order to deprotonate charged particles, break down salt bridges and disrupt weak bonds (hydrogen bonds, electrostatic bonds and hydrophobic bonds) in cell membranes, high pressure in the range of 100–1000 megapascals was applied at temperatures of 5–35°C. This, in turn, helps promote cell permeability and improves the extractability of target compounds [43]. The permeability of the cells increases, and the amount of target molecules that may be extracted increases as the pressure is increased to the critical threshold level. Because the high pressure used in this method increases cell permeability, a large amount of valuable chemicals may be extracted from the cells [44].

Li et al. (2017) conducted research on the process of extracting astaxanthin from the shell powder of Penaeus vannamei Boone (Whiteleg shrimp) using high pressures (0.1–600 MPa); solvents, including ethanol, dichloromethane and acetone; solvent-to-sample ratios ranging from 10 $mL.g^{-1}$ to 50 $mL.g^{-1}$, and treatment times ranging from 0 min to 20 min. In comparison to other extraction methods like pressurized hot ethanol extraction, sunflower oil extraction, ethanol mixture extraction, enzymatic treatment and DPCD, the efficiency of the astaxanthin extraction was highest (72 $\mu g.g^{-1}$) when HHP conditions were optimized by applying 200 MPa with an ethanol solvent to dried shells at a ratio of 20 $mL.g^{-1}$ for 5 min [45]. Extracting astaxanthin with HHP resulted in a pigment with strong antioxidant properties. When the pressure was raised during the extraction process, the rate of mass transfer rose, and the cellular membranes were disrupted, both of which contributed to the rise in permeability. In this way, the extraction solvent was able to more easily enter the cells [46]. In accordance with the Pascal theory and the equilibrium of pressure both outside and inside of the cells, an optimal time has been identified. The maximal extraction yield might be readily reached when the solvent diffusion speed reaches the level needed to achieve equilibrium. If the storage period is extended, the content of astaxanthin, free astaxanthin and structure may be changed. The extraction yield would therefore drop. Compared to microwave-assisted extraction (MAE) and conventional heating extraction, the HHP treatment was shorter by 11 min and 39 min, respectively [47].

The HHP was also used on tilapia fish waste in order to extract bioactive peptides from protein by enzymatic hydrolysis. Increased peptide bond exposure at high pressures facilitates enzymatic proteolysis and subsequent fractionation. Protein hydrolysis at high pressure is essential for producing antioxidant peptides in protein hydrolysates and increasing the digestibility of the hydrolyzed protein [48]. Hemker et al. (2020) employed a central composite design to study the interaction effects of pressure (38–462 MPa) and time (6–35 min) in HHP on bioactive peptide extraction from tilapia by-product protein hydrolysates [49]. Thirty-five minutes of extraction at 250 MPa yielded 5.7 mg.mL^{-1} of soluble protein and a 23% trichloroacetic acid-solubility index. It is possible that an increase in pressure would lead to a rise in the concentration of soluble proteins, as this would cause enzymes to be activated and protein structures to be unfolded.

In addition, HHP has found use in the maritime sector, where it has been employed in the recovery of omega-3 PUFAs from fish waste. Omega-3 polyunsaturated fatty acids are being studied for their potential to prevent cardiovascular and inflammatory autoimmune diseases [50]. Environmental parameters, including the constitution, pH and packing materials, as well as extrinsic factors, like pressure endurance, levels of heating and pressure, decompression duration, processing conditions and initial product temperature, all have an impact on the extraction of omega-3 PUFAs [51].

HHP is an effective invention that has been utilized to extract bioactive substances from marine by-products, including astaxanthin, bioactive peptides and PUFAs. Compared to other cutting-edge technologies and traditional approaches, it demonstrated a greater extraction yield. The starting material's chemical and physical qualities must also be taken into account, as the holding period is dependent on the sample's porosity. Pressure and holding times are other significant elements. Thus, more focused research and study are needed to identify the ideal circumstances before applying HHP on an industrial scale.

7.3.2 Pulsed electric field

Pulsed electric field (PEF) is one example of a nonthermal processing technique that has the potential to find widespread application in the marine animal processing sector. PEF technique outperforms heat-based conventional extraction techniques because of its relatively high efficiency, fast response time, shorter processing time, less energy and temperature costs and high flexibility [52]. PEF technique uses high voltages and brief electric pulses to damage cell membranes in samples of marine animal by-products that are sandwiched between two electrodes. The process of releasing intracellular substances, known as electroporation, happens when the number of pores is increased. As a consequence, the great extraction efficiency of value-added chemicals is made possible [53]. The primary use of PEF processing is dependent on a number of variables, including electric strength, the number and duration of electric pulses, width, cell, shape and frequency characteristics. Additionally, the starting material's chemical and physical characteristics as well as the equipment's capability, may have an impact on the PEF working conditions [52].

Marine by-products are effectively valorized via PEF treatment, which has the advantages of speeding up the extraction process, increasing extraction yield and being ecologically benign. According to a recent study, PEF technology was used to extract antioxidants from the by-products (gills, bones and heads) of sea bass and sea bream [54]. After the PEF process was performed with a voltage of 7,000 V, 20 s pulse width, frequency of 10 Hz, 100 pulses and 1.40 kV.cm^{-1} electric field by a semiconductor-based positive Marx modulator (Epulsus-PM1-10), the efficiency of antioxidant extract from heads, bones and gills were enhanced significantly. However, the research only yielded crude antioxidant extracts, and no efforts were made to purify further or identify them.

Another impressive use of PEF technology is in the marine by-products processing industry. Because of its high effectiveness and low application temperature requirements for the extraction of naturally occurring chemicals with added value, this technology has been frequently employed to recover bioactive compounds from marine by-products [55]. Gulzar et al. (2020) employed PEF technology to pre-treat the cephalothorax of *L. vannamei* at 16 kV.cm^{-1} voltage and 240 pulses were selected depending on the lipid extraction efficiency and the rate of oxidation [56]. Afterwards, researchers used an ultrasonic processor set to run in continuous mode at an amplitude of 80% (0.12 $W.cm^{-3}$) to extract the solvent using hexane, isopropanol and ultrasound-assisted extraction (UAE) for 25 min. The extraction efficiency of fatty acids and the content of carotenoids was improved when using the UAE or PEF methods following solvent extraction as compared to the usual solvent extraction procedure using isopropanol and hexane. Since more electroporation occurred, better solvent transport into the samples was achieved, and matrix holes were created at higher voltages and higher pulse numbers, leading to a higher extraction yield of lipids [57]. Therefore, a synergistic effect was found for lipid extraction by combining UAE with turbulence; the solvent was better able to penetrate the matrix pores created by the PEF. PEF treatment reduced oxidation levels in carotenoids and UAE improved extraction performance. In light of the reduced oxidation and hydrolysis that resulted from the PEF and UAE treatments, its potential for use in lipid extraction applications is evident [56].

He et al. (2019) reported using a high-intensity PEF technique to extract collagen from the bones of marine animals [58]. With eight pulses and a 20 $kV.cm^{-1}$ electric field, the highest collagen content was attained. Gavahian et al. (2017) used the same starting material—marine food bone—to generate calcium and chondroitin sulphate by applying a strong electric field [59]. When it comes to extracting molecules with added value, PEF is incredibly fast and effective. In this regard, Li et al. (2016) developed PEF enzymatic-assisted extraction to facilitate the extraction of protein from abalone viscera [60]. A treatment length of 600 s, an intensity strength of 20 $kV.cm^{-1}$ and a material-to-solvent ratio of 1:4 all contributed to a 39.99% extraction yield. Another research suggested utilizing PEF treatment to recover protein hydrolysates from by-products of marine animals. In comparison to hydrolysates prepared using standard enzymatic extraction techniques, the PEF-extracted hydrolysates had improved solubility, emulsification and hydrolysis characteristics [61].

PEF treatment has demonstrated advantages in a number of processes used to handle marine by-products, and it offers a possible means of disrupting biological cells in food matrices without having any unfavourable side effects [62]. PEF treatment is suitable for processing both liquid and solid forms of heat-sensitive foods since it does not involve the use of potentially harmful solvents. It is a waste-free method that improves mass transfer through cell rupture. However, since electrode degradation and the migration of electrodes into target samples may happen, PEF treatment has remained difficult for dependable equipment at the industrial level up to this point [63]. It is necessary to do more research into the application of PEF treatment for the commercialization of marine by-products for use in a wider variety of industries.

7.3.3 Dense phase carbon dioxide

The dense phase carbon dioxide (DPCD) approach, which extracts value-added components from marine by-products using supercritical extraction techniques and a fluid solvent, is another possible nonthermal technology. In the DPCD method of working with pressurized circumstances, solvents are the primary tool for separating the solutes of liquid and solid matrices at temperatures and pressures close to the critical point. Solvents induce transitory combinations between gases and liquids under these receptive circumstances, eventually assisting in the extraction of the target chemicals from the food matrix. In the DPCD process, safe CO_2 is by far the most common and widely used solvent, particularly for food-related uses. The DPCD method has garnered a great deal of interest as a result of its prompt extraction rate, low toxicity, lower viscosity, greater diffusivity and minimal surface tension, in addition to its moderate processing parameters (pressure of 7.38 MPa and temperature of 31.1°C). This method is also less expensive than other conventional techniques [64]. Because of DPCD's elevated diffusion coefficient and low viscosity, CO_2 may quickly pass through the pores of heterogeneous matrices. The extraction period, extraction temperature and pressure, CO_2 flow rate and density and concentration of the solvent are a few variables that influence DPCD efficiency [65]. The modification of the pressure and temperature, which in turn affects the density, is necessary for achieving DPCD selectivity. On the other hand, the DPCD selectivity may be altered with the use of a cosolvent, which works by either increasing or lowering the polarity of the solvent.

Recently, lipid constituents derived from waste products from a wide variety of marine sources were recovered with DPCD. For example, it is believed that the viscera of the common carp fish (*Cyprinus carpio L.*) that is expelled as waste contains both monounsaturated fatty acids (MUFAs) and PUFAs [64]. Waste from common carp, which includes fillets, roe and viscera, contains a significant quantity of raw material that may be utilized for the separation of omega-3 and omega-6 PUFAs. Therefore, the DPCD method was utilized in order to extract lipid constituents from the viscera of common carp fish. This was accomplished by optimizing the lipids' extraction using the DPCD method and reducing the degradation of PUFAs throughout the processing stage [66]. Using multiple DPCD procedures at several pressures and temperatures (45–75°C), the Rastrelliger kanagurta by-product (skin) was found as a potential source of oil (20–35 MPa). Over the course of 6 hours, 2 mL.min^{-1} of CO_2 and 0.4 mL.min^{-1} of ethanol (as a cosolvent) were continually added to the by-product. The skin was continuously exposed to carbon dioxide (at a 2 mL.min^{-1} rate) and pressure (5 min) to extract oil. The by-product was exposed to constant pressure for 2 hours while being soaked in CO_2 in a vessel and then removed for 3 hours. For the soaking method, the sample from the vessel is submerged in CO_2 for 10 hours before being removed for another 6 hours [67]. As a consequence, the DPCD procedures using the soaking, pressure swing, continuous and cosolvent approaches, respectively, yielded extraction rates of 52.8%, 52.3%, 24.7% and 53.2%.

Similarly, DPCD has seen considerable usage in the extraction of oil from marine fauna and by-products under mild settings of 25 MPa and 39.85°C. Omega-3-PUFAs, notably EPA and DHA, are of critical relevance in marine sectors and marine oil is a significant source of these nutrients [68]. DPCD may be used to extract oil from shrimp (Farfantepenaeus paulensis) manufacturing by-products. EPA and DHA, omega-3 and omega-6 unsaturated fatty acids, are particularly plentiful in marine oil [69]. Carotenoids, including astaxanthin, are abundant in the waste from freeze-drying red-spotted shrimp. High levels of protein (49%) and ash (27%) and relatively low levels of lipids (4.9%) are typical for these by-products (head, shells and tails). To this end, the yield of astaxanthin from by-products was studied under varying conditions of moderate pressure (200–400 bar) and temperature (40–60°C). Under DPCD treatment conditions of 43°C and 370 bar, astaxanthin recovery was 39%. Although there are a number of elements that have a significant impact on the yield of substantial components extracted using DPCD methods, the conditions of pressure and temperature are believed to be the most critical ones [70].

Due to its favourable processing conditions, fast rate of extraction, low toxicity, low viscosity, elevated diffusivity and minimal surface tension, the DPCD technique has attained a lot of attention in the marine processing industry. This method is therefore preferred to other conventional approaches in terms of solvent consumption and heat costs [64]. In addition, this method of digesting lipids is free of the hazards posed by heavy metals and inorganic salts, making it both safe and effective. When it comes to industrial applications, however, the expensive cost of equipment and commercial viability are both factors that must be considered. There is a degree of selectivity in using CO_2 due to the inability to remove polar molecules, but consistent functioning needs a lot of power and a constant concentration of carbon dioxide [71].

7.3.4 Membrane technology

A useful method for recovering, concentrating, separating, or extracting physiologically active chemicals is membrane technology [72]. A semipermeable membrane serves as a barrier in this process, allowing biomolecules to pass across it only when they meet certain membrane molecular weight cut-offs (MWCOs). According to molecular weight, separation is accomplished; molecules that are smaller than the membrane pores may flow through the membrane with ease, but big ones cannot. In comparison to other nonthermal technologies, these membranes, which are generally composed of polymer materials (such as polysulfone, polyamides, polyethersulfone and composite fluoropolymer) display solvent permeability and selective separation properties [73–75]. In these technologies, membrane-based techniques such as microfiltration (MF), nanofiltration (NF), ultrafiltration (UF) and reverse osmosis (RO), developed in the previous several decades, are employed to recover essential bioactive compounds from marine by-products [76].

As a result, bioactive compounds, such as peptides, proteins, chitooligosaccharides and PUFAs, with radical scavenging properties and desirable anticoagulant, antihypertensive and antimicrobial properties have been extracted from marine-related by-products using membrane bioreactor technology [77]. In total, 2.5 kg of protein were extracted from one tonne of processed mackerel blood fluids by utilizing 10-kDa and 3-kDa MWCO filtration. These blood fluids had a high concentration of both essential and nonessential amino acids [78].

It was hypothesized by Ghalamara et al. (2020) that UF membranes might be used to sustainably produce protein peptide-enriched fractions from marine by-products (codfish blood) (50 kDa). For this reason, protein hydrolysates from yellowfin tuna viscera were fractionated utilizing UF membranes of varying MWCOs due to the beneficial antibacterial and antioxidant capabilities of protein peptides against pathogenic microbes (3 kDa, 10 kDa and 30 kDa). It was possible to generate four distinct types of permeates, each of which comprised protein hydrolysates with varied molecular size fractions (3 kDa, 3–10 kDa, 10–30 kDa and > 30 kDa), strong antioxidant and antibacterial properties and prospective applications in the pharmaceutical and nutraceutical sectors [79].

In this context, cod by-products were treated to valorize hydrolysates containing collagen (74%) by integrated membrane ultrafiltration. These hydrolysates have the potential to find use in the cosmetic and pharmaceutical sectors [80]. In their study on the fractionation of protein and astaxanthin from shrimp cooking wastewater, Amado et al. (2016) made use of a succession of UF membranes with MWCOs of 300, 100 and 30. According to the findings, the two phases of sequential membrane UF (MWCOs of 100 kDa and 300 kDa), which were used to separate the protein, exhibited the greatest protein recoveries (90% and 75%, respectively) [81]. Interestingly, the aggregation of high-molecular-weight proteins, which was also achieved during UF employing a 300-kDa membrane, caused ultrafiltration membranes to display the greatest astaxanthin retention rate at 300 kDa. Enzymatic hydrolysis of marine products was also used to conduct membrane separation of valuable components. Saidi et al. (2018) recovered tuna protein by-products using UF (> 4 kDa) and NF (4 MW > 1 kDa; 1 kDa) membrane processes with enzymatic hydrolysis. The NF retentate fraction (1–4 kDa) was shown to have the maximum superoxide radical activity, and four new antioxidant peptides were produced. These peptides might be employed as food additives and in the development of nutritional goods. In a similar manner, the viscera of cuttlefish contain a number of beneficial protein hydrolysates that are likewise capable of being created as by-products of enzymatic hydrolysis. The majority of this procedure is carried out utilizing an ultrafiltration membrane with an MWCO of 4 kDa, which is able to recover greater than 70% of the bioactive peptides [82]. It is possible that other membrane technologies may be employed to extract substantial amounts of valuable substances from marine by-products. For instance, UF membranes (5 kDa and 20 kDa) were used in an electro-separation process to improve the efficacy and yield of peptides, particularly arginine and lysine, during their extraction from snow crab coproducts [83].

In addition, membrane processing technology, which has wide commercial use in the pharmaceutical and cosmetic industries, has been employed as a potential processing strategy that has been used to extract key components from marine by-products. Additionally, various membrane technologies, notably electro membrane, NF and UF, are largely utilized in conjunction with enzymes to hydrolyze bioactive peptides and proteins from marine by-products. Membrane technologies are more financially viable than conventional ones since they don't need additional agents or harmful substances but instead use environmentally benign and industrially sustainable manufacturing techniques [76]. Fouling and concentration polarization, both of which may be caused by bacteria as well as inorganic and organic substances, can have a negative impact on the effectiveness of membrane processing [84].

7.3.5 Hurdle Technologies

Nonthermal technologies offer numerous applications, yet the majority of these technologies have drawbacks. However, these disadvantages can be addressed by improving the extraction effectiveness and the quality of the target compounds by combining the use of numerous nonthermal approaches.

One of the most recent lines of investigation focused on combining the use of UAE technology with that of other methods in order to improve both the purity of the bioactive compounds and the efficiency of their extraction. By using isoelectric solubilization precipitation (ISP) on marine by-products for the best protein extraction, Alvarez et al. (2018) investigated the relevance of UAE. ISP encourages protein extraction through pH variations and a number of variables, including the raw materials employed and the extraction parameters, such as temperature, time and pH. These had an influence on the extraction yields [85]. In particular, a 10-min soak at 60% amplitude in 0.1 M NaOH will recover 94% of the total protein. When compared to conventional ISP, the rate of recovered protein was greater when utilizing an ultrasonic bath with a lower amplitude (20%). More than 95% of the total protein available in marine by-products was extracted using UAE in an alkaline-based extraction. Therefore, the use of UAE in conjunction with ISP is an excellent green extraction procedure for the extraction of high quantities of protein from marine by-products, while simultaneously reducing the amount of water and reagent that is consumed in the process.

Using supercritical CO_2, Riera (2004) studied the impact of ultrasound on oil extraction [86]. At 280 bar and 55°C, when SFE was supplemented by ultrasound, the yield of oil in the final extraction stage surprisingly rose by about 20%, resulting in a yield that was equivalent to that obtained by SFE alone. Enzymatic hydrolysis and ultrafiltration methods were combined, according to Tonon et al. (2016), to recover protein hydrolysates from shrimp effluent. The degree of hydrolysis and antioxidant activity were 235.56 mol TE/L and 16.07%, respectively, under the ideal circumstances (a pH of 9.0, an E/S ratio of 0.1% and a temperature of 75°C) [87]. Additionally, by combining enzymatic hydrolysis and ultrafiltration, the extraction of protein hydrolysates from sardine by-products was improved. pH 1.5 and 48°C were the ideal conditions for enzymatic hydrolysis, which considerably increased the protein hydrolysate output at a good rate with a rejection rate of around 90% and a concentration factor of 2.5 [88]. Enzyme-linked ultrafiltration and nanofiltration were used to test the antioxidant activity of a hydrolysate of tuna by-products and its peptide fractions. The enzymes used in the by-product hydrolysis and the ultrafiltration and nanofiltration membranes used for the three-way fractionation were both crucial. The hydrolysate of tuna by-products showed more iron-chelating activity than the hydrolysate of the peptide fraction. For this reason, the combination of enzymatic hydrolysis and membrane technology is favoured for extracting valuable components from marine waste [89].

7.4 COMMERCIAL ASPECTS

Several nonthermal technologies, including HHP, PEF, DPCD and UAE, have been commercially implemented at the industrial scale for a variety of applications. These technologies have been used for many reasons. Nonthermal methods have demonstrated benefits in terms of water and energy conservation, in addition to having little influence on the surrounding environment [4].

Fruits, vegetables, wine, meat, meat products (including seafood), dairy and egg products, grains and more have all been subjected to HHP treatment for commercial use. This treatment has been used to inactivate microbes and enzymes, modify microstructures, develop products, extend shelf lives and alter physicochemical properties. It is vital to include a processing time of less than 20 min when calculating the cost-effectiveness of HHP treatment, together with pressure and temperature [90]. The energy input for treating cans with HHP might be reduced from 300 $kJ.kg^{-1}$ to 270 $kJ.kg^{-1}$, yielding a greater than 10% energy savings. A litre or kilogram of a product that has been HHP-treated typically costs between \$ 0.05 and \$ 0.5, which is less than a similar product that has been thermally processed [91].

Genesis juice, the first commercially successful PEF-treated product in the United States, was said to have reduced energy consumption by traditional thermal technology by 95%, gas use by 100% due to the removal of thermal processing and electricity consumption by up to 18% [92]. Toepfl (2011) was able to effectively translate the PEF technology's laboratory processing settings to an industrial scale, with average powers of 5, 30 and 80 kW enabling a capacity of up to 10,000 $L.h^{-1}$. The overall cost of the procedure was calculated to be between 1–2 $€.t^{-1}$ for cell disintegration and 0.01–0.02 $€.L^{-1}$ for the preservation of liquid media. However, because solid samples cannot be pumped, PEF technology is not appropriate for them [93].

Environmental and economic benefits occurred from the use of DPCD treatment with moderate temperature and pressure to extract bioactive chemicals from seafood by-products as opposed to traditional chemical procedures, including solvents, alkalis and acids [94]. Factors including temperature and pressure have a significant role in DPCD, which has the potential to enhance the solubility of bioactive chemicals. However, since bioactive chemicals might be damaged by high temperatures, high pressure is preferable for enhancing solubility. By incorporating polar modifiers or cosolvents, DPCD can be used for both polar and nonpolar compounds. Better amino acid recoveries were seen in krill by-product residues following DPCD treatment than in the raw krill by-product residues following hydrolysis, demonstrating the efficiency and sustainability of DPCD. More research is needed to find the best proportions of solvents, create continuous processes, establish operating parameters that ensure product quality and increase production to a commercial level [4].

The UAE has made great strides in the extraction process, and the result is a green and cost-effective method that has many potential uses, including those in the food, pharmaceutical and bioenergy industries. The sample parameters, intrinsic medium properties—such as dissolved gases, solvent, temperature and external pressure— and the physical parameters of power, frequency, intensity, shape and size of ultrasonic reactors are all significant UAE parameters. The Soxhlet process needed 8 kWh and 6 kWh for heating when standard techniques were employed to extract fat and oil, and the cost of energy was correlated with mixing and heating maceration. However, the UAE only needed 0.25 kWh. Additionally, solid materials were used to determine carbon dioxide emissions. For the Soxhlet, maceration and UAE processes, the quantities were 6,400 g CO_2/100 g, 3,600 g CO_2/100 g and 200 g CO_2/100 g samples, respectively. This showed that the UAE technique is a green technology. At Euphytos, GMC and Giotti enterprises, food extraction and medicinal additives were utilized on a commercial scale [4].

The delay in industrializing nonthermal technology is due to equipment development, processing parameters control and regulatory approval. Since diverse starting materials are employed, more research is necessary to improve equipment settings, especially when procedures are applied to the industrial scale from the laboratory size. Additionally, integrating nonthermal approaches may address some of their drawbacks, increase extraction effectiveness and retain quality prospects to enhance the sustainability of marine industrial output, which is ultimately advantageous from an economic and environmental standpoint.

7.5 CHALLENGES AND FUTURE PERSPECTIVES

Processed marine by-products are utilized in a wide variety of sectors, and the scope of their commercial use is growing each year. On the other hand, their viability as bioactive chemicals and the nutraceutical potential of their products have received little attention. The study of bioactive compounds for their potential nutraceutical value is a burgeoning area of research, and the utilization of marine processing wastes represents a novel strategy for the development of further commercial applications. A small number of the bioactivities that can be derived from isolated chemicals have been uncovered as of yet, and further study is required to create ways that can utilize these bioactivities for the improvement of human health.

It is also necessary to do more research on the stability of bioactive compounds that have been identified under a variety of conditions. In order to have a successful commercial extraction process, it is necessary first to optimize the nonthermal extraction processes.

7.6 CONCLUSIONS

As a result of significant changes in the western lifestyle, there is rising concern about chronic illnesses, including obesity, diabetes, hypertension, hypercholesterolaemia, cancer and cardiovascular disorders. By-products of aquaculture are produced in enormous numbers and because they include bioactive components that may be utilized as food or health supplements, they can be recycled commercially. For a sustainable industry to avoid or reduce its negative environmental effects, waste utilization in the marine food processing sector must be improved. Accordingly, fish by-products are a fantastic source of protein and omega-3 PUFAs, which have been specifically examined in relation to Atlantic salmon or rainbow trout. Fish protein hydrolysate (FPH), a source of amino acids and peptides with high digestibility, quick absorption and significant biological activity, is produced by chemical, enzymatic and microbiological hydrolysis of processing by-products. Eicosapentaenoic (EPA) and docosahexaenoic (DHA) acids, two omega-3 PUFAs derived from fish waste, have been shown to reduce postprandial triacylglycerol levels, blood pressure, platelet aggregation and the inflammatory response. By-products from crustaceans may also be utilized to make carotenoids, which have significant biological activity, and chitosan, which has antioxidant and antibacterial action, for the food and pharmaceutical sectors. Similarly, it was observed that bones are a good source of collagen and gelatine.

Various extraction processes have been explored by researchers, and among these, nonthermal methods have been the focus of significant study and practical application. Extraction efficiency, processing time, toxicity, environmental friendliness and solvent usage were all areas where nonthermal technology excelled compared to conventional extraction techniques. Therefore, nonthermal approaches hold great potential as instruments for extracting valuable molecules from marine by-products and bringing them closer to their potential uses. This chapter reviewed current research that examined the impact of PEF, DPCD, HHP, membrane, UAE, enzyme-assisted extraction and hurdle technologies on the recovery of bioactive chemicals and the quality of the resulting products. Nonthermal technologies are becoming increasingly applicable at the industrial scale as a result of recent advancements in material science, the creation of energy transducers and the discovery of efficient enzyme compounds. This is largely due to the growing concern for both the quality of the environment and the reliability of manufactured goods. Nonthermal methods have shown promise for industrializing the processing of marine by-products, but their use has thus far been limited to the laboratory. Given the economic advantages and environmental implications, it is imperative that greater effort be made to solve obstacles during scale-up in order to effectively deploy the technology on its own or to obtain a high yield with little harm to physiochemical qualities.

ACKNOWLEDGEMENT

The study is supported by the Indian National Academy of Engineering (INAE/121/AKF/22), Gurgaon, India. The authors are solely responsible for all of the opinions, results and conclusions expressed in this study; INAE's viewpoints are not necessarily reflected in any of these aspects.

REFERENCES

1. P. Duarah, D. Haldar, M.K. Purkait, Technological advancement in the synthesis and applications of lignin-based nanoparticles derived from agro-industrial waste residues: A review, *Int. J. Biol. Macromol.* 163 (2020) 1828–1843.

2. D. Haldar, P. Duarah, M.K. Purkait, Chapter 16 - Progress in the synthesis and applications of polymeric nanomaterials derived from waste lignocellulosic biomass, in: D. Giannakoudakis, L. Meili, I. Anastopoulos (Eds.), *Advanced Materials for Sustainable Environmental Remediation, Elsevier2022*, pp. 419–433.
3. S. Ulagesan, S. Krishnan, T.J. Nam, Y.H. Choi, A review of bioactive compounds in oyster shell and tissues, *Front. Bioeng. Biotechnol.* 10 (2022) 913839.
4. A. Ali, S. Wei, Z. Liu, X. Fan, Q. Sun, Q. Xia, S. Liu, J. Hao, C. Deng, Non-thermal processing technologies for the recovery of bioactive compounds from marine by-products, *LWT* 147 (2021) 111549.
5. S.-K. Kim, E. Mendis, Bioactive compounds from marine processing byproducts: A review, *Food Res. Int.* 39(4) (2006) 383–393.
6. F.A. Khawli, F.J. Martí-Quijal, E. Ferrer, M.-J. Ruiz, H. Berrada, M. Gavahian, F.J. Barba, B. de la Fuente, *Aquaculture and Its By-products as a Source of Nutrients and Bioactive Compounds, Advances in Food and Nutrition Research*, Elsevier, 2020, pp. 1–33.
7. A.M. Shabbirahmed, D. Haldar, P. Dey, A.K. Patel, R.R. Singhania, C.-D. Dong, M.K. Purkait, Sugarcane bagasse into value-added products: A review, *Environ. Sci. Pollut. Res.* 29 (2022). https://doi.org/10.1007/s11356-022-21889-1
8. B. Debnath, D. Haldar, M.K. Purkait, Potential and sustainable utilization of tea waste: A review on present status and future trends, *J. Environ. Chem. Eng.* 9(5) (2021) 106179.
9. T.N. Baite, B. Mandal, M.K. Purkait, Ultrasound assisted extraction of gallic acid from Ficus auriculata leaves using green solvent, *Food Bioprod. Process.* 128 (2021) 1–11.
10. S.-K. Kim, P.-J. Park, H.-P. Yang, S.-S. Han, Subacute toxicity of chitosan oligosaccharide in Sprague-Dawley rats, *Arzneimittelforschung* 51(09) (2001) 769–774.
11. Y.-J. Jeon, F. Shahidi, S.-K. Kim, Preparation of chitin and chitosan oligomers and their applications in physiological functional foods, *Food Rev. Int.* 16(2) (2000) 159–176.
12. Y.-J. Jeon, S.-K. Kim, Continuous production of chitooligosaccharides using a dual reactor system, *Process Biochemistry* 35(6) (2000) 623–632.
13. O. Kanauchi, K. Deuchi, Y. Imasato, M. Shizukuishi, E. Kobayashi, Mechanism for the inhibition of fat digestion by chitosan and for the synergistic effect of ascorbate, *Biosci. Biotechnol. Biochem.* 59(5) (1995) 786–790.
14. M. Sugano, K. Yoshida, M. Hashimoto, K. Enomoto, S. Hirano, Hypocholesterolemic activity of partially hydrolyzed chitosan in rats, Advances in chitin and chitosan 11(2) (1992) 472–478.
15. I. Ikeda, M. Sugano, K. Yoshida, E. Sasaki, Y. Iwamoto, K. Hatano, Effects of chitosan hydrolyzates on lipid absorption and on serum and liver lipid concentration in rats, *J. Agric. Food Chem.* 41(3) (1993) 431–435.
16. G.I. Howling, P.W. Dettmar, P.A. Goddard, F.C. Hampson, M. Dornish, E.J. Wood, The effect of chitin and chitosan on fibroblast-populated collagen lattice contraction, *Biotechnol. Appl. Biochem.* 36(3) (2002) 247–253.
17. H.-G. Byun, Y.-T. Kim, P.-J. Park, X. Lin, S.-K. Kim, Chitooligosaccharides as a novel β-secretase inhibitor, *Carbohydr. Polym.* 61(2) (2005) 198–202.
18. T. Sun, W. Xie, P. Xu, Antioxidant activity of graft chitosan derivatives, *Macromol. Biosci.* 3(6) (2003) 320–323.
19. M. Hamdi, R. Nasri, N. Dridi, H. Moussa, L. Ashour, M. Nasri, Improvement of the quality and the shelf life of reduced-nitrites turkey meat sausages incorporated with carotenoproteins from blue crabs shells, *Food Control* 91 (2018) 148–159.
20. M. Hamdi, R. Nasri, N. Dridi, S. Li, M. Nasri, Development of novel high-selective extraction approach of carotenoproteins from blue crab (Portunus segnis) shells, contribution to the qualitative analysis of bioactive compounds by HR-ESI-MS, *Food Chem.* 302 (2020) 125334.
21. Z. Guo, F. Zhao, H. Chen, M. Tu, S. Tao, Z. Wang, C. Wu, S. He, M. Du, Heat treatments of peptides from oyster (Crassostrea gigas) and the impact on their digestibility and angiotensin I converting enzyme inhibitory activity, *Food Sci. Biotechnol.* 29(7) (2020) 961–967.
22. K. Sadeghi, K. Park, J. Seo, Oyster shell disposal: Potential as a novel ecofriendly antimicrobial agent for packaging: A mini review, *Korean J. Food Sci. Technol.* 25(2) (2019) 57–62.
23. M. Bonnard, *Identification of Valuable Compounds from the Shell of the Edible Oyster Crassostrea Gigas*, Université Montpellier, 2020.
24. R. Chaturvedi, P.K. Singha, S. Dey, Water soluble bioactives of nacre mediate antioxidant activity and osteoblast differentiation, *PLoS One* 8(12) (2013) e84584.
25. G. Atlan, O. Delattre, S. Berland, A. LeFaou, G. Nabias, D. Cot, E. Lopez, Interface between bone and nacre implants in sheep, *Biomaterials* 20(11) (1999) 1017–1022.

26. A. Flausse, C. Henrionnet, M. Dossot, D. Dumas, S. Hupont, A. Pinzano, D. Mainard, L. Galois, J. Magdalou, E. Lopez, Osteogenic differentiation of human bone marrow mesenchymal stem cells in hydrogel containing nacre powder, *J Biomed Mater Res A*. 101(11) (2013) 3211–3218.
27. D.W. Green, H.-J. Kwon, H.-S. Jung, Osteogenic potency of nacre on human mesenchymal stem cells, *Molecules and Cells* 38(3) (2015) 267.
28. X. Feng, S. Jiang, F. Zhang, R. Wang, T. Zhang, Y. Zhao, M. Zeng, Extraction and characterization of matrix protein from pacific oyster (Crassostrea gigs) shell and its anti-osteoporosis properties in vitro and in vivo, *Food Funct* 12(19) (2021) 9066–9076.
29. T. Latire, F. Legendre, M. Bouyoucef, F. Marin, F. Carreiras, M. Rigot-Jolivet, J.-M. Lebel, P. Galéra, A. Serpentini, Shell extracts of the edible mussel and oyster induce an enhancement of the catabolic pathway of human skin fibroblasts, in vitro, *Cytotechnology* 69(5) (2017) 815–829.
30. M.C. Gómez-Guillén, J. Turnay, M. Fernández-Dıaz, N. Ulmo, M.A. Lizarbe, P. Montero, Structural and physical properties of gelatin extracted from different marine species: A comparative study, *Food Hydrocoll.* 16(1) (2002) 25–34.
31. S.-K. Kim, H.-G. Byun, E.-H. Lee, Optimum extraction conditions of gelatin from fish skins and its physical properties, *Appl. Chem. Eng.* 5(3) (1994) 547–559.
32. T. Nakagawa, T. Tagawa, Ultrastructural study of direct bone formation induced by BMPs-collagen complex implanted into an ectopic site, *Oral Dis.* 6(3) (2000) 172–179.
33. G.-H. KIM, Y.-J. JEON, H.-G. BYUN, Y.-S. LEE, E.-H. LEE, S.-K. KIM, Effect of calcium compounds from oyster shell bouind fish skin gelatin peptide in calcium deficient rats, Korean *J Fish Aquat Sci.* 31(2) (1998) 149–159.
34. T. Larsen, S.H. Thilsted, K. Kongsbak, M. Hansen, Whole small fish as a rich calcium source, *Br. J. Nutr.* 83(2) (2000) 191–196.
35. S.S. Jensen, M. Aaboe, E.M. Pinholt, E. Hjørting-Hansen, F. Melsen, I. Ruyter, Tissue reaction and material characteristics of four bone substitutes, *Int J Oral Maxillofac Implants.* 11(1) (1996).
36. S. Chantachum, S. Benjakul, N. Sriwirat, Separation and quality of fish oil from precooked and non-precooked tuna heads, *Food Chem.* 69(3) (2000) 289–294.
37. K. Hidajat, C. Ching, M. Rao, Preparative-scale liquid chromatographic separation of ω-3 fatty acids from fish oil sources, *J. Chromatogr. A* 702(1–2) (1995) 215–221.
38. C. Von Schacky, n– 3 fatty acids and the prevention of coronary atherosclerosis, *The American Journal of Clinical Nutrition* 71(1) Supplement (2000) 224s–227s.
39. P.M. Kris-Etherton, W.S. Harris, L.J. Appel, Omega-3 fatty acids and cardiovascular disease: New recommendations from the American Heart Association, *Arteriosclerosis, Thrombosis, and Vascular Biology* 23(2) (2003) 151–152.
40. J.H. Christensen, E. Korup, J. Aarøe, E. Toft, J. Møller, K. Rasmussen, J. Dyerberg, E.B. Schmidt, Fish consumption, n-3 fatty acids in cell membranes, and heart rate variability in survivors of myocardial infarction with left ventricular dysfunction, *Am. J. Card.* 79(12) (1997) 1670–1673.
41. H.G. BYUN, P.J. PARK, N.I. SUNG, S.K. KIM, Purification and characterization of a serine proteinase from the tuna pyloric caeca, *J. Food Biochem.* 26(6) (2002) 479–494.
42. S.K. KIM, P.J. PARK, H.G. BYUN, J.Y. JE, S.H. MOON, S.H. KIM, Recovery of fish bone from hoki (Johnius belengeri) frame using a proteolytic enzyme isolated from mackerel intestine, *J. Food Biochem.* 27(3) (2003) 255–266.
43. T. Tsironi, D. Houhoula, P. Taoukis, Hurdle technology for fish preservation, *Aquac. Fish.* 5(2) (2020) 65–71.
44. Z. Shouqin, Z. Junjie, W. Changzhen, Novel high pressure extraction technology, *Int. J. Pharm.* 278(2) (2004) 471–474.
45. J. Li, W. Sun, H.S. Ramaswamy, Y. Yu, S. Zhu, J. Wang, H. Li, High pressure extraction of astaxanthin from shrimp waste (Penaeus Vannamei Boone): Effect on yield and antioxidant activity, *J. Food Process Eng.* 40(2) (2017) e12353.
46. X. Jun, S. Deji, L. Ye, Z. Rui, Micromechanism of ultrahigh pressure extraction of active ingredients from green tea leaves, *Food Control* 22(8) (2011) 1473–1476.
47. X. Guo, D. Han, H. Xi, L. Rao, X. Liao, X. Hu, J. Wu, Extraction of pectin from navel orange peel assisted by ultra-high pressure, microwave or traditional heating: A comparison, *Carbohydr. Polym.* 88(2) (2012) 441–448.
48. C. Yu, Y. Cha, F. Wu, W. Fan, X. Xu, M. Du, Effects of ball-milling treatment on mussel (Mytilus edulis) protein: Structure, functional properties and in vitro digestibility, *Int. J. Food Sci.* 53(3) (2018) 683–691.
49. A.K. Hemker, L.T. Nguyen, M. Karwe, D. Salvi, Effects of pressure-assisted enzymatic hydrolysis on functional and bioactive properties of tilapia (Oreochromis niloticus) by-product protein hydrolysates, *Lwt* 122 (2020) 109003.

50. A. Monteiro, D. Paquincha, F. Martins, R.P. Queirós, J.A. Saraiva, J. Švarc-Gajić, N. Nastić, C. Delerue-Matos, A.P. Carvalho, Liquid by-products from fish canning industry as sustainable sources of ω3 lipids, *J. Environ. Manage.* 219 (2018) 9–17.
51. A. Marciniak, S. Suwal, N. Naderi, Y. Pouliot, A. Doyen, Enhancing enzymatic hydrolysis of food proteins and production of bioactive peptides using high hydrostatic pressure technology, *Trends Food Sci Technol.* 80 (2018) 187–198.
52. B. Gómez, P.E. Munekata, M. Gavahian, F.J. Barba, F.J. Martí-Quijal, T. Bolumar, P.C.B. Campagnol, I. Tomasevic, J.M. Lorenzo, Application of pulsed electric fields in meat and fish processing industries: An overview, *Food Res. Int.* 123 (2019) 95–105.
53. M. López-Pedrouso, J.M. Lorenzo, C. Zapata, D. Franco, *Proteins and Amino Acids, Innovative Thermal and non-thermal Processing, Bioaccessibility and Bioavailability of Nutrients and Bioactive Compounds*, Elsevier, 2019, pp. 139–169.
54. D. Franco, P.E. Munekata, R. Agregán, R. Bermúdez, M. López-Pedrouso, M. Pateiro, J.M. Lorenzo, Application of pulsed electric fields for obtaining antioxidant extracts from fish residues, *Antioxidants* 9(2) (2020) 90.
55. J. Xi, Z. Li, Y. Fan, Recent advances in continuous extraction of bioactive ingredients from food-processing wastes by pulsed electric fields, *Crit Rev Food Sci Nutr.* 61(10) (2021) 1738–1750.
56. S. Gulzar, S. Benjakul, Impact of pulsed electric field pretreatment on yield and quality of lipid extracted from cephalothorax of Pacific white shrimp (Litopenaeus vannamei) by ultrasound-assisted process, *Int. J. Food Sci.* 55(2) (2020) 619–630.
57. J.R. Sarkis, N. Boussetta, I.C. Tessaro, L.D.F. Marczak, E. Vorobiev, Application of pulsed electric fields and high voltage electrical discharges for oil extraction from sesame seeds, *J. Food Eng.* 153 (2015) 20–27.
58. G. He, X. Yan, X. Wang, Y. Wang, Extraction and structural characterization of collagen from fishbone by high intensity pulsed electric fields, *J. Food Process Eng.* 42(6) (2019) e13214.
59. M. Gavahian, R. Farhoosh, K. Javidnia, F. Shahidi, M.-T. Golmakani, A. Farahnaky, Effects of Electrolyte Concentration and Ultrasound Pretreatment on Ohmic-Assisted Hydrodistillation of Essential Oils from Mentha piperita L, *International J. Food Eng.* 13(10) (2017).
60. M. Li, J. Lin, J. Chen, T. Fang, Pulsed electric field-assisted enzymatic extraction of protein from abalone (Haliotis discus hannai Ino) viscera, *J. Food Process Eng.* 39(6) (2016) 702–710.
61. G. He, Y. Yin, X. Yan, Y. Wang, Semi-bionic extraction of effective ingredient from fishbone by high intensity pulsed electric fields, *J. Food Process Eng.* 40(2) (2017) e12392.
62. E. Puertolas, M. Koubaa, F.J. Barba, An overview of the impact of electrotechnologies for the recovery of oil and high-value compounds from vegetable oil industry: Energy and economic cost implications, *Food Res. Int.* 80 (2016) 19–26.
63. C. Mannozzi, T. Fauster, K. Haas, U. Tylewicz, S. Romani, M. Dalla Rosa, H. Jaeger, Role of thermal and electric field effects during the pre-treatment of fruit and vegetable mash by pulsed electric fields (PEF) and ohmic heating (OH), *Innov. Food Sci. Emerg. Technol.* 48 (2018) 131–137.
64. S. Kuvendziev, K. Lisichkov, Z. Zeković, M. Marinkovski, Z.H. Musliu, Supercritical fluid extraction of fish oil from common carp (Cyprinus carpio L.) tissues, *J Supercrit Fluids* 133 (2018) 528–534.
65. M. Plaza, I. Rodríguez-Meizoso, Advanced extraction processes to obtain bioactives from marine foods, *Bioactive Compounds from Marine Foods: Plant and Animal Sources* (2013) 343–371. https://doi.org/10.1002/9781118412893.ch16
66. C. Prieto, L. Calvo, The encapsulation of low viscosity omega-3 rich fish oil in polycaprolactone by supercritical fluid extraction of emulsions, *J Supercrit Fluids* 128 (2017) 227–234.
67. F. Sahena, I. Zaidul, S. Jinap, M. Jahurul, A. Khatib, N. Norulaini, Extraction of fish oil from the skin of Indian mackerel using supercritical fluids, *J. Food Eng.* 99(1) (2010) 63–69.
68. N. Rubio-Rodríguez, M. Sara, S. Beltrán, I. Jaime, M.T. Sanz, J. Rovira, Supercritical fluid extraction of fish oil from fish by-products: A comparison with other extraction methods, *J. Food Eng.* 109(2) (2012) 238–248.
69. M. Sprague, J.R. Dick, D.R. Tocher, Impact of sustainable feeds on omega-3 long-chain fatty acid levels in farmed Atlantic salmon, 2006–2015, *Sci. Rep.* 6(1) (2016) 1–9.
70. A.P. Sánchez-Camargo, M.Â.A. Meireles, B.L.F. Lopes, F.A. Cabral, Proximate composition and extraction of carotenoids and lipids from Brazilian redspotted shrimp waste (Farfantepenaeus paulensis), *J. Food Eng.* 102(1) (2011) 87–93.
71. K. Ivanovs, D. Blumberga, Extraction of fish oil using green extraction methods: A short review, *Energy Procedia* 128 (2017) 477–483.
72. V.L. Dhadge, M. Changmai, M.K. Purkait, Purification of catechins from Camellia sinensis using membrane cell, *Food Bioprod. Process.* 117 (2019) 203–212.

73. C.M. Galanakis, Recovery of high added-value components from food wastes: Conventional, emerging technologies and commercialized applications, *Trends Food Sci Technol.* 26(2) (2012) 68–87.
74. M.K. Purkait, R. Singh, D. Haldar, P. Mondal, *Thermal Induced Membrane Separation Processes*, Elsevier, 2020.
75. S. Sinha Ray, H. Singh Bakshi, R. Dangayach, R. Singh, C.K. Deb, M. Ganesapillai, S.-S. Chen, M.K. Purkait, Recent developments in nanomaterials-modified membranes for improved membrane distillation performance, *Membranes* 10(7) (2020) 140.
76. R. Castro-Muñoz, G. Boczkaj, E. Gontarek, A. Cassano, V. Fíla, Membrane technologies assisting plant-based and agro-food by-products processing: A comprehensive review, *Trends Food Sci Technol.* 95 (2020) 219–232.
77. S.-K. Kim, M. Senevirathne, Membrane bioreactor technology for the development of functional materials from sea-food processing wastes and their potential health benefits, *Membranes* 1(4) (2011) 327–344.
78. M. Hayes, M. Gallagher, Processing and recovery of valuable components from pelagic blood-water waste streams: A review and recommendations, *J. Clean. Prod.* 215 (2019) 410–422.
79. R. Abejón, A. Abejón, M.-P. Belleville, A. Garea, A. Irabien, J. Sanchez-Marcano, Water recovery and reuse in the fractionation of protein hydrolysate by ultrafiltration and nanofiltration membranes, *Chem. Eng. Trans.* 52 (2016) 283–288.
80. I. Krasnova, G. Semenov, N.Y. Zarubin, Modern technologies for using fish wastes in the production of collagen hydrolysates and functional beverages, *IOP Conference Series: Earth and Environmental Science*, IOP Publishing, 2020, p. 062030.
81. I.R. Amado, M.P. González, M.A. Murado, J.A. Vázquez, Shrimp wastewater as a source of astaxanthin and bioactive peptides, *J. Chem. Technol. Biotechnol.* 91(3) (2016) 793–805.
82. E. Soufi-Kechaou, M. Derouiniot-Chaplin, R.B. Amar, P. Jaouen, J.-P. Berge, Recovery of valuable marine compounds from cuttlefish by-product hydrolysates: Combination of enzyme bioreactor and membrane technologies: Fractionation of cuttlefish protein hydrolysates by ultrafiltration: Impact on peptidic populations, *Comptes Rendus Chimie* 20(9–10) (2017) 975–985.
83. S. Suwal, C. Roblet, J. Amiot, A. Doyen, L. Beaulieu, J. Legault, L. Bazinet, Recovery of valuable peptides from marine protein hydrolysate by electrodialysis with ultrafiltration membrane: Impact of ionic strength, *Food Res. Int.* 65 (2014) 407–415.
84. S.A. Ilame, S.V. Singh, Application of membrane separation in fruit and vegetable juice processing: A review, *Crit Rev Food Sci Nutr.* 55(7) (2015) 964–987.
85. C. Álvarez, P. Lélu, S.A. Lynch, B.K. Tiwari, Optimised protein recovery from mackerel whole fish by using sequential acid/alkaline isoelectric solubilization precipitation (ISP) extraction assisted by ultrasound, *LWT* 88 (2018) 210–216.
86. E. Riera, Y. Golas, A. Blanco, J. Gallego, M. Blasco, A. Mulet, Mass transfer enhancement in supercritical fluids extraction by means of power ultrasound, *Ultrason Sonochem* 11(3–4) (2004) 241–244.
87. R.V. Tonon, B.A. dos Santos, C.C. Couto, C. Mellinger-Silva, A.I.S. Brígida, L.M. Cabral, Coupling of ultrafiltration and enzymatic hydrolysis aiming at valorizing shrimp wastewater, *Food Chem.* 198 (2016) 20–27.
88. M. Benhabiles, N. Abdi, N. Drouiche, H. Lounici, A. Pauss, M. Goosen, N. Mameri, Fish protein hydrolysate production from sardine solid waste by crude pepsin enzymatic hydrolysis in a bioreactor coupled to an ultrafiltration unit, *Mater. Sci. Eng. C.* 32(4) (2012) 922–928.
89. S. Saidi, A. Deratani, M.-P. Belleville, R.B. Amar, Production and fractionation of tuna by-product protein hydrolysate by ultrafiltration and nanofiltration: Impact on interesting peptides fractions and nutritional properties, *Food Res. Int.* 65 (2014) 453–461.
90. D. Bermúdez-Aguirre, G.V. Barbosa-Cánovas, An update on high hydrostatic pressure, from the laboratory to industrial applications, *Food Eng. Rev.* 3(1) (2011) 44–61.
91. N. Rastogi, K. Raghavarao, V. Balasubramaniam, K. Niranjan, D. Knorr, Opportunities and challenges in high pressure processing of foods, *Crit Rev Food Sci Nutr.* 47(1) (2007) 69–112.
92. R. Pereira, A. Vicente, Environmental impact of novel thermal and non-thermal technologies in food processing, *Food Res. Int.* 43(7) (2010) 1936–1943.
93. S. Toepfl, A. Mathys, V. Heinz, D. Knorr, Potential of high hydrostatic pressure and pulsed electric fields for energy efficient and environmentally friendly food processing, *Food Rev. Int.* 22(4) (2006) 405–423.
94. S. Ahmadkelayeh, K. Hawboldt, Extraction of lipids and astaxanthin from crustacean by-products: A review on supercritical CO2 extraction, *Trends Food Sci Technol.* 103 (2020) 94–108.

8 Extraction of bioactive compounds from tea, coffee and wine processing waste

8.1 INTRODUCTION

Beyond climate change, the world's biggest problems right now include the sharp rise in energy demand, the production and consumption of food and other resources that are not sustainable and the production of anthropogenic waste. By 2050, around 9.6 billion people are expected to inhabit the globe, using up the equivalent of 1.6 a planet's worth of resources and producing a significant quantity of waste. The generation of the massive amount of food waste is becoming a global concern, as the world needs to feed the 9.6 billion people by 2050 in a way that promotes economic growth and eases environmental pressure. A recent study has focused on the sustainable usage of waste as a value-added product due to increased environmental awareness and severe legislation [1]. Nevertheless, the majority of these wastes are frequently burned or dumped in landfills without adequate treatment, which has substantial environmental consequences by polluting water, air and soil. "Waste to riches" is a recent, quickly expanding idea for environmentally friendly, sustainable growth. In the context of the circular bioeconomy, food wastes are recognized as sources for the recovery of bioactive chemicals. A sustainable method of managing these wastes that will also strengthen the circular economy is the recovery of value-added bioactive compounds, such as phenolic compounds, dietary fibres, polysaccharides, polyphenols, catechins and flavonoids, among others, and their subsequent use in various applications [2].

One of the best sources of naturally occurring bioactive substances that are incredibly good for human health is tea and its by-products. Tea waste is made up of a variety of chemical substances, including natural polymers with strong antioxidant activity and nontoxic tea polyphenols. Catechins are the main antioxidants found in tea leaves, particularly epigallocatechin gallate (EGCG), which is regarded as essential for tea's exceptional antioxidant capacity [3]. Alkaloids, tannins and saponins can all be found in tea waste thanks to phytochemical analysis. In addition to their antibacterial, antioxidant, antidiabetic, antihypertensive, anti-inflammatory and immunoregulatory properties, these bioactive chemicals recovered from tea waste also have other health-promoting properties [4]. Many by-products are produced during the manufacturing of coffee as well. The primary by-products of this stage are the parchment, silverskin, mucilage, pulp and coffee husk. Spent ground coffee is another by-product of the commercial manufacture of soluble coffee. With regard to their utilization in food and pharmaceutical research, the presence of phenolic acids, flavonoids, caffeine and chlorogenic acid in coffee and its by-products has drawn considerable attention, particularly when taking into account the bioactive potential of such compounds. Anti-inflammatory, antibacterial and antioxidant properties are among the health-promoting effects of such bioactive chemicals recovered from coffee waste, in addition to the prevention of type 2 diabetes and Parkinson's disease [5]. Additionally, one of the most important agricultural activities conducted globally is the production of wine. Both the red and white winemaking processes produce large amounts of solid organic waste and by-products that are extraordinarily rich in bioactive components and need to be properly extracted and utilized. The main by-products of the winemaking business are grape pomace, grape marc and grape stalks. They are an intricate matrix of bioactive substances, including polyphenols, flavonoids, procyanidins, dietary fibres, anthocyanins and phenolic compounds. Such

DOI: 10.1201/9781003315469-8

bioactive substances recovered from winery waste have antihyperglycemic, antihyperlipidemic and anticancer properties in addition to their ability to prevent coronary artery disease. As such, effective utilization of such nonalcoholic and alcoholic beverage waste has gained significant interest amongst the research community in recent times [6].

This chapter elaborately highlights the presence and recovery of different bioactive compounds from both nonalcoholic (tea and coffee) and alcoholic beverage (wine) waste. Evaluation of the present scenario of tea, coffee and wine production has also been covered in details. Besides the health-promoting benefits and proper utilization of bioactive compounds recovered from tea, coffee and wine processing, waste in various applications (wastewater treatment and soil composting), food and polymer-based research are also extensively discussed and summarized.

8.2 PRESENT STATUS OF TEA, COFFEE AND BEVERAGE PRODUCTS

The estimated value of the tea market (worldwide) in 2020 was around US$ 210 billion, and by 2025, it is expected to reach over US$ 318 billion. In 2018, there were over 5.8 million metric tonnes of tea produced worldwide, and there were roughly 1.8 million metric tonnes exported. China, India, Kenya, Indonesia and Sri Lanka are the top tea-producing nations in the world, with China leading the pack, followed by India and Kenya. China produces around 42% of the world's total tea generation, while India makes up roughly 24%. According to the International Tea Committee's (ITC) 2020 Annual Bulletin of Statistics, 6.1 million tonnes of tea were produced worldwide in 2019. The Food and Agricultural Organization (FAO) predicts that during the next ten years, the total output of black tea would expand at a rate of 2.2% annually, reaching 4.4 million tonnes in 2027 [7]. By 2027, it is predicted that global green tea production will increase at a higher yearly growth rate of 7.6% with a total of 3.5 million tonnes. Therefore, it is expected that green tea output would quadruple in 2027, rising from 1.6 to 3.8 million tonnes, with China serving as the main driving force. Therefore, this growth in global tea production, which is anticipated to expand at the same rate over the next ten years in order to meet the growing demand, would lead to the generation of a vast amount of tea waste globally. After China, India is the second-highest tea manufacturer. About 5,79,000 acres of land are used for tea cultivation in India, where an estimated 8,57,000 tonnes of tea are produced annually, or 27.4% of all tea produced globally. Following processing, the aforementioned facilities collectively create about 1,90,400 tonnes of waste tea [8]. During the fiscal year 2020–2021, India produced over 25 million kg of discarded tea. All of the tea factories in India produce huge amounts of waste tea, including tea leaves, buds and stems, that have been abandoned, but the majority of this waste is not properly disposed of in accordance with the standards established by the Tea Board of India. These wastes can pollute the earth, water and air if they are not properly disposed of. The number of purchasers for tea waste in India is lower than the enormous amounts of waste produced. The majority of tea waste is disposed of as waste, causing a lot of environmental problems. Consequently, the potential and sustainable use of tea waste has received considerable attention in recent times.

The coffee industry also produces a lot of waste and by-products, both on farms and in commercial locations, because it is the most traded and consumed commodity [9]. Over 11 million hectares of coffee are grown worldwide. A total of 4, 54,722 ha of coffee is farmed in India, of which 2, 28,910 ha are Arabica and 2, 25,812 ha are Robusta. 9.96 million metric tonnes (MMT) of coffee, made up of 5.7 MMT of Arabica and 4.26 MMT of Robusta, are produced worldwide. Arabica production is 0.1 MMT and Robusta production is 0.22 MMT in the Indian scenario. Coffee production generates 25 MMT of solid waste on average, including coffee husks, pulp, used coffee grounds and damaged coffee beans, all of which are significant sources of pollution and a threat to the environment. Wet processing is used to produce 10% of Robusta coffee and 90% of Arabica coffee. 16.46 MMT of wet pulp are produced globally during wet processing, with 0.08 MMT produced in India. Husks from coffee beans and faulty beans of low grade are the solid waste produced during dry processing [10]. During the curing of coffee, approximately 4.4 MMT of cherry husk and 13.89 MMT of parchment husk are generated worldwide. India contributes 0.21 MMT of cherry husk and 0.27 MMT of parchment husk to this total. India's overall coffee exports to China have climbed from around 307 metric tonnes in 2014 to 547 metric tonnes in 2018, although India exported about 632 MT to China in 2017 and saw a compound annual growth rate of 11.5%. China is India's potential market for coffee. A sizeable portion of India's overall coffee exports is made up of soluble coffee. All of the coffee processing facilities in India produce a lot of waste coffee, including abandoned husk, pulp and silverskin as well as used coffee grounds, the majority of which are not disposed of correctly [5]. These wastes can pollute the air, soil and water if they are not properly disposed. Only a small fraction of coffee waste is purchased and used in relation to the enormous amounts of waste generated, and the remainder is discharged as waste, leading to several environmental problems. Additionally, research has shown that for every tonne of clean Arabica and Robusta coffee processed, respectively, 12,000 L to 60,000 L and 18,000 L to 70,000 L of effluent and pollution are produced. Every year, India produces between 1.8 and 2 million cubic metres of wastewater. High levels of biodegradable organic materials are dispersed and suspended in this extremely acidic wastewater [11]. Consequently, it has become crucial to explore the potential for and effectiveness of using coffee waste to recover valuable constituents.

A total of 260 million hectolitres (Mhl) of wine were produced worldwide (Mhl) in the year 2020. This is a 1% increase from 2019, which equals nearly 3 Mhl. This is barely below the average for the long term. Since 2000, we have experienced three extremely successful years, with 290–295 Mhl in 2018, 2013 and 2004. The lowest year during this time period was 2017, with an MHL below 250; 2002, with an MHL of only 255, came in second. With 49.1 Mhl, Italy continues to be the largest wine-producing nation. With 46.6 Mhl and 40.7 Mhl, respectively, France and Spain come in second and third, accounting for 46% of global production. In 2020, the United States produced 22.8 Mhl of wine, ranking fourth in the world. Wine output in the European Union (EU) increased by a sizeable 8% from 2019 to 165 Mhl in 2020 [12]. Accordingly, 63% of the wine consumed worldwide is produced in the EU. A total of 71% of the wine produced worldwide in 2000 came from different EU nations. Around 57% of the total wine manufacturing took place in the EU in 2020, and by 2030, the rest of the world is expected to surpass the EU in wine production. On an area of roughly 6,000 acres, Indian wine output was increased to 16.8 million litres in 2019. This is a 5% increase in productivity compared to 2018, which also had a healthy harvest. Despite the country's expanding wine industry, less than 1% of the world's wine is produced in India. The majority of India's wine consumption, according to Wine Intelligence, occurs in urban areas [13]. India bought 5.2 million litres of wine worth $27.4 million in 2018. In addition, throughout the winemaking process, the main solid waste products are grape stalks after destemming, followed by grape pomace after pressing and wine lees after fermentation and storage. According to reports, grape pomace makes up between 15% and 20% of the by-products produced during the winemaking process. During the winemaking process, grape stems and seeds make up around 2.5% to 7.5% and 3% to 6%, respectively, while yeast lees make up about 3.5% to 8.5%. Such waste might have an adverse influence on the ecosystem if it is not effectively discharged or used. Therefore, priority has been placed on the employment of various extraction techniques for the efficient valorization of such waste [6].

8.3 EXTRACTION OF BIOACTIVE COMPOUNDS

8.3.1 Tea processing waste

The most popular and widely used nonalcoholic beverage across the globe is tea. An estimated 16–18 billion cups of tea are consumed around the world on a daily basis. Only packaged drinking water was consumed more per person globally in 2018 (approximately 35 litres), and it was anticipated that by 2021, that number would rise to 37.5 litres. Around 6.3 million metric tonnes of tea were consumed globally in 2020, and by 2025, that number is projected to increase to 7.4 million metric tonnes. A lot of solid waste is produced as a result of large-scale tea production, which increases the need for recycling waste into many new applications [14].

Tea waste is a lignocellulosic biomass made up of proteins, tannins, lignin, hemicellulose and cellulose. Tea contains a variety of biologically active ingredients, including methylxanthines, alkaloids (caffeine, theophylline and theobromine), polyphenols (catechins, flavonoids and proanthocyanidins), terpenoids, pigments, amino acids and polysaccharides. Tea waste typically has nearly the same amounts of comparable constituents to regular tea. In addition to being high in cellulose, hemicellulose and lignin, pruned tea leaves, branches and residues are also high in bioactive substances, viz. water-insoluble proteins, polysaccharides and polyphenols [15]. Tea can be divided into 3 primary categories, viz. black, oolong and green tea, based on the manufacturing process and chemical composition. Catechins are present in considerable amounts in green tea. Polyphenols make up nearly 32% of the dry weight of green tea. The grade of the tea leaves, processing conditions and geographic location all affect the composition of green tea. Flavonol, theanine, flavan-3-ol, caffeine, vitamin C and phenolic acids are some of the main ingredients. Green tea contains more catechins than the other two forms of tea combined. Green tea consists of four principle catechins, viz. epicatechin-3-O-gallate (ECG), epigallocatechin (EGC), epigallocatechin-3-O-gallate (EGCG) and epicatechin (EC). Oolong tea contains polyphenols such as tannins, catechins, flavonol glycosides and flavonols, as well as polysaccharides, minerals, proteins, organic acids, fragrances and vitamins. It also contains caffeine, theobromine and theophylline as purine alkaloids. Major flavonols found in oolong tea include quercetin, myricetin and kaempferol [16]. On the other hand, condensed tannins, lignin, hemicellulose, cellulose, various structural proteins, along with insoluble and soluble proteins, as well as soluble polysaccharides are all present in black tea. Caffeine makes up close to 2–5% of black tea's dry weight. About 30–42% of the dry weight of black tea is made up of polyphenols. Only 20–30% of the total flavonoids are catechins, which are present in black tea. Generally speaking, tea waste is rich in polyphenolic compounds that have remarkable health-promoting properties. Consequently, tea waste can be used as a significant source for the creation of such medically significant compounds as polyphenols and polysaccharides [17]. Natural polyphenolic substances like flavonols, flavanols and other flavonoids are abundant in tea. Consequently, a sizeable number of these beneficial components are also present in the waste produced during the preparation of tea. According to research by Abdeltaif et al. (2018), there are roughly 45.75 mg of catechin per gramme of spent black tea in the processing waste from black tea [18]. Secondary metabolites known as tea polyphenols (TPs) are made up of one or more hydroxyl group and one aromatic ring. According to Sui et al. (2019), tea manufacturing waste can be activated by steam explosion to yield a sizeable amount of bioactive polyphenols [19]. Due to their remarkable health benefits, polyphenols are regarded as the most significant components of tea waste. Flavonoids and non-flavonoids are two general subcategories of polyphenols. Most of the polyphenols are dihydroflavanols, which are reduced to generate flavanols or flavan-3-ols, sometimes referred to as catechins. The majority of tea's biological effects are linked to catechins. The antioxidant properties of flavonoids help lower blood pressure and stop the spread of cancer cells. Amongst the black, oolong and green teas, the flavonoids present in black tea showed the strongest anticancer and antihypertensive properties, which were linked to its flavonol components such as patuletin, kaempferol and quercetin. The anticarcinogenic, antimutagenic and antioxidant properties along with cardiovascular disease prevention by black, oolong and green teas are mostly attributed to their catechin content [20]. Bioactive substances called polysaccharides are also found in the buds and leaves of tea plants. It has been discovered that tea polysaccharides provide a number of health advantages. Extracted from black, oolong and green teas, polysaccharide conjugates have been shown to have significant health advantages, including anti-obesity, antioxidant, anti-fatigue, anti-tumour, anti-skin ageing, antidiabetic and antiadhesion properties against pathogenic bacteria. The antioxidant qualities of polysaccharides derived from the leaves of green tea were examined by Li et al. (2019) [21]. The polysaccharides showed an encouraging ability to scavenge free radicals, which are responsible for oxidative damage in both human and animal bodies. Additionally, the polysaccharides demonstrated increased antioxidant capacity on chicken and were able to greatly increase their body weight. In light of this, the study came to the conclusion that the polysaccharides recovered from green tea can be employed as natural antioxidants and may find value in the feed sector.

The inhibition of certain enzymes, genetic material modification and interactions with cell membranes all play a role in the antibacterial mechanism of TPs. TPs have a broad range of antimicrobial activity to control food deterioration. The antibacterial activity of TPs against several bacteria, including Staphylococcus aureus, Escherichia coli, Helicobacter pylori, Bacillus, Clostridium and Streptococcus, has been the subject of numerous research. TPs also have antiviral and antifungal properties in addition to their antibacterial properties [22]. By attaching to receptor sites or altering the characteristics of the viral membranes, EGCG has demonstrated antiviral efficacy against adenoviruses, enteroviruses and influenza. The opportunistic fungus Candida albicans, which causes vaginal and oral infections, has been shown to be susceptible to the antifungal effects of EGCG. Moreover, Trichophyton mentagrophytes, T. rubrum, T. violaceum and T. tonsurans have also been treated with EGCG [23]. A schematic diagram representing the molecular mechanism of TPs for preventing various types of diseases is shown in Figure 8.1. The extensive usage of TPs in the food packaging sector is made possible by their distinctive antibacterial action. This is due to the fact that microbes are primarily to blame for the majority of food deterioration. Antimicrobial substances that may actively react to packaging conditions are thus necessary. A growing body of research has revealed that polyphenols not only slow down the ageing process but can also be used to prevent and treat diseases like cancer, cardiovascular, cerebrovascular and neurological disorders. The effect of antioxidants is reflected by a number of mechanisms, including the suppression of lipid peroxidation, reduction in oxidation through the chelation of metal ions, enhancement in the activity of antioxidant enzymes and scavenging of free radicals along with other nutrients. Rather than being actively involved in a plant's growth and development, polyphenols are secondary metabolites that provide the plant with distinctive qualities. The phenolic structures of TPs are what give them their natural antioxidant properties [24]. Free radicals are scavenged by TPs, which also restrict the

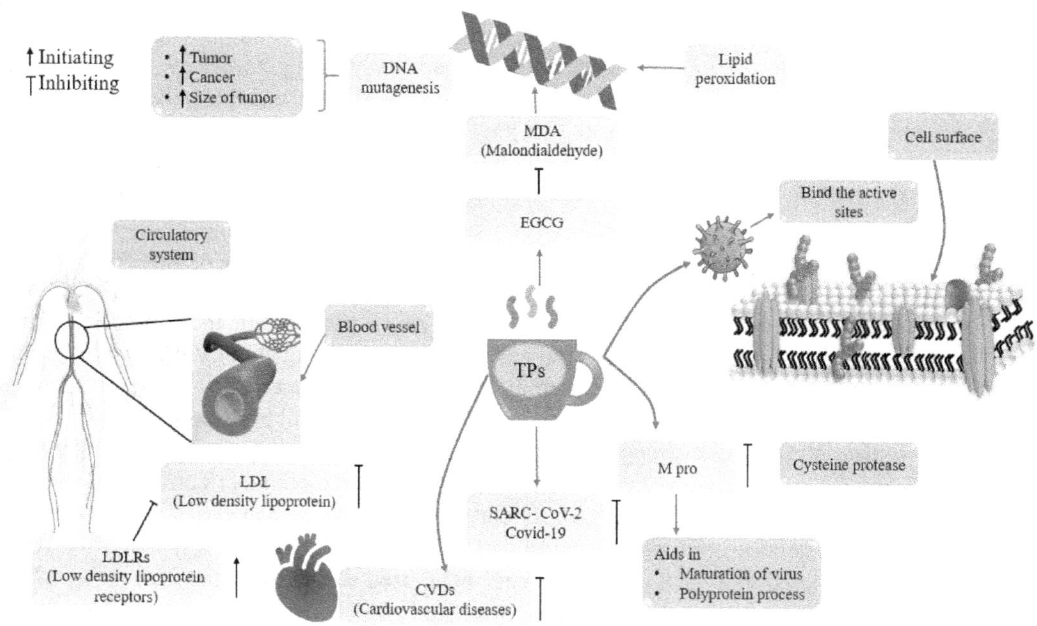

FIGURE 8.1 Molecular mechanism of tea polyphenols with biological activities [Reproduced with permission from Dai et al. (2022) [24]].

production of reactive oxygen species (ROS). The phenolic hydroxy groups on the polyphenol rings of TPs are what are thought to be responsible for their antioxidant activity. According to Li, et al. (2019), the radical chelating ability follows the EC<ECG<EGC<EGCG trend, while the theaflavins follow the following trend TF2<TF<TF1 [25]. TPs have a wider range of applications due to their antioxidant characteristics. Free radicals are eliminated when TPs and ROS interact to create relatively stable phenolic oxygen radicals. This is essential for stopping the accumulation of ROS, which is known to cause inflammation and neurodegenerative disorders. The lone electron on the oxygen atom of the phenolic hydroxyl group is conjugated by the electron on the benzene ring in TP. The solitary electron that is prone to moving toward the benzene ring reduces the activity of the hydrogen-oxygen bond in the phenolic hydroxyl group. As a result, the phenolic hydroxyl group's hydrogen activity increases. In summary, the environment, the stability of the phenolic oxygen radicals and the quantity of hydroxyl groups in the structure all have a role in how well TPs scavenge ROS. The human body is additionally shielded by TPs against oxidase and antioxidant enzymes. Xanthin oxidase, lipoxidase, NADPH oxidase and cyclic oxidase can all be inhibited by TPs [26]. Additionally, TPs suppress the activity of the enzymes to lessen the creation of oxygen free radicals. Additionally, polyphenols can affect the way that antioxidant defence enzymes work. Heavy metals pose a major hazard to the environment, human health and animal health since they are very poisonous even at very low quantities. Because of the rapid expansion of industries, the amounts of heavy metals released into wastewater effluents, including cadmium (Cd), chromium (Cr), manganese (Mn), copper (Cu), lead (Pb) and mercury (Hg) among others have dramatically risen. The development of inexpensive, effective and bio-based adsorbents for heavy metal removal has become a key focus of study in the recent era since adsorption is the most straightforward, affordable and promising way to remove various contaminants from the aquatic environment [9]. While Cherdchoo et al. (2019) found a greater Cr(VI) removal effectiveness of over 95% utilizing mixed tea waste [27], Celebi (2020) tested the adsorption of Cr(VI) utilizing waste from black, green and rooibos tea and got high removal efficacy (90%, 85% and 71%, respectively) [28]. We can therefore draw the conclusion that mixed tea waste exhibits better adsorption behaviour than single type of tea waste. Green biodegradable polymer composites with good performance and fewer environmental problems can be made by using natural fibres as a reinforcing agent in polymer matrices. Natural fibres often have high specific properties, are affordable and hydrophobic, but their mechanical strength is substantially lower. In contrast, composites made of synthetic fibres have good mechanical strength but are less recyclable. Therefore, integrating both fibres in a single polymer matrix has benefits since the resulting hybrid composite will have the greatest qualities of both fibres. As such, a hybrid polymer composite made of tea waste, sisal and glass fibres was created in an epoxy resin polymer matrix [24]. To get rid of hydrophilic lignin and hemicellulose, sisal fibres and tea waste were first given a 5 wt% NaOH treatment. They were then mixed with glass fibres in an epoxy matrix. The composites' mechanical qualities were assessed using different weight ratios of sisal and tea waste fibres. The composite with 10 wt% tea waste and 20 wt% sisal had the highest tensile and flexural strengths, measuring 75.5 MPa and 220 MPa, respectively, while the composite with 15 wt% tea waste and 15 wt% sisal had the best impact strength, measuring 96.2 kJ/m^2. 20 wt% tea and 5 wt% sisal fibre composition had a strong sound absorption coefficient of 0.9. Thus, as compared to a neat glass fibre composite, the hybridization with tea waste and sisal fibres improved the impact strength, flexural strength, tensile strength and absorption ability of the polymer composite by 6.5%, 21.6%, 41.0% and 37%, respectively. The hybrid composite can therefore be effectively employed in vehicle parts like front and rear bumpers, dashboards, brake pedals, seat bases, aircraft interior panelling, speakers, sound-proof materials and furniture [29]. TPs are frequently employed in the construction of functional food packaging. TPs have been used successfully for fruit preservation, according to several studies. Zhang et al. (2016) who studied various physicochemical and functional characteristics of Chinese winter jujube under ambient storage conditions incorporated TPs in an edible coating prepared from alginate [30]. Winter jujube's shelf life was found to be extended by the coating's ability to lower respiration rates and retain total phenol, ascorbic acid and

chlorophyll contents. It has been demonstrated that an assembly of chitosan and green TPs in alginate coated on PET sheets has antibacterial and antioxidant qualities, extending the shelf life of recently cut peaches. To increase the shelf life of packaged strawberries, Lan et al. (2019) created an antimicrobial composite film utilizing PVA and TPs in various ratios [31]. During the fabrication process, it was discovered that TP had a significant impact on the size and shape of fibres. At 25°C and a relative humidity of 75%, the ideal 8:2 PVA/TP film was discovered to maintain the firmness of strawberries and prevent microbial incursion into the fruit. The inclusion of TPs enhanced the mechanical characteristics of the film by weakening the hydrogen bonds of PVA, assisting in the creation of new bonds and decreasing the solubility of the film. Additionally, TPs are proven to prolong the shelf life of vegetables and enhance their physical beauty. In another study, novel films were prepared by Chen et al. (2022) [32]. Along with tea polyphenols, xanthan gum and hydroxypropyl methyl cellulose were used. Results demonstrated that TP actively improved the packed food's quality and shelf life. Additionally, TPs improved the mechanical performance and antibacterial qualities of packaging materials. They concluded that TPs are excellent for packaging veggies like freshly cut bell peppers. In order to preserve beet leaves, Fernandez et al. (2018) combined various preservation techniques while determining the antibacterial properties of nisin and green tea extract [33]. In order to further prevent nutrient depletion, modified atmosphere packaging and a modified starting atmosphere were used. The beet leaves were found to be better preserved using the combination of approaches. The effects of nisin and green tea extracts high in catechins and natamycin have been examined separately and in combination. The findings demonstrated that the beet leaves were greatly reduced in microbial load by green tea extract.

8.3.2 Coffee processing waste

Coffee is one of the most consumed nonalcoholic beverages in the world. 169.34 million bags of coffee were produced worldwide in 2019–2020. Brazil is currently the world's greatest manufacturer, generating around 60.55 million bags a year, while the US reports the highest consumption of coffee. Coffea arabica, which accounts for 75% of global output and Coffea canephora, which accounts for 25% of global production, are the two coffee species that are economically exploited. The annual output of coffee has increased, which has led to an equal growth in the amount of solid waste produced. As a result, the search for biological and technical uses for this waste material has become more intense [5].

Phenolic compounds are identified by the presence of one or more hydroxy substituents linked to at least one aromatic ring. The phenolic constituents are classified into lignans, stilbenes, flavonoids and phenolic acids, based on their structural differences. Phenolic acids are categorized into hydroxybenzoic acids (vanillic acids, syringic, protocatechuic, gallic, ellagic and hydroxybenzoic) and hydroxycinnamic acids (synaptic acids, ferulic, caffeic, p-coumaric, chlorogenic and cinnamic). On the other hand, flavonoids are categorized into anthocyanidins, flavanones, tannins, flavones, isoflavones, flavonols and flavanonols. Chlorogenic acid is the principle bioactive compound found in coffee and coffee by-products. The chemical structure of chlorogenic acid derivatives is shown in Figure 8.2. Because the pulp is high in p-coumaric acid, ferulic acid, caffeic acid and chlorogenic acids, its phenolic chemical makeup resembles that of green coffee [34]. Flavonoids (quercitrin, rutin, kaempferol and quercetin) and chlorogenic acid have been identified in the silverskin. Most hydroxycinnamic acids discovered in coffee are chlorogenic acid and its derivatives. Caffeic, p-coumaric, ferulic and sinapic acids are the most prevalent hydroxycinnamates found in coffee, and they combine with quinic acid to generate esters [35]. The most prevalent soluble ester phenolic constituent is 5-caffeoylquinic acid (dicaffeoylquinic acid and caffeoylquinic acid are denoted as n-diCQA and n-CQA respectively); on the other hand, 3,4-O-dicaffeoylquinic acid, 3-O-caffeoylquinic acid, 3,5-O-dicaffeoylquinic acid, 4-O-caffeoylquinic acid, 4,5-O-dicafeoylquinic acid, 5-O-feruloylquinic acid and 4-O-feruloylquinic acid have also been identified in coffee husk [36]. These phenolic compounds play a crucial role in the flavour and distinctive bitterness of coffee,

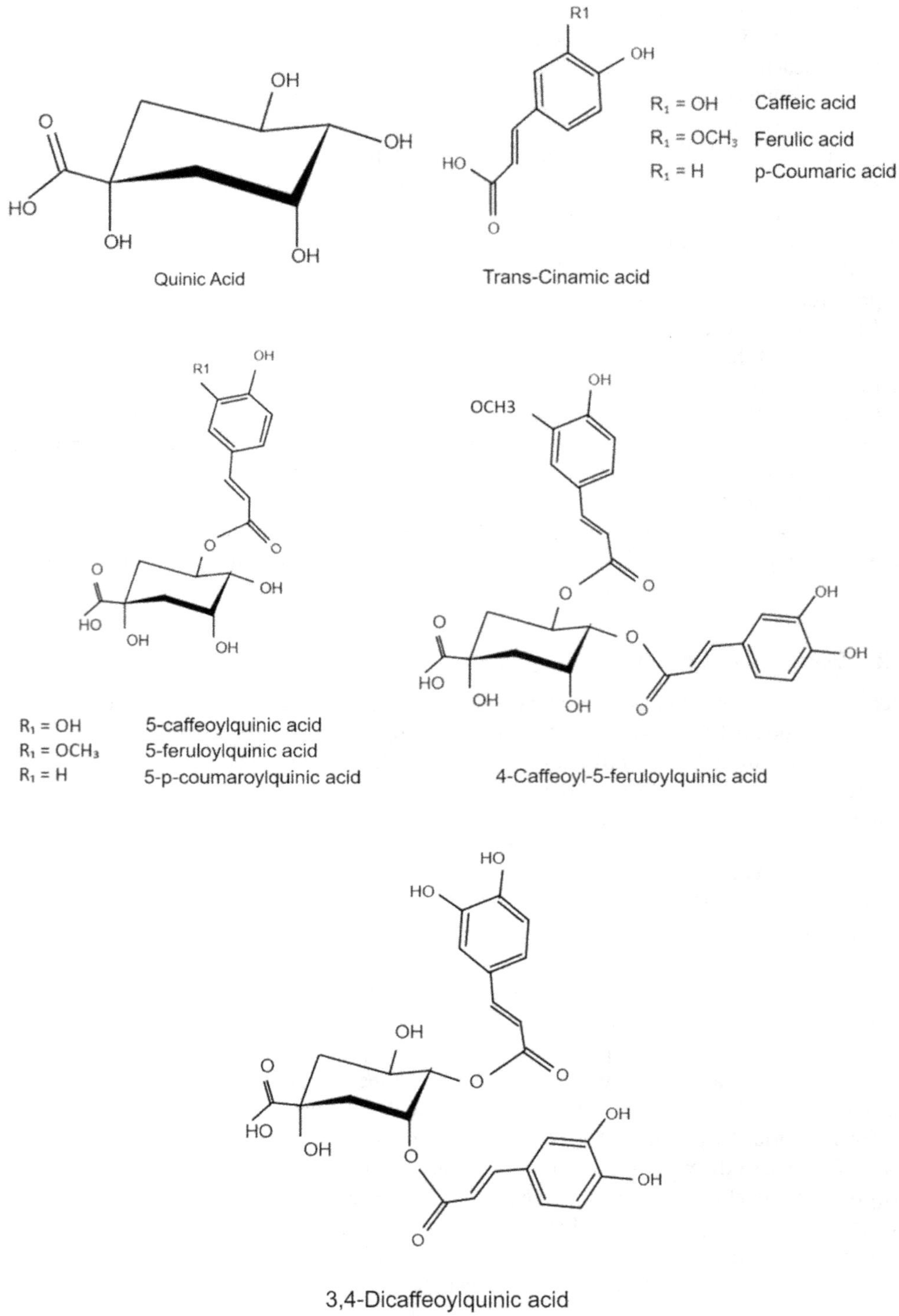

FIGURE 8.2 Chemical structure of chlorogenic acid derivatives in coffee waste [Reproduced with permission from Bondam et al. (2022) [5]].

but they are also significant because of their positive impacts on health, including their antioxidant, anti-inflammatory, antibacterial and antiradical capabilities. There have been various positive effects associated with chlorogenic acid, viz. anti-inflammatory, anticancer and antioxidant activities. Chlorogenic acid in coffee waste may offer significant advantage in terms of neurological illnesses, like ischaemic stroke, according to Mikami and Yamazawa (2015) [37]. Arauzo et al. (2020) used hydrothermal delignification (HDL) with ultrasound assistance to extract polyphenols from used coffee grounds [38]. To determine the concentration of total phenolic compounds, the authors utilized methanol, water and methanol:water ratio of 1:1 as solvents, followed by keeping the extracts at 40°C for 30 min. The optimal settings produced 20250 mg GAE/kg of dry sample from HDL at 190°C for 60 min while using methanol. In order to recover the phenolic compounds from coffee pulp, Kieu Tran, et al. (2020) utilized ultrasonication (150 W) technique at 40°C for 30 min [39]. Both hot air and vacuum drying temperatures were also adjusted. The sample that produced the greatest result was vacuum-dried at 85°C (14810 mg GAE/kg), however there was no discernible difference between samples that were dried at 110°C. The concentrations of extracted caffeine and chlorogenic acid were also assessed by the authors. They discovered that there was no discernible variation in the extracted caffeine concentrations over a wide range of drying conditions and operating temperature, however the concentration of chlorogenic acid was higher following vacuum drying at 85°C (3415 mg/kg). Furthermore, the amount of chlorogenic acid, caffeine and total phenolics in the silverskins derived from coffee beans was assessed by Barbosa-Pereira et al. (2018) [40]. The pre-treatment was carried out by the authors using 3.5% ethanol as a solvent in a pulsed electric field (PEF) using a high-voltage generator operating at 100 A and 14 kV. One of the silverskin types showed a significant level of total phenolic content at 12,870 mg GAE/kg and another variety yielded 365 mg/kg and 6405 mg/kg of chlorogenic acid and caffeine, respectively. The total phenolic constituent achieved via PEF increased by 83% as compared to the usual approach. Coffee pulp and coffee husk were employed for solid state fermentation as substrates with Penicillium purpurogenum by Garcia et al. (2015), who discovered that they are excellent substrates for the growth of this microbe [41]. The extraction of more polyphenols was made possible by the fermentation process. Particularly, fermentation at 33°C and humidity levels of 50–75% produced substantial polyphenol concentrations. Additionally, the pulp and husk of the coffee bean yielded 22,780 mg/kg and 132,450 mg/kg of chlorogenic acid and 4,350 mg/kg and 28,150 mg/kg of caffeic acid, respectively.

Antimicrobials either eliminate or stop the growth of microorganisms. They are essential in managing bacterial infections and have a variety of action methods and classifications (artificial and natural). However, there is a growing interest in natural antimicrobial compounds due to the rise in bacterial resistance and the hunt for more natural foods. As a result, the presence of active dual-purpose chemicals in coffee by-products promotes their usage as functional additives in culinary applications. Bacteria like Escherichia coli (E. coli), Staphylococcus aureus (S. aureus), Pseudomonas aeruginosa (P. aeruginosa), Bacillus subtilis (B. subtilis) and Enterococcus faecalis (E. faecalis) are susceptible to the antibacterial properties of coffee pulp [5]. Jim'enez-Zamora et al. (2015) investigated the antibacterial properties of coffee melanoidins, spent coffee grounds and extracts from silverskin [42]. When examined independently, the antibacterial activity of silverskin extracts and spent coffee grounds did not exhibit any appreciable impact against microbes. The melanoidin extract did, however, significantly blocked several microbe mixtures. A coffee pulp aqueous extract's capacity to kill bacteria was assessed by Duangjai et al. (2016) [43]. By using the agar well diffusion method, the authors assessed the antibacterial activity and calculated the minimal bactericidal concentration (MBC) and minimal inhibitory concentration (MIC) for E. coli, S. aureus, P. aeruginosa and S. epidermidis, thereby obtaining inhibition zones of 10 mm, 12 mm, 10 mm and 16 mm as well as MIC values of 36.8 mg/mL, 4.72 mg/mL, 77 mg/mL and 38.9 mg/mL and MBC value of 300mg/mL for all the bacteria. The antioxidant property of coffee waste is of particular interest to the research community. Phenolic constituents are antioxidants, which are naturally found in foods or can be incorporated into foods in small amounts to prevent the oxidation of nutritious substances and the formation of free radicals. Because it has been hypothesized that some antioxidants may be hazardous, the utilization of simulated antioxidants in the food processing sector is currently under scrutiny; as a result, efforts are being made to find natural alternatives that have antioxidant qualities. Due to their potential as antioxidants, phytochemical components found in coffee by-products are utilized in the food business [44]. For instance, roasted coffee extracts have been shown by Lin et al. (2015) to be an efficient antioxidant (natural) in beef, prolonging its shelf life even in the presence of chloride salts [45].

From the food technology standpoint, the coffee processing waste materials (husk, pulp, parchment and silverskin) are intriguing since they include a variety of bioactive substances and nutrients that may have biological benefits. In the food and pharmaceutical sectors, for instance, these substances can be employed as preservatives, antioxidants, colours and natural flavours. Bioactive substances having useful qualities, such as anti-inflammatory, neuroprotective, antibacterial and anticancer activities, are present in coffee extracts and their by-products. These characteristics point to the potential use of these extracts as ingredients and food supplements, as well as in cosmetics. There are many goods available for purchase on the market that are made with ingredients derived from coffee extracts or their by-products [46]. For instance, the Pectcof company created a product called Dutch Gum that is made from coffee pulp and has emulsifying and stabilizing capabilities; it is offered as an ingredient to the food and beverage sector. Additionally, Aqia Nutrition has created a line of goods called AQIA coffee. Ten products from this line are based on green coffee and cherry coffee. Green and cherry coffee oils, which are produced by cold pressing coffee seeds, are among the goods marketed. There are investigations into the creation of foods and beverages using coffee by-products as components in addition to the items already on the market [5]. For instance, Martinez-Saez et al. (2014) created an antioxidant beverage based on extracts from silverskin that contains caffeine and chlorogenic acid [47]. These ingredients have acceptable sensory qualities and are present in physiologically active concentrations for the regulation of body weight. Later, the hyaluronidase inhibitory impact was examined in vivo by Rodrigues et al. (2016) [48]. Twenty volunteers participated in this analysis using a coffee silverskin cream. After some time of use, the skin's firmness and moisture were assessed by contrasting the silverskin-containing cream with a hyaluronic acid control. Coffee silverskin has been demonstrated to be a potent component, producing outcomes comparable to hyaluronic acid and offering skin hydration and firmness. Iriondo-DeHond et al. (2016) examined the anti-ageing potential of silverskin extracts in the face of oxidative agents' accelerated ageing [49]. The scientists came to the conclusion that silverskin extracts can be used as components in cosmetic goods and hypothesized that the anti-ageing qualities of silverskin may be due to its complex variety of antioxidants that work in concert. The possible impact of different constituents isolated from C. arabica beans against UV-B incited skin damage were assessed by Cho et al. (2017) [50]. The anti-wrinkle properties of pyrocatechol, 3,4,5-tricaffeoyl quinic acid and chlorogenic acid were confirmed by the authors. These findings suggest that some phenolic compounds can be employed as shields against UV-induced skin ageing. Choi et al. (2016) assessed the protective effects of ethanolic extracts (42.58 mg/mL) and the oil fraction (547.32 mg/mL of caffeine and 119.25 mg/mL of chlorogenic acid) of used coffee grounds [51]. Both extracts significantly decreased UV-B-induced wrinkle formation when applied. In addition, the combined application of both fractions reduced the impacts of UV-B exposure, such as erythema development in 48%, transdermal water loss in 27%, epidermal thickness in 40% and wrinkle area by over 35%. These findings point to the possibility of using high-caffeine formulations as anti-photoaging products.

8.3.3 Winery processing waste

Due to their importance to the wine industry, grapes are regarded as a crop with a high economic value. In terms of global fruit production in 2019, grapes come in at number five with 77.14 MMT. In contrast, China is expected to produce over 10,800 thousand metric tonnes of grapes in 2019 and 2020, followed by Turkey and India, which will produce 1,950 and 3,000 thousand metric tonnes, respectively. Grapes, berries and other fruits can be used to make wine, which is a fermented product. The majority of the regions that produce wine are found in Europe, America, Australia and South Africa. In addition to wine, the winery process also produces a number of by-products, including grape pomace, grape marc and wine lees. If vineyard wastes are not properly disposed of, they may be deemed hazardous; as a result, therefore, emphasis should be given on effective recovery of valuable compounds from them, with subsequent applications [13].

There is a significant volume of grape pomace produced throughout the vinification process. Nutrients like carbohydrates, fibre, minerals and vitamins are found in grape pomace. Among the nutrients, dietary fibre is observed to be present in large concentrations. According to several studies, grape pomace contains up to 70% total dietary fibre, with insoluble dietary fibres including cellulose and hemicellulose making up 26–78% of the material. Contrarily, 9–11% of pomace is made up of water-soluble dietary fibre (DF), which includes gums, pectins and β-glucans among others. The capacity of pomace fibre to ferment in the colon, releasing short chain fatty acids that function as prebiotics, is associated to its physiological health advantages [52]. Pinheiro et al. (2009) found that the vitamin C content in grapes' edible component ranged from 4.90 mg/100g to 26.25 mg/100g, whereas that of the grape pomace was found to be between 4.90 mg/100g and 11.25 mg/100g [53]. In addition to these nutrients, grape pomace is a possible source of a number of bioactive chemicals, which are non-nutrient substances. Phenolic compounds are the most significant of these. Gonzalez-Centeno et al. (2013) assert that even after processing, the phenolic chemicals found in grapes will still be present in the pomace [54]. Compared to other fruit and vegetable waste, these are present in greater quantities. In general, phenolic compounds have an aromatic ring with hydroxyl groups in their structure, either one (phenolic acid) or several (polyphenols). Polyphenols are further divided into non-flavonoids and flavonoids. Non-flavonoids consist of hydroxycinnamic acid and hydroxybenzoic acid along with their derivatives, viz. stilbenes and phenolic alcohols among others, while flavonoids consist of anthoxanthins (isoflavone, flavan-3-ol, flavanone, flavanol), leucoxanthins, anthocyanins and flavonoidal alkaloids. The chemical structure of flavonoids and non-flavonoids are shown in Figure 8.3. Grape pomace is a potential source of resveratrol, procyanidin dimers and trimers, epicatechin, and catechins. Furthermore, the amount of anthocyanins found in grape pomace varies between 131.15 mg/100g and 1,790.20 mg/100g. These results are significantly greater than the residues from guava (3.2 g/100g) and acerola (8.4 g/100g) in terms of anthocyanin concentration [55]. Red grape bagasse includes anthocyanin levels of 390–940 mg/100g, according to Rockenbach et al. (2011) [56]. Procyanidins (1,051 mg/100 g), flavonols (190–6,119 mg/100 g), catechin (0.379–155 mg/100 g), proanthocyanidins (245–1,575 mg catechin equivalents (CTE)/ 100 g), epicatechin (45.55–118 mg/100g) and tannins (2,875–31,555 mg CTE/100 g) and hydroxycinnamic acids (271 mg/100 g) are the most prevalent phenolic constituents present in winery waste. Iora et al. (2015) found that grape pomace is a very strong source of resveratrol (0.62–6.35 mg/100g) and resveratrol gallic acid (4.65–388 mg/100g), both of which are present in significant amounts [57]. The aforementioned values have been reported by several writers in earlier investigations. These phenolic chemicals are in charge of the pomace extracts' potential antioxidant activity, which supports health. Many polyphenol chemicals, such as anthocyanidins, which are responsible for the red, blue, violet, orange and purple colours, are known for their role in the pigmentation of plant cells. The major qualities of wines, such as colour and flavour, are provided by polyphenols, which are of enormous importance to the wine business. Studies have concentrated on the sensory qualities of polyphenols as well as their health-related advantages. Actually, a number of in vitro studies have demonstrated that ingesting polyphenols lowers the risk of degenerative diseases (such as cardiovascular diseases) by preventing the oxidation of low-density lipoproteins in vitro. Additionally, polyphenols have anti-inflammatory, anti-ulcer and anticarcinogenic characteristics [13]. The technology known as pulsed ohmic heating (POH) has been used in the production of grape pomace extract. POH uses mild electric fields and temperatures along with thermal treatments to extract chemicals from products. POH damages cell membranes, promoting the recovery of phenolic constituents from red grape pomace. The following circumstances have been used to

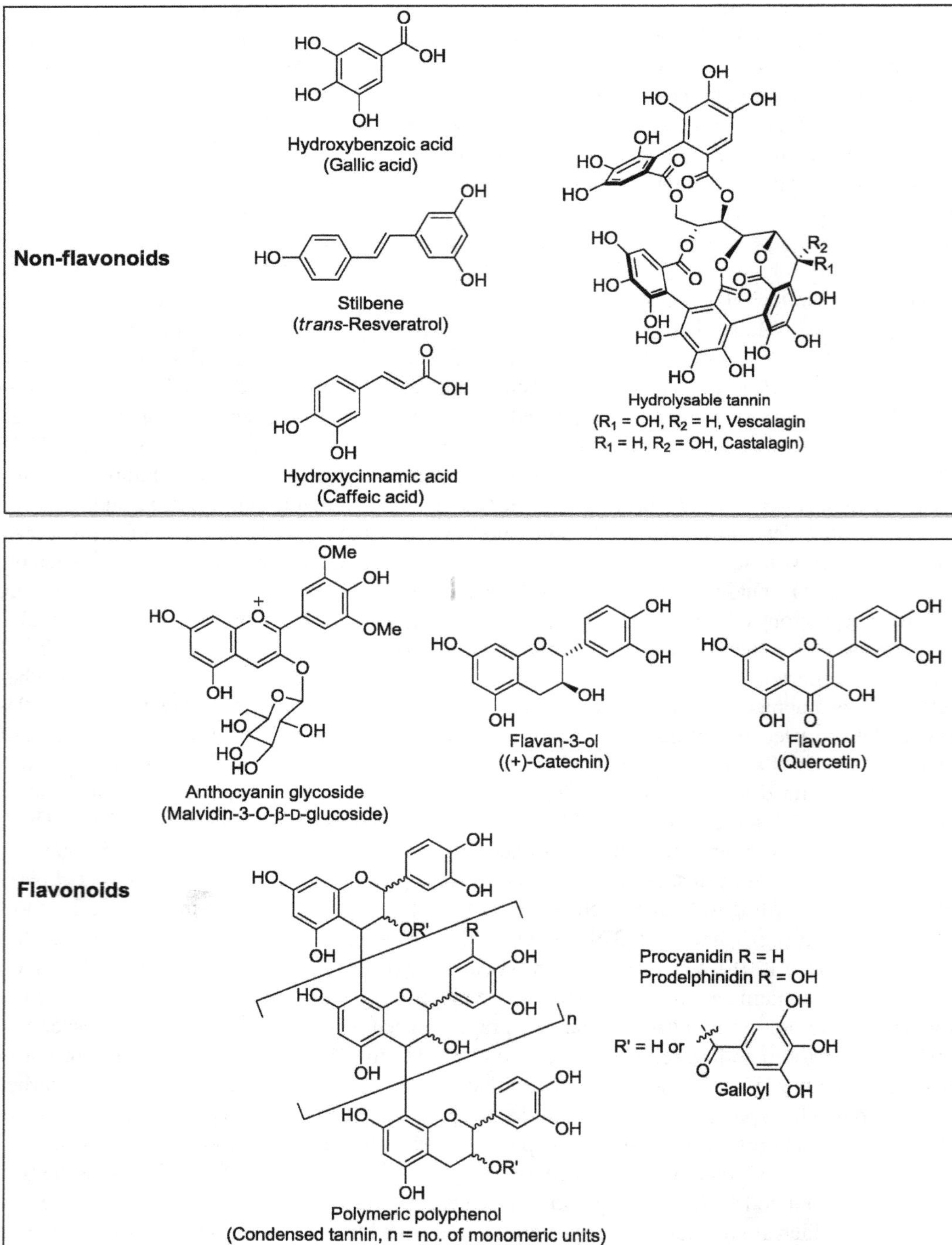

FIGURE 8.3 Chemical structure of flavonoids and non-flavonoids in winery waste [Reproduced with permission from Muhlack et al. (2018) [12]].

study POH's impact on grape pomace: 0–50% ethanol in water and an electric field strength of 100–800 V/cm. The findings indicated that denaturation of the cell membrane improved the extraction of phenolic constituents. The extraction of phenolic compounds was shown to be directly correlated with the electric field strength. The most successful POH treatment was discovered to be

500°C diffusion temperature and 30% ethanol. POH was found to accelerate the kinetics of the phenolic compounds when employed as a pre-treatment, avoiding the need for high temperatures during the extraction process. Because of this, it has been assumed that POH may one day be useful for valorising fruit and vegetable pomace without the usage of hydroalcoholic solvents [58]. Ferri et al. (2020) sought to improve and validate the valorisation of grape agro-waste to generate bioactive compounds and novel materials [59]. For this experiment, pressurized liquid extraction (PLE) and solvent-based methods were employed. Up to 47.5 g GAE/kg dw of total phenols were present in solvent-based techniques, whereas 79 g GAE/kg dw was extracted using PLE. However beneficial the use of elevated temperatures in PLE may be, it may also occasionally prove to be counterproductive, particularly when dealing with thermolabile materials, like anthocyanins, that break down at relatively low temperatures. Thus, when optimizing the process, these elements must be taken into account. In order to recover polyphenols from grape waste, pulse electric field technology has primarily been used. The amount of anthocyanins in red grape waste rose by 60% when PEF was used as a pre-treatment for 1 minute at 30°C along with a traditional thermal extraction for 1 hour at 65°C. A total of 10% more polyphenols were extracted from white grape skins after PEF treatment at 20°C compared to untreated samples. Bioactive chemicals isolated from grape pomace have been found to exert anticancer effects in in vitro and in vivo testing [60]. The bioactive contents found in grape pomace display cancer protection through a number of different methods. Angiogenesis, invasion and metastasis are a few metabolic pathways that are known to be inhibited by the polyphenol in grape pomace, as well as proteases and phase I and II drug-metabolizing enzymes. Additionally, they change cell-cycle checkpoints and apoptosis as well as disrupt receptor-mediated actions. A cell line investigation verified that grape seed proanthocyanidins inhibited cancer cell invasion in a dose-dependent manner for doses of 0, 10, 20 and 40 μg/mL. It was determined that this effect against human cutaneous HNSCC cells was caused by EGFR targeting and reversing the epithelial-to-mesenchymal transition process [61]. Pérez-Ortiz et al. (2019) showed anti-proliferative effects of grape pomace extract on fibroblasts and colon cancer cell lines (Caco-2, HT-29) at various doses of 5–250 μg/mL [62]. By increasing Ptg2 in Caco-2 cells and decreasing Myc gene expression in HT-29, the study showed that grape seed extract has anti-tumour properties. Grape seed extract's non-anthocyanin component has demonstrated potential activity against colorectal cancer cells. Cardiovascular disease (CVD) is primarily characterized by oxidative stress, which is also regarded as the primary disease marker for therapeutic interventions. Additionally, it was shown that oxidative stress indicators, particularly malondialdehyde, were downregulated while serum antioxidant levels were increased [63]. When used in concentrations between 0.25 mg/100g and 2 mg/100g, the heart-protective effects of pomace obtained from various grape cultivars, viz. Syrah, Marselan, Sauvignon and Cabernet, were also demonstrated [64]. A condition known as hyperglycaemia occurs when there is an excessive flow of sugar through the blood. Type 2 diabetes is linked to this issue. Inhibiting intestinal α-glucosidase, an enzyme that aids in the digestion and absorption of carbohydrates, is a treatment option. At a concentration of 10 μg/mL, red and white grape pomace extract reduced the activity of the α-glucosidase enzyme in yeast cells by 63% and 43%, respectively. Similar outcomes against the rat intestinal α-glucosidase enzyme were noted, with inhibitory activities of 47% (red grape pomace) and 39% (white grape pomace). On the other hand, STZ-induced mice that were given red grape pomace extract at a dose of 40 mg/100 g had their postprandial hyperglycaemia under control [65].

The removal and/or recovery of heavy metal pollutants from industrial effluent, which would otherwise result in hazardous pollution or eutrophication, has been recognized as a potential use for agricultural wastes like grape marc. Functional groups found in proteins or phenolics, such as carboxyl, amino or hydroxyl, cause metal ion binding. The effectiveness of different residual materials, such as grape marc, for the adsorption of pesticides was examined by Rodriguez-Cruz et al. (2012), who discovered that adding organic residues to soil improved retention of the tested hydrophobic pesticides [66]. Although the adsorption capacity was reduced due to changes in the organic carbon levels with incubation time, the same impact was seen in soils that had been incubated with the organic residue for 12 months prior to pesticide adsorption. Previous research examined the adsorption of metal ions, such as Cd(II) and Pb(II), onto dried grape marc (without further pre-treatment). With an equilibrium contact period of five minutes, batch adsorption studies for Cd(II) and Pb(II) showed maximum adsorption at pH 7.0 and 3.0, respectively. According to Langmuir's investigations, Cd(II) and Pb(II) have adsorption capacities of 0.479 mol/kg and 0.204 mol/kg, respectively, which are greater than other biomass adsorbents in the case of Pb and much lower in the case of Cd [67].

Chand et al. (2009) looked into the adsorption of Cr(IV) from aqueous solution by grape marc [68]. In batch-wise testing with regulated particle sizes of 100–150 mm under controlled temperature and pH, an adsorption gel made from grape marc by cross-linking with concentrated sulphuric acid was assessed. In contrast to pH 1 and 2, where adsorption declined with contact time, pH > 3 saw an increase in adsorption with time until equilibrium was reached and a constant level of adsorption was achieved. At pH 4, there was the greatest amount of adsorption, with a Cr(IV) adsorption capacity of 1.91 mol/kg that was comparable to that of biomasses such as sugarcane bagasse and chitosan. This method might produce a viable adsorbent for industrial use because wastewaters containing chromium are frequently acidic. In a novel study, Perez-Ameneiro et al. (2014) specifically examined the adsorption of micronutrients from winery wastewaters, including phosphate, ammonium, nitrate, potassium, magnesium and sulphate, utilizing grape marc entrapped in calcium alginate beads [69]. Three months of biodegradation of grape marc were followed by immobilization using calcium chloride and sodium alginate. With more than 50% of Mg, P and K adsorbed and more than 97% of NH_4 and NO_3 adsorbed, adsorption experiments utilizing a factorial design with adsorbent:wastewater ratio, contact time and agitation rate as the parameters revealed exceptionally high adsorption rates. When employed as fertilizer afterwards, the adsorption of these micronutrients may improve the quality of the adsorbent. Although the biochemical use of marc is frequently used to manufacture alcohol, it is also used to make a variety of other products, including biofertilizers, utilizing both fresh and spent marc [12]. In the process of composting, mesophilic temperatures and aerobic microbial activity break down organic material, which is subsequently stabilized by thermophilic temperatures. Pathogens are removed from the finished product, enabling it to be used in soil applications, thanks to thermophilic microbial activity and the heat that results from that action. Due to its high nutritional and organic content, marc is often utilized as a soil conditioner. In terms of chemical properties, soil conditioners based on marc are comparable to compost made from other organic waste, with the exception of the higher calcium content of marc as a result of winemaking conditions [70]. Hungria et al. (2017) used a pilot-scale dynamic respirometer operated in aerobic settings with monitoring of physicochemical, respirometric and olfactometric parameters to study the co-composting of grape marc with the organic fraction of municipal solid waste (OFMSW) at a 50:50 w/w ratio [71]. It was discovered that combining grape marc with OFMSW neutralized the acidity of the grape marc while creating a final composted product rich in nitrogen and phosphorus, improving its potential for reuse as an organic fertilizer. Under comparison to composting OFMSW alone in the same circumstances, it was reported that co-composting grape marc with OFMSW reduced odour emissions. In another study, Achmon et al. (2016) examined the ability of tomato pomace, red and white grape marc and biosolarization to inactivate soil pests [72]. In accordance with soil disinfestation, soil treatment with tomato pomace and white grape marc elevated soil temperature noticeably in aerobic soil conditions and decreased pH in anaerobic soil conditions. Red grape marc, however, did not cause these alterations; instead, it was found that unwanted soil methanogenesis was occurring, indicating that the area is less appropriate for biosolarization.

8.4 CONCLUSION

Agro-industrial wastes have received special attention over the past two decades, both to lessen the harm they cause to the environment and to explore the possibility of using them to create goods with a high added value. A possible alternative to using agro-industrial waste as an ingredient in the food and polymer industries is the recovery of value-added chemicals from both nonalcoholic and alcoholic beverages. The waste generated during the manufacturing of tea, coffee and wine has also been shown in this chapter to include a significant number of bioactives, which are ultimately responsible for significant health-promoting effects. The bioactive compounds recovered from the tea waste include polyphenols, catechin, polysaccharides, flavonoids and phenolic compounds among others which play important roles in promoting antimicrobial, antioxidant, antidiabetic and immunoregulatory activities. Coffee waste also contains considerable amounts of phytochemicals, viz. phenolic compounds, caffeine, chlorogenic acid, hydroxycinnamates and flavonoids, among others, which show significant health benefits like anti-inflammatory, antimicrobial and antioxidant activities, along with the prevention of type 2 diabetes. Furthermore, winery waste exhibited anti-hyperglycemic activity, antihyperlipidemic activity and anticancer activity along with the promotion of gut health due to the presence of high-value bioactives such as procyanidins, dietary fibres, anthocyanins, polyphenols and proanthocyanidins, among others. In view of the above concluding remarks, it can be said that this chapter emphasized the effective recovery of high-value phytochemicals from the generated tea, coffee and wine processing waste via different extraction methods along with their uses in numerous applications including wastewater treatment, soil composting, as well as food and polymer-based research.

ACKNOWLEDGEMENT

The study is supported by the Indian National Academy of Engineering (INAE/121/AKF/22), Gurgaon, India. The authors are solely responsible for all of the opinions, results and conclusions expressed in this study; INAE's viewpoints are not necessarily reflected in any of these aspects.

REFERENCES

1. S. Dahiya, A.N. Kumar, J. Shanthi Sravan, S. Chatterjee, O. Sarkar, S.V. Mohan, Food waste biorefinery: Sustainable strategy for circular bioeconomy, *Bioresour. Technol.* 248 (2018) 2–12. https://doi.org/10.1016/j.biortech.2017.07.176.
2. A. Zabaniotou, P. Kamaterou, Food waste valorization advocating Circular Bioeconomy - A critical review of potentialities and perspectives of spent coffee grounds biorefinery, *J. Clean. Prod.* 211 (2019) 1553–1566. https://doi.org/10.1016/j.jclepro.2018.11.230.
3. G. Aboagye, B. Tuah, E. Bansah, C. Tettey, G. Hunkpe, Comparative evaluation of antioxidant properties of lemongrass and other tea brands, *Sci. African.* 11 (2021) e00718. https://doi.org/10.1016/j.sciaf.2021.e00718.
4. A. Shang, J. Li, D.D. Zhou, R.Y. Gan, H. Bin Li, Molecular mechanisms underlying health benefits of tea compounds, *Free Radic. Biol. Med.* 172 (2021) 181–200. https://doi.org/10.1016/j.freeradbiomed.2021.06.006.
5. A.F. Bondam, D. Diolinda da Silveira, J. Pozzada dos Santos, J.F. Hoffmann, Phenolic compounds from coffee by-products: Extraction and application in the food and pharmaceutical industries, *Trends Food Sci. Technol.* 123 (2022) 172–186. https://doi.org/10.1016/j.tifs.2022.03.013.
6. F.J. Barba, Z. Zhu, M. Koubaa, A.S. Sant'Ana, V. Orlien, Green alternative methods for the extraction of antioxidant bioactive compounds from winery wastes and by-products: A review, *Trends Food Sci. Technol.* 49 (2016) 96–109. https://doi.org/10.1016/j.tifs.2016.01.006.
7. V. Basumatary, R. Saikia, R. Narzari, N. Bordoloi, L. Gogoi, D. Sut, N. Bhuyan, R. Kataki, Tea factory waste as a feedstock for thermo-chemical conversion to biofuel and biomaterial, *Mater. Today Proc.* 5 (2018) 23413–23422. https://doi.org/10.1016/j.matpr.2018.11.081.
8. A. Chowdhury, S. Sarkar, A. Chowdhury, S. Bardhan, P. Mandal, M. Chowdhury, Tea waste management: A case study from West Bengal, India, *Indian J. Sci. Technol.* 9 (2016) 11–19. https://doi.org/10.17485/ijst/2016/v9i42/89790.
9. B. Debnath, D. Haldar, M.K. Purkait, Potential and sustainable utilization of tea waste: A review on present status and future trends, *J. Environ. Chem. Eng.* 9 (2021) 106179. https://doi.org/10.1016/j.jece.2021.106179.

10. F.M. DaMatta, Ecophysiological constraints on the production of shaded and unshaded coffee: A review, F. Crop. *Res.* 86 (2004) 99–114. https://doi.org/10.1016/j.fcr.2003.09.001.
11. R.C. Campos, V.R.A. Pinto, L.F. Melo, S.J.S.S. da Rocha, J.S. Coimbra, New sustainable perspectives for "Coffee Wastewater" and other by-products: A critical review, Futur. *Foods.* 4 (2021) 100058. https://doi.org/10.1016/j.fufo.2021.100058.
12. R.A. Muhlack, R. Potumarthi, D.W. Jeffery, Sustainable wineries through waste valorisation: A review of grape marc utilisation for value-added products, *Waste Manag.* 72 (2018) 99–118. https://doi.org/10.1016/j.wasman.2017.11.011.
13. A.K. Chakka, A.S. Babu, Bioactive compounds of winery by-products: Extraction techniques and their potential health benefits, *Appl. Food Res.* 2 (2022) 100058. https://doi.org/10.1016/j.afres.2022.100058.
14. Q. Xu, Y. Yang, K. Hu, J. Chen, S.N. Djomo, X. Yang, M.T. Knudsen, Economic, environmental, and emergy analysis of China's green tea production, Sustain. Prod. Consum. 28 (2021) 269–280. https://doi.org/10.1016/j.spc.2021.04.019.
15. S. Guo, M. Kumar Awasthi, Y. Wang, P. Xu, Current understanding in conversion and application of tea waste biomass: A review, *Bioresour. Technol.* 338 (2021) 125530. https://doi.org/10.1016/j.biortech.2021.125530.
16. R. Kiyama, Estrogenic biological activity and underlying molecular mechanisms of green tea constituents, *Trends Food Sci. Technol.* 95 (2020) 247–260. https://doi.org/10.1016/j.tifs.2019.11.014.
17. S. Hussain, K.P. Anjali, S.T. Hassan, P.B. Dwivedi, Waste tea as a novel adsorbent: A review, *Appl. Water Sci.* 8 (2018) 1–16. https://doi.org/10.1007/s13201-018-0824-5.
18. S.A. Abdeltaif, K.A. Sirelkhatim, A.B. Hassan, Estimation of phenolic and flavonoid compounds and antioxidant activity of spent coffee and black tea (Processing) waste for potential recovery and reuse in Sudan, *Recycling.* 3 (2018) 12–18. https://doi.org/10.3390/recycling3020027.
19. W. Sui, Y. Xiao, R. Liu, T. Wu, M. Zhang, Steam explosion modification on tea waste to enhance bioactive compounds' extractability and antioxidant capacity of extracts, *J. Food Eng.* 261 (2019) 51–59. https://doi.org/10.1016/j.jfoodeng.2019.03.015.
20. G. Serdar, E. Demir, M. Sökmen, Recycling of tea waste: Simple and effective separation of caffeine and catechins by microwave assisted extraction (MAE), *Int. J. Second. Metab.* 4 (2017) 78–78. https://doi.org/10.21448/ijsm.288226.
21. X. Li, S. Chen, J.E. Li, N. Wang, X. Liu, Q. An, X.M. Ye, Z.T. Zhao, M. Zhao, Y. Han, K.H. Ouyang, W.J. Wang, Chemical composition and antioxidant activities of polysaccharides from yingshan cloud mist tea, *Oxid. Med. Cell. Longev.* 19 (2019) 22–27. https://doi.org/10.1155/2019/1915967.
22. L. Bouarab Chibane, P. Degraeve, H. Ferhout, J. Bouajila, N. Oulahal, Plant antimicrobial polyphenols as potential natural food preservatives, *J. Sci. Food Agric.* 99 (2019) 1457–1474. https://doi.org/10.1002/jsfa.9357.
23. S. Bansal, S. Choudhary, M. Sharma, S.S. Kumar, S. Lohan, V. Bhardwaj, N. Syan, S. Jyoti, Tea: A native source of antimicrobial agents, *Food Res. Int.* 53 (2013) 568–584. https://doi.org/10.1016/j.foodres.2013.01.032.
24. J. Dai, D.E. Sameen, Y. Zeng, S. Li, W. Qin, Y. Liu, An overview of tea polyphenols as bioactive agents for food packaging applications, *LWT.* 167 (2022) 113845. https://doi.org/10.1016/j.lwt.2022.113845.
25. Y. Li, X. Jiang, J. Hao, Y. Zhang, R. Huang, Tea polyphenols: The application in oral microorganism infectious diseases control, *Arch. Oral Biol.* 102 (2019) 74–82. https://doi.org/10.1016/j.archoralbio.2019.03.027.
26. U. Čakar, M. Čolović, D. Milenković, B. Medić, D. Krstić, A. Petrović, B. Đorđević, Protective effects of fruit wines against hydrogen peroxide—Induced oxidative stress in rat synaptosomes, *Agronomy.* 11 (2021) 1–14. https://doi.org/10.3390/agronomy11071414.
27. W. Cherdchoo, S. Nithettham, J. Charoenpanich, Removal of Cr(VI) from synthetic wastewater by adsorption onto coffee ground and mixed waste tea, *Chemosphere.* 221 (2019) 758–767. https://doi.org/10.1016/j.chemosphere.2019.01.100.
28. H. Çelebi, Recovery of detox tea wastes: Usage as a lignocellulosic adsorbent in Cr6+ adsorption, *J. Environ. Chem. Eng.* 8 (2020) 104310. https://doi.org/10.1016/j.jece.2020.104310.
29. L. Prabhu, V. Krishnaraj, S. Gokulkumar, S. Sathish, M. Ramesh, Mechanical, chemical and acoustical behavior of sisal - Tea waste - Glass fiber reinforced epoxy based hybrid polymer composites, *Mater. Today Proc.* 16 (2019) 653–660. https://doi.org/10.1016/j.matpr.2019.05.142.
30. L. Zhang, P. Chen, G. Gu, Q. Wu, W. Yao, Novel synthesis and photocatalytic performance of Ce 1 - X Zr x O 2 /silica fiber, *Appl. Surf. Sci.* 382 (2016) 155–161. https://doi.org/10.1016/j.apsusc.2016.04.122.

31. W. Lan, R. Zhang, S. Ahmed, W. Qin, Y. Liu, Effects of various antimicrobial polyvinyl alcohol/tea polyphenol composite films on the shelf life of packaged strawberries, *Lwt.* 113 (2019) 108297. https://doi.org/10.1016/j.lwt.2019.108297.
32. J. Chen, M. Zheng, K.B. Tan, J. Lin, M. Chen, Y. Zhu, Development of xanthan gum/hydroxypropyl methyl cellulose composite films incorporating tea polyphenol and its application on fresh-cut green bell peppers preservation, *Int. J. Biol. Macromol.* 211 (2022) 198–206. https://doi.org/10.1016/j.ijbiomac.2022.05.043.
33. M.V. Fernandez, R.J. Jagus, M.V. Agüero, Natural antimicrobials for beet leaves preservation: In vitro and in vivo determination of effectiveness, *J. Food Sci. Technol.* 55 (2018) 3665–3674. https://doi.org/10.1007/s13197-018-3295-7.
34. A.S.G. Costa, R.C. Alves, A.F. Vinha, E. Costa, C.S.G. Costa, M.A. Nunes, A.A. Almeida, A. Santos-Silva, M.B.P.P. Oliveira, Nutritional, chemical and antioxidant/pro-oxidant profiles of silverskin, a coffee roasting by-product, *Food Chem.* 267 (2018) 28–35. https://doi.org/10.1016/j.foodchem.2017.03.106.
35. L. Castaldo, G. Graziani, A. Gaspari, L. Izzo, C. Luz, J. Mañes, M. Rubino, G. Meca, A. Ritieni, Study of the chemical components, bioactivity and antifungal properties of the coffee husk, *J. Food Res.* 7 (2018) 43. https://doi.org/10.5539/jfr.v7n4p43.
36. M. de O. Silva, J.N.B. Honfoga, L.L. de Medeiros, M.S. Madruga, T.K.A. Bezerra, Obtaining bioactive compounds from the coffee husk (Coffea arabica L.) using different extraction methods, *Molecules.* 26 (2020) 31–38. https://doi.org/10.3390/molecules26010046.
37. Y. Mikami, T. Yamazawa, Chlorogenic acid, a polyphenol in coffee, protects neurons against glutamate neurotoxicity, *Life Sci.* 139 (2015) 69–74. https://doi.org/10.1016/j.lfs.2015.08.005.
38. P.J. Arauzo, M. Lucian, L. Du, M.P. Olszewski, L. Fiori, A. Kruse, Improving the recovery of phenolic compounds from spent coffee grounds by using hydrothermal delignification coupled with ultrasound assisted extraction, *Biomass and Bioenergy.* 139 (2020) 105616. https://doi.org/10.1016/j.biombioe.2020.105616.
39. T.M. Kieu Tran, T. Kirkman, M. Nguyen, Q. Van Vuong, Effects of drying on physical properties, phenolic compounds and antioxidant capacity of Robusta wet coffee pulp (Coffea canephora), *Heliyon.* 6 (2020) 11–19. https://doi.org/10.1016/j.heliyon.2020.e04498.
40. L. Barbosa-Pereira, A. Guglielmetti, G. Zeppa, Pulsed electric field assisted extraction of bioactive compounds from cocoa bean shell and coffee silverskin, *Food Bioprocess Technol.* 11 (2018) 818–835. https://doi.org/10.1007/s11947-017-2045-6.
41. L.R.P. García, C.R. Biasetto, A.R. Araujo, V.L. del Bianchi, Enhanced extraction of phenolic compounds from coffee industry's residues through solid state fermentation by Penicillium purpurogenum, *Food Sci. Technol.* 35 (2015) 704–711. https://doi.org/10.1590/1678-457X.6834.
42. A. Jiménez-Zamora, S. Pastoriza, J.A. Rufián-Henares, Revalorization of coffee by-products. Prebiotic, antimicrobial and antioxidant properties, *LWT - Food Sci. Technol.* 61 (2015) 12–18. https://doi.org/10.1016/j.lwt.2014.11.031.
43. A. Duangjai, N. Suphrom, J. Wungrath, A. Ontawong, N. Nuengchamnong, A. Yosboonruang, Comparison of antioxidant, antimicrobial activities and chemical profiles of three coffee (Coffea arabica L.) pulp aqueous extracts, *Integr. Med. Res.* 5 (2016) 324–331. https://doi.org/10.1016/j.imr.2016.09.001.
44. T.A. Hashimoto, F. Caporaso, C. Toto, L. Were, Antioxidant capacity and sensory impact of coffee added to ground pork, *Eur. Food Res. Technol.* 245 (2019) 977–986. https://doi.org/10.1007/s00217-018-3200-7.
45. C. Lin, C. Toto, L. Were, Antioxidant effectiveness of ground roasted coffee in raw ground top round beef with added sodium chloride, *LWT - Food Sci. Technol.* 60 (2015) 29–35. https://doi.org/10.1016/j.lwt.2014.08.010.
46. A. Farah, J. de P. Lima, Consumption of chlorogenic acids through coffee and health implications, *Beverages.* 5 (2019) 124–129. https://doi.org/10.3390/beverages5010011.
47. N. Martinez-Saez, M. Ullate, M.A. Martin-Cabrejas, P. Martorell, S. Genovés, D. Ramon, M.D. Del Castillo, A novel antioxidant beverage for body weight control based on coffee silverskin, *Food Chem.* 150 (2014) 227–234. https://doi.org/10.1016/j.foodchem.2013.10.100.
48. F. Rodrigues, R. Matias, M. Ferreira, M.H. Amaral, M.B.P.P. Oliveira, In vitro and in vivo comparative study of cosmetic ingredients Coffee silverskin and hyaluronic acid, *Exp. Dermatol.* 25 (2016) 572–574. https://doi.org/10.1111/exd.13010.
49. A. Iriondo-DeHond, P. Martorell, S. Genovés, D. Ramón, K. Stamatakis, M. Fresno, A. Molina, M.D. Del Castillo, Coffee silverskin extract protects against accelerated aging caused by oxidative agents, *Molecules.* 21 (2016) 1–14. https://doi.org/10.3390/molecules21060721.

50. Y.H. Cho, A. Bahuguna, H.H. Kim, D. in Kim, H.J. Kim, J.M. Yu, H.G. Jung, J.Y. Jang, J.H. Kwak, G.H. Park, O. jun Kwon, Y.J. Cho, J.Y. An, C. Jo, S.C. Kang, B.J. An, Potential effect of compounds isolated from Coffea arabica against UV-B induced skin damage by protecting fibroblast cells, *J. Photochem. Photobiol. B Biol.* 174 (2017) 323–332. https://doi.org/10.1016/j.jphotobiol.2017.08.015.
51. H.S. Choi, E.D. Park, Y. Park, S.H. Han, K.B. Hong, H.J. Suh, Topical application of spent coffee ground extracts protects skin from ultraviolet B-induced photoaging in hairless mice, *Photochem. Photobiol. Sci.* 15 (2016) 779–790. https://doi.org/10.1039/c6pp00045b.
52. A. Nayak, B. Bhushan, A. Rosales, L.R. Turienzo, J.L. Cortina, Valorisation potential of Cabernet grape pomace for the recovery of polyphenols: Process intensification, optimisation and study of kinetics, *Food Bioprod. Process.* 109 (2018) 74–85. https://doi.org/10.1016/j.fbp.2018.03.004.
53. É.S. Pinheiro, J.M.C. da Costa, E. Clemente, P.H.S. Machado, G.A. Maia, Physical chemical and mineral stability of grape juice obtained by steam extraction, *Rev. Cienc. Agron.* 40 (2009) 373–380. https://doi.org/10.1016/j.rciag.2009.113905.
54. M.R. González-Centeno, M. Jourdes, A. Femenia, S. Simal, C. Rosselló, P.L. Teissedre, Characterization of polyphenols and antioxidant potential of white grape pomace byproducts (Vitis vinifera L.), *J. Agric. Food Chem.* 61 (2013) 11579–11587. https://doi.org/10.1021/jf403168k.
55. M. Fanzone, F. Zamora, V. Jofré, M. Assof, C. Gómez-Cordovés, Á. Peña-Neira, Phenolic characterisation of red wines from different grape varieties cultivated in Mendoza province (Argentina), *J. Sci. Food Agric.* 92 (2012) 704–718. https://doi.org/10.1002/jsfa.4638.
56. I.I. Rockenbach, E. Rodrigues, L.V. Gonzaga, V. Caliari, M.I. Genovese, A.E.D.S.S. Gonalves, R. Fett, Phenolic compounds content and antioxidant activity in pomace from selected red grapes (Vitis vinifera L. and Vitis labrusca L.) widely produced in Brazil, *Food Chem.* 127 (2011) 174–179. https://doi.org/10.1016/j.foodchem.2010.12.137.
57. S.R.F. Iora, G.M. Maciel, A.A.F. Zielinski, M. V da Silva, P.V. de A. Pontes, C.W.I. Haminiuk, D. Granato, Evaluation of the bioactive compounds and the antioxidant capacity of grape pomace, *Int. J. Food Sci. Technol.* 50 (2015) 62–69. https://doi.org/10.1111/ijfs.12583.
58. A.R. Fontana, A. Antoniolli, R. Bottini, Grape pomace as a sustainable source of bioactive compounds: Extraction, characterization, and biotechnological applications of phenolics, *J. Agric. Food Chem.* 61 (2013) 8987–9003. https://doi.org/10.1021/jf402586f.
59. M. Ferri, M. Vannini, M. Ehrnell, L. Eliasson, E. Xanthakis, S. Monari, L. Sisti, P. Marchese, A. Celli, A. Tassoni, From winery waste to bioactive compounds and new polymeric biocomposites: A contribution to the circular economy concept, *J. Adv. Res.* 24 (2020) 1–11. https://doi.org/10.1016/j.jare.2020.02.015.
60. H. Wijngaard, M.B. Hossain, D.K. Rai, N. Brunton, Techniques to extract bioactive compounds from food by-products of plant origin, *Food Res. Int.* 46 (2012) 505–513. https://doi.org/10.1016/j.foodres.2011.09.027.
61. J.K. Kundu, Y.J. Surh, Cancer chemopreventive and therapeutic potential of resveratrol: Mechanistic perspectives, *Cancer Lett.* 269 (2008) 243–261. https://doi.org/10.1016/j.canlet.2008.03.057.
62. J.M. Pérez-Ortiz, L.F. Alguacil, E. Salas, I. Hermosín-Gutiérrez, S. Gómez-Alonso, C. González-Martín, Antiproliferative and cytotoxic effects of grape pomace and grape seed extracts on colorectal cancer cell lines, *Food Sci. Nutr.* 7 (2019) 2948–2957. https://doi.org/10.1002/fsn3.1150.
63. Ş.S. Balea, A.E. Pârvu, N. Pop, F.Z. Marín, A. Andreicuţ, M. Pârvu, Phytochemical profiling, antioxidant and cardioprotective properties of pinot noir cultivar pomace extracts, *Farmacia.* 66 (2018) 432–441. https://doi.org/10.31925/farmacia.2018.3.7.
64. S. Chacar, J. Hajal, Y. Saliba, P. Bois, N. Louka, R.G. Maroun, J.-F. Faivre, N. Fares, Long-term intake of phenolic compounds attenuates age-related cardiac remodeling, Aging Cell. 18 (2019) 12894. https://doi.org/10.1111/acel.12894.
65. Shelly Hogan; Lei Zhang, Antioxidant rich grape pomace extract suppresses postprandial hyperglycemia in diabetic mice by specifically inhibiting alpha-glucosidase, *Nutr. Metab.* 5 (2010) 1–9. https://doi.org/10.4236/nutmet.2010.48a022.
66. M.S. Rodríguez-Cruz, E. Herrero-Hernández, J.M. Ordax, J.M. Marín-Benito, K. Draoui, M.J. Sánchez-Martín, Adsorption of pesticides by sewage sludge, grape marc, spent mushroom substrate and by amended soils, *Int. J. Environ. Anal. Chem.* 92 (2012) 933–948. https://doi.org/10.1080/03067319.2011.609933.
67. N. V. Farinella, G.D. Matos, M.A.Z. Arruda, Grape bagasse as a potential biosorbent of metals in effluent treatments, *Bioresour. Technol.* 98 (2007) 1940–1946. https://doi.org/10.1016/j.biortech.2006.07.043.

68. R. Chand, K. Narimura, H. Kawakita, K. Ohto, T. Watari, K. Inoue, Grape waste as a biosorbent for removing Cr(VI) from aqueous solution, *J. Hazard. Mater.* 163 (2009) 245–250. https://doi.org/10.1016/j.jhazmat.2008.06.084.
69. M. Perez-Ameneiro, X. Vecino, L. Vega, R. Devesa-Rey, J.M. Cruz, A.B. Moldes, Elimination of micronutrients from winery wastewater using entrapped grape marc in alginate beads, *CYTA - J. Food.* 12 (2014) 73–79. https://doi.org/10.1080/19476337.2013.797923.
70. V.K. Sharma, M. Canditelli, F. Fortuna, G. Cornacchia, Processing of urban and agro-industrial residues by aerobic composting: Review, *Energy Convers. Manag.* 38 (1997) 453–478. https://doi.org/10.1016/S0196-8904(96)00068-4.
71. J. Hungría, M.C. Gutiérrez, J.A. Siles, M.A. Martín, Advantages and drawbacks of OFMSW and winery waste co-composting at pilot scale, *J. Clean. Prod.* 164 (2017) 1050–1057. https://doi.org/10.1016/j.jclepro.2017.07.029.
72. Y. Achmon, D.R. Harrold, J.T. Claypool, J.J. Stapleton, J.S. VanderGheynst, C.W. Simmons, Assessment of tomato and wine processing solid wastes as soil amendments for biosolarization, *Waste Manag.* 48 (2016) 156–164. https://doi.org/10.1016/j.wasman.2015.10.022.

9 Commercial aspects of bioactive compounds extracted from food waste

9.1 INTRODUCTION

The exponential rise in population has resulted in an equally exponential rise in the demand for all kinds of food and energy. Industrialization and modernization of the agricultural and food-based industries were made possible by the ever-increasing demand for these sectors' output, leading to more variety in the sectors' output and an increase in both the quality and quantity of their output [1]. The massive volumes of agro-industrial food waste that have resulted from this growth are typically dumped into landfills without much thought, the exorbitant expense of treatment procedures being the main reason for this. The amount of wasted food that is produced on a global scale was estimated to be between 1.3 and 1.4 billion tonnes and it is anticipated that this number may increase to as much as 2.6 billion tonnes by the year 2025 [2]. The studies produced by the Food and Agriculture Organization (FAO) of the United Nations (UN) made it abundantly evident that vegetable waste was contributing to "carbon footprint," and fruit waste was contributing to "blue water hotspots" [3, 4]. According to the estimates of the United Nations Environment Programme, the amount of money that is lost due to the improper disposal of food is around US$ 400 billion [5]. These concerns compelled the implementation of sustainable and environmentally compatible methods for the maximum utilization of agro-industrial food waste as a valuable resource to produce bioproducts, like bioactive compounds [3, 6, 7] rather than carelessly discharging it into the environment, which can lead to severe air, land and water pollution.

Figure 9.1 shows various routes for the valorization of food waste which is categorized from high grade to low grade. It's crucial to take into account the environmental effect of the production process as a whole, but there's a lot of untapped potential in using food wastes as precursors for the synthesis of pharmaceutical and bioactive compounds because they're cheap and renewable. In the most recent few decades, there has been a rise in interest in the circular bioeconomy concept, which emphasizes the maximum utilization of the wastes generated from primary production processes as cost-effective resources for the production of secondary products. This concept was initially introduced to bring attention to the maximum utilization of waste produced from primary production processes. It is frequently difficult to employ a single manufacturing technique that will be successful for the purpose of completely using food waste due to the heterogeneity of food waste and the complicated biochemical makeup of the waste itself. Due to the difficulties of the situation, an integrated manufacturing method has been developed, which allows for the creation of numerous goods using a channelized waste valorization approach [8].

Consumers' interest in natural and high-quality meals appears to be growing, as seen by recent developments in the food business and the ongoing quest for healthier products. Furthermore, the global health issue brought on by COVID-19 altered existing consumer attitudes, perceptions and behavioural patterns towards lessening food waste and caring more about the food goods they eat. Therefore, there has been a rise in interest in natural bioactive compounds over the course of the past year, particularly as a result of consumers becoming more informed about the evidence that consuming a nutritious and well-balanced diet has a positive impact on health; consequently, consumers all over the world have become more health-conscious [10]. In addition, agro-food waste

DOI: 10.1201/9781003315469-9

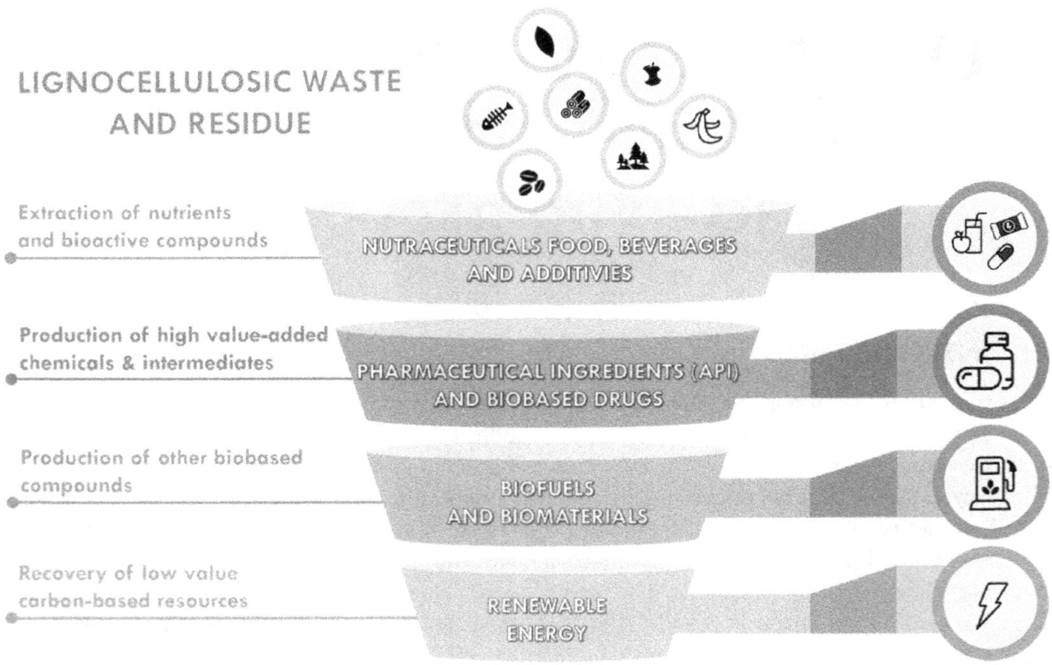

FIGURE 9.1 Several distinct avenues for the valorization of food waste [Reproduced with permission from Espro et al. (2021) [9]].

is a rich source of various bioactive compounds, the exact composition of which varies according to the type of waste being disposed of (i.e., whether it is fruit and vegetable, dairy, meat and fish, grain, root, tuber, or oilseed waste) [11, 12]. These bioactive compounds may be recovered from food waste to create unique functional food items; food waste is both inexpensive and renewable. There has been a rise in the variety of nutrients, nutritional supplements and functional ingredients available on the market during the past several decades. Approximately $210 billion will be spent on the worldwide sale of nutraceuticals, food antioxidants and nutritional supplements in 2026. Europe is dedicated to gathering clinical data about the safety and health advantages of functional goods, whereas Asia and North America are the world's largest consumer markets for nutraceuticals and nutritional supplements. More stable, useful and user-friendly food additives that may be applied to a wide variety of food items are gaining popularity in the food business [3].

In this chapter, we will attempt to examine the current literature on the topic of food waste's potential economic viability as a substrate for the production of medicines, cosmetics, energy, food packaging, etc. This chapter provides an in-depth review of the latest developments in production strategy and prospects.

9.2 UTILIZATION OF BIOACTIVE COMPOUNDS IN VARIOUS INDUSTRIES

9.2.1 Pharmaceutical industry

The nutraceutical and pharmaceutical sectors play an essential role in our economy and have made significant contributions to enhancing human well-being. Creating active pharmaceutical ingredients (APIs) is likely at the heart of every pharmaceutical company's operations. As the world's population ages and chronic illnesses proliferate, the worldwide API market is expected to expand from €158.9 billion in 2020 to €210.6 billion in 2025, at a compound annual growth rate (CAGR) of 5.8% (in the areas of diabetes and cancer in particular). Nutraceutical products (which include nutritional supplements, functional/pharmaceutical foods, drinks and personal care items) also have a booming global market that had a value of 324.2 billion EUR in 2014 and is estimated to increase up to 613.7 billion EUR by the year 2027 (at a CAGR of 8.3%) [9]. By employing food loss, food waste and lignocellulosic agricultural/forestry residues as starting materials for the manufacture of APIs and bioactive compounds, it is possible to significantly increase the sustainability of the pharmaceutical and nutraceutical sectors. The use of lignocellulosic materials has already been widely embraced for the manufacture of fine chemicals, energy (biogas, hydrogen) and fuels (ethanol, biodiesel). The extraction of physiologically active substances (such as anthocyanins, polyphenols and flavonoids) for nutraceuticals from food waste has the potential to be done effectively [13, 14].

Two primary factors will accelerate the development of novel, environmentally friendly processes for the synthesis of bioactive chemicals, APIs and pharmaceutical intermediates from biomass resources. Such high-value-added compounds can offer innovative "greener" techniques for the pharmaceutical and nutraceutical sectors and would greatly improve the overall economic viability of a biorefinery.

Over the past 20 years, the "12 Green Chemistry Principles" have slowly but steadily filtered through the pharmaceutical sector, becoming well-established either by incorporation into specific corporate policies or through advocacy by top organizations. But they have not been equally incorporated into the everyday routines of all pharmaceutical industry scientists i.e., medicinal chemists and process chemists [14, 15]. While medicinal chemists have recently begun to implement Green Chemistry principles for the identification of new bioactive compounds for drug discovery as well as development pipeline, great strides have been made in applying these principles to the redesign of API production methods as well as for the advancement of synthetic sequences and pilot plant processes. Additionally, the majority of the efforts are being focused on three strategies: waste minimization, avoiding use of toxic and/or hazardous reagents/solvents and biocatalysis. There has been an increasing focus on the utilization of renewable biomass as precursors only in the past ten years.

Similarly, the nutraceutical sector views the use of food waste for bioactive ingredients as a promising path towards lowering the price of formulations of nutraceuticals and utilizing fewer synthetic chemicals.

Compared to fossil-based resources, using food waste and food loss to make medications has real benefits. In terms of short- and long-term pricing, health, environmental and security issues, locally obtained feedstocks appear to be more advantageous. For instance, Bristol Myers Squibb utilized bio-based materials back in 2004 when it harvested paclitaxel (the active ingredient in Taxol) effectively from plant cell cultures instead of using an 11-step process.

An Italian firm known as Herbal & Antioxidant Derivatives (H&AD) is capitalizing on the global demand for bergamot by-products to sell dietary supplements and nutraceuticals. The essential oil found in the bergamot (Citrus bergamia) fruit peel is commonly used in the perfume business, whereas the rest of the fruit (the pulp and juice) is thrown away. As bergamot waste may cause soil pollution if not adequately destroyed or valorized, H&AD rethought its potential uses by capitalizing on the high polyphenol content. Bergacyn is a nutraceutical compound developed by H&AD and protected by a patent for a novel technology comprising ultrafiltration and absorptive extraction. Clinical studies in patients have substantiated the advantages of Bergacyn. Notably, in order to further the source of nutraceuticals from recycled raw materials, H&AD formed a relationship with the US natural product business DolCas biotech LLC in 2019 [16].

Another joint venture with Swisse Wellness, the Fight Food Waste Cooperative Research Center, Viridi Innovation Pty Ltd, Austeng and the Swinburne University of Technology intends to transform leftover grapes (skins and seeds) from the Australian wine industry into nutraceutical components. Most vineyard trash ends up as harmful soil because of the high polyphenol concentration. Two hundred fifty tonnes of mostly Pinot marc will be used in this initiative to make grape seed extract for the 2020 harvest that will be sold in Asia. Polyphenol extraction from grape waste, by a novel method using microwave and ultrasound, followed by incubation with fungal enzymes (Basidiomycete and Ascomycete), has been patented. It is expected that by 2021, the validation at the laboratory level will have been completed successfully, and an industrial-scale pilot plant constructed [9].

The strategy used by the Danish biotech firm Kaffe Bueno is another illustration of food waste and food loss exploitation for the manufacture of nutraceuticals. Used coffee grounds (SCGs) are upcycled by Kaffe Bueno into useful and active components for nutraceuticals, functional meals and cosmetics [17]. SCGs are a beneficial source of fatty acids and antioxidants with several health advantages. In order to create a lipid fraction, Kaffe Bueno built a bio-preservation system and improved a supercritical CO_2 extraction. The resultant defatted coffee grounds still contain significant amounts of insoluble dietary fibre (4/9 necessary amino acids) and protein. Currently, there are three products available on the market (Kaffe Bueno Oil, Kafflour and Kaffibre), which are used as food additives as well as ingredients in nutraceuticals and cosmetics. In order to commercialize its coffee oil under the brand name Koffee'Up, Kaffe Bueno teamed up with Givaudan, a world leader in beauty products with headquarters in Switzerland, in June 2020. It's significant to note that Kaffe Bueno wants to process 1,200 tonnes of SCG annually by 2026.

In the year 2000, Repolar Pharmaceuticals also had success in preparing products based on bioactive compounds extracted from food waste. A multidisciplinary team created a resin-based solution from Norwegian spruce trees (Picea abies) that were injured by occurrences like lightning strikes [18]. The product, Abilar, was inspired by a traditional Norwegian medicine used to cure wounds. Antimicrobial and anti-inflammatory properties, as well as the promotion of re-epithelialization, have been attributed to the resin, which is a complex mixture of different acids (such as abietic, pimaric, dehydroabietic, neoabietic, palustric acids, etc.) and lignans (such as p-hydroxycinnamic acid, pino- and larici-matairesinol). The resin ointment Abilar was introduced by Repolar in 2008. Since its introduction to the market, Abilar sales have been rather constant [19].

Recently, cashew nut shell liquid (CNSL) from food waste has been used in the synthesis of physiologically active phenolic compounds, which might lead to the development of novel pharmaceuticals. In spite of their many biological actions, CNSL phenolic lipids (8–10) on their own are not potent enough for any practical therapeutic use at this time. Therefore, it is worthwhile to explore strategies that try to synthetically elaborate and diversify towards more complex physiologically active compounds, which may then be tuned further into therapeutic candidates [20].

Since many of the world's top CNSL producers are located in tropical and subtropical regions (South Asia, Africa and South America), this provides a significant potential for the development of treatments for neglected tropical diseases (NTDs). In order to achieve synergy and higher potency for hybrid medications, Bolognesi and colleagues argued for the use of a medicinal chemistry technique to create the semi-synthetic derivatives of CNSL. Specifically, a trypanocidal profile was achieved by fusing the chemical structures of CNSL derivatives and 2-phenoxy-1,4-naphthoquinone. To join CNSL phenols with 2-bromo-1,4-naphtoquinones, a simple, accessible and scalable approach based on a nucleophilic substitution was created [21]. This could potentially make the entire production process locally affordable, especially given how inexpensive and readily available the initial material is. The synthetic compounds exhibited fast trypanocidal action in the low micromolar range, suppressed parasite multiplication and exhibited no observable toxicity on human cell lines. They appeared to have a mechanism of action involving mitochondrial deterioration.

These studies highlight the fact that the pharmaceutical sectors have a significant potential for commercializing food waste. It is clear that the pharmaceutical industry has a lot of room for food waste commercialization. There has to be a shift in the pharmaceutical sectors towards the utilization of feedstocks derived from food waste. There are a number of scientific and technological hurdles that need to be overcome in this area, including developing an efficient transportation system, an efficient catalyst for the desired transformation, a low energy separation process and a continuous process [9].

9.2.2 Cosmetic industry

Cosmetic is any product intended to be rubbed, poured, sprinkled, sprayed on, infused into, or used on any area of the human body for washing, grooming, encouraging attractiveness, or modifying the look. Despite the lack of a formal definition for "cosmeceuticals" in the law, the term is commonly used to describe items that fall between the cosmetics and pharmaceutical industries. The creation of new components and products for cosmeceutical and pharmaceuticals relies heavily on plant sources. Because they may improve health and attractiveness using substances that affect the skin's biological texture and function and give protection against degenerative skin disorders, cosmeceuticals have found widespread use in the personal care business. By supplying essential nutrients, they can enhance the skin's tone, texture and glow while minimizing fine lines and wrinkles.

Formulations of herbal cosmetics that employ diverse cosmetic elements of natural origin to enhance the state of the skin are also referred to as natural cosmetics. Cosmetic compounds of plant origin are often promise to prevent any adverse effects on the human body. These substances are meant to improve both the health and beauty of the skin while simultaneously offering a certain outcome. It is essential to have an understanding of the physiognomy of the skin as well as the elements that are responsible for changes in the skin in order to be able to comprehend the application of plant extracts. The skin is the biggest organ in the human body and its primary role is to defend the body from pathogens, ionizing radiation and a wide variety of allergens and irritants. Bioactive chemicals can be released from polyphenol-rich plant extracts, entering the skin, bypassing the *stratum corneum* barrier and permeating the epidermis and dermis [22]. The *stratum corneum* is the most superficial layer of the epidermis and the outermost layer of the skin. It is a heterogeneous layer of the epidermis and selectively permeable, as seen in Figure 9.2. Its primary role is to protect the skin from dryness and environmental stress. In addition, the *stratum corneum* stores a suitable quantity of water, which is necessary for the functioning of the skin. A breach in the skin barrier of this layer is the root cause of both decreased skin hydration and increased rates of water loss [23].

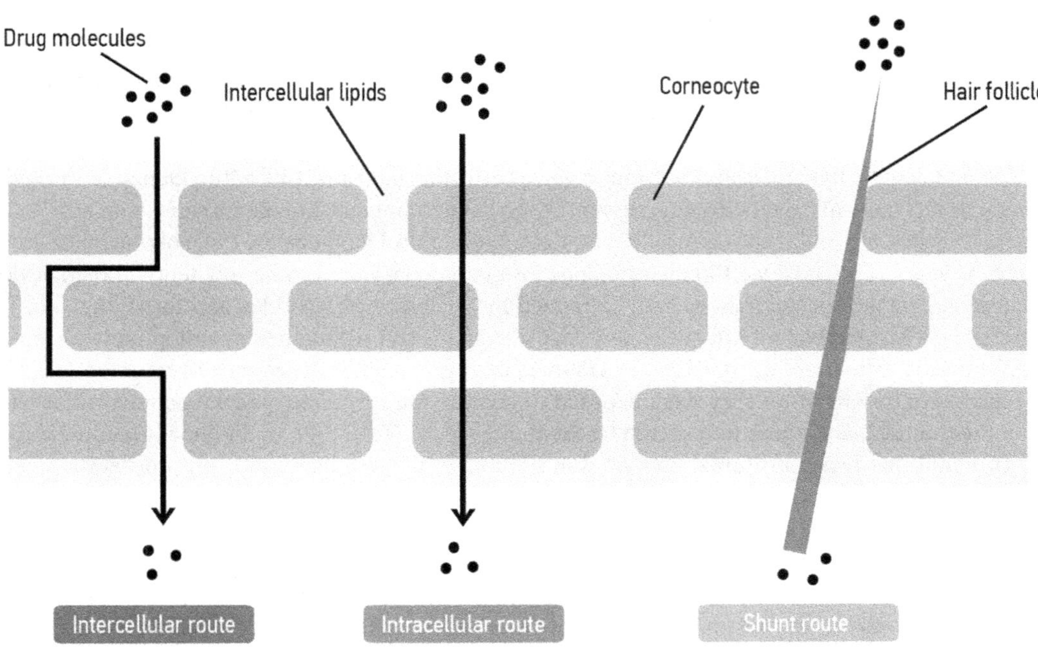

FIGURE 9.2 *Stratum corneum* pathways for product penetration [Reproduced with permission from Zoric et al. (2022) [22]].

The formulation of the cosmetic has a role not only in the molecular qualities of the specific component, such as its molecular weight and lipophilicity but also in the release of active compounds from cosmetic products and subsequent skin penetration. The formulations for cosmetics need to be physically, chemically and microbiologically stable in order to guarantee that the active ingredient will remain unchanged and will be delivered to the appropriate skin layers [24].

Extracting bioactive components from food waste can be a viable option for the cosmetic industry. It is evident that antioxidants are particularly abundant in food by-products. The functional advantage that may be offered by the extracts of food by-products to skin is their antioxidant activity. The antioxidant capabilities of several substances are used in the current generation of cosmetic treatments to combat wrinkles. By minimizing the biochemical effects of oxidation, such extracts and substances are becoming prospects for reducing the ageing process of skin. Generation of free radicals, the highly reactive molecules containing unpaired electrons, is the key factor in skin ageing. When the equilibrium between the synthesis and removal of reactive sulphur species (RSS), reactive nitrogen species (RNS) and reactive oxygen species (ROS) is disturbed, oxidative stress results [25]. Pro-oxidants, such as ROS, RNS and RSS, primarily target proteins, DNA and RNA molecules, carbohydrates and lipids. These reactive species may cause both internal damage (during metabolism) and external damage (via different oxidative stressors, like ultraviolet radiation). In fact, as we age, the generation of free radicals rises while endogenous defences decline, hastening the ageing process. Antioxidant defence is one of the functions of the skin, in which different antioxidant molecules and enzymes actively interact with the radicals to stop them from reaching their target. The antioxidant compounds possess the ability to bind free radicals, prevent many skin illnesses and reduce the ageing process of the skin. Antioxidants applied topically, such as vitamin C or E, polyphenolic compounds and coenzyme Q10, can improve the natural defence mechanism of skin, shielding it from ROS's damaging effects and oxidative damage. Phenolic chemicals have a role in the process of defence against harmful oxidative damage, preventing disorders linked to oxidative stress. This may lead to more obvious indications of youthful, healthy skin. Researchers have been captivated by the powers of the well-known bioflavonoids, quercetin, rutin and hesperidin, in preventing and reversing wrinkles, lessening the onset of age spots and combating spider veins [26]. Erythema, acne, psoriasis, cutaneous vasculitis, irritating contact dermatitis, cancer, photoaging and allergy are all skin illnesses that have been linked to reactive oxygen species and UV irradiation. Redness, itching, swelling and blistering are all symptoms of dermatitis, a sort of skin inflammation that can take many forms [27]. The most powerful contact dermatitis inhibitors are phenolics and terpenoids. Many of these chemicals have anti-inflammatory effects and can alleviate skin inflammation and contact hypersensitivity through nonspecific mechanisms, like antioxidants, or through particular mechanisms, including suppression of mediators implicated in the immune response [28]. Due to a variety of special anti-aging characteristics, resveratrol has been the topic of much research in recent years. The most well-known polyphenol that is considered to be a potent antioxidant to enhance the texture of the skin as well as reduce the visibility of wrinkles and fortifying skin is resveratrol. This polyphenol affects nuclear factor kappa B (NF-B), matrix metalloproteinases and MAP kinases, which are all involved in UV-mediated photoaging. Resveratrol prevented HaCaT cell mortality in vitro after they were exposed to sodium nitroprusside, which is a donor of nitric oxide free radical, according to research by Bastianetto et al. 2010 [29]. In addition, Giardina et al. (2010) found that resveratrol treatment of skin fibroblasts in vitro increased cell proliferation and decreased collagenase activity in a dose-dependent manner [30]. Procyanidins are a class of biopolymers that are made up of chains of catechin or epicatechin monomer units. Proanthocyanidins are antioxidants that also help maintain collagen and elastin, two vital proteins in connective tissue. Due to the high concentration of polyunsaturated fatty acids in olive by-products, notably linoleic acid, they seem to be the most intriguing in terms of fatty acids. Ceramides make up the majority of the skin's lipid-enriched component (approximately 50% of its weight), together with cholesterol and free fatty acids, according to Schure et al. 1991 [31]. These substances specifically affect the lipid layer that makes up the skin's surface, preserving the skin's hydration, suppleness and elasticity as well as its protective barrier. In reality, ROS alters the membrane's phospholipids by peroxidizing their polyunsaturated fatty acids, causing cell damage and final cell death [32]. By acting as a permeability barrier, signalling keratinocytes to control epidermal homeostasis, inhibiting the lipids in sebum generated by sebaceous glands and encouraging the acidity of the stratum corneum, fatty acids serve a variety of roles in the epidermis. Additionally, as people age, their skin loses its ability to retain moisture and emollience. For this reason, using enhanced skin care products is crucial [33].

The cosmetics industry places a premium on finding cellulose and its derivative products in olive wastes. Owing to their molecular makeup, these compounds have the potential to serve as active ingredients in cosmetic formulations by enhancing the products' hydration, viscosity, emulsion, oil-holding capacity, texture, sensory qualities, oxidative stability and shelf life [34]. In contrast, squalene has been shown to provide a number of benefits to the skin tissues, including antioxidant qualities at the epidermal level against sun radiation and functioning as a biological filter of singlet oxygen [35]. In addition, squalene may function as a sink for highly lipophilic xenobiotics, facilitating the organism's removal of these substances. Other intriguing qualities of squalene encourage their usage as constituents in dermo-protective ointments and other cosmetic formulations as moisturizers or emollients, in addition to those already mentioned [36]. Caffeine, which is abundant in coffee by-products, is another intriguing substance for cosmetic goods. Due to its strong biological activity and capacity to penetrate the skin barrier, it is being utilized more frequently in cosmetics [37]. Bresciani et al. (2014) assert that coffee silverskin has a significant quantity of caffeine that is comparable to that found in coffee beans [38]. Through phosphodiesterase activity inhibition, this alkaloid promotes the breakdown of lipids during lipolysis and possesses strong antioxidant effects. Some claims for anti-cellulite products rest on the idea that the formulation's bioactive ingredient is absorbed by the skin and distributed throughout the dermal and subcutaneous layers, where it inhibits the growth of fat cells and aids in weight loss [37]. Caffeine also has anti-ageing effects, which is a bonus. The extracts from two of the most prevalent coffee species have been shown to improve facial skin strength, resilience and suppleness by protecting against UV damage [39]. While Coffea robusta contains a high concentration of caffeine and chlorogenic acid, which limit photodamage and minimize skin roughness, wrinkle development and the appearance of crow's feet, Coffea arabica seed oil greatly increases the creation of collagen and elastin [39]. As was already noted, minerals are another intriguing component found in food by-products. As is widely known, minerals are a crucial element of natural moisturizing factor (NMF), which generates the osmotic force that draws water to corneocytes and other stratum corneum cells [40]. According to Rodrigues et al., NMF mostly consists of amino acids (about 40%), mineral ions (about 18%), lactate (about 12%), sugars (about 8%), urea (about 7%) and water-soluble ions (about 8.5%) and is responsible for the hydration, rigidity and pH of the stratum corneum [41]. Due to different hormonal considerations, changes in skin function and ageing occur at varying speeds. The skin gradually loses elasticity as it ages, increasing its extensibility, making it more brittle and vulnerable to injury. In turn, this increases the chance of skin damage (such as lacerations, rips, ulcerations and bruises) and impairs wound healing. The advantages and disadvantages of oestrogen replacement therapy (ERT) have become key areas of study since the average woman in a developed country spends around one-third of her life following the start of menopause. ERT enhances skin collagen content while maintaining thickness, which slows the ageing process. With ERT, skin moisture content improves while levels of sebum, acid mucopolysaccharides, hyaluronic acid and perhaps barrier function of stratum corneum are maintained. One of the main causes of skin changes brought on by ageing is the gradual sparing, disorganization and atrophy of collagen. Numerous studies show that menopause causes oestrogen insufficiency, and research conducted over the past 60 years shows a strong relationship between skin thickness, oestrogen concentration and skin collagen. Topical oestrogen therapy might significantly enhance the look of the skin while preserving dermal collagen [42]. Dermal fibroblasts, epidermal keratinocytes and melanocytes, as well as the hair follicles, skin extensions and the sebaceous gland, are all significantly influenced by oestrogen, which has far-reaching physiological impacts on the skin [42]. Numerous studies have documented the variations experienced by postmenopausal women, providing a foundation for many of oestrogen's effects on human skin that are based on the alterations noticed in this demographic. It is worth noting that skin thickness varies during the menstrual cycle, being at its thinnest as the cycle begins when the levels of oestrogen and progesterone are the lowest and increasing as oestrogen levels rise [42]. Oestrogen enhances vascularization and has actions at many dermal tissue depths in skin tissue. Oestrogen compounds have been shown to alleviate the signs of skin ageing in perimenopausal women, including an increase in epidermal hydration, thickness as well as elasticity of skin and the reduction of wrinkles. These effects were achieved alongside an increase in collagen content and quality and a boost in vascularization [43]. Due to the fact that lycopene has double conjugated bonds, which allow it to absorb light at long wavelengths of the visible spectrum, it can be said that it may lessen the negative effects of UV light on skin and protect against the long-term consequences of sun exposure (cancer) [44]. In addition, earlier research conducted in vitro and in vivo demonstrated that lycopene plays a helpful function in the treatment of chronic illnesses such as cardiovascular disease, atherosclerosis, cancer and neurological disorders [45]. With the knowledge that the formation of free radicals causes damage to DNA, collagen and elastin—all of which are essential for the elasticity of the skin as well as its natural process of renewal—future topical formulations can be developed through the incorporation of tomato by-products [43].

Although the cosmetic business is rich in fascinating chemicals derived from food by-products, few topical formulations, including these active components, have been created. However, by-products of coffee and Cannabis sativa are employed successfully in topical formulations. The silverskin from coffee is one instance. It has been extensively studied how coffee silverskin has a high antioxidant capacity, particularly because of chlorogenic acid and other phenolic components. Additionally, the amount of caffeine is considerable and comparable to that of coffee beans [38]. These substances are thought to offer in vivo defence against free radical harm. Coffee silverskin, like coffee beans, includes a variety of classes of health-promoting substances, including phenolics, diterpenes, xanthines and vitamin precursors [46]. In several monolayers of skin cell lines, Rodrigues et al. assessed the cell viability and cytotoxicity of coffee silverskin. Cytotoxicity was not noticed. Additionally, the beauty industry created a variety of in vitro three-dimensional model tests, such as the Human Corneal Epithelial Model (SkinEthicTM HCE) and the Reconstructed Human Epidermis Test (EpiSkinTM), to evaluate possible skin or eye irritants. These tests were conducted on three distinct coffee silverskin extracts by the same study team. Additionally, the histopathology of the animals following extract application was examined. The in vitro outcomes showed that extracts were not led as irritants, and the histological studies showed that they had no impact on the structural integrity of either model. After the epidermal test, the quantities of caffeine, 5-hydroxymethyl furfural and chlorogenic acid were measured, demonstrating that coffee beans contain comparable levels of caffeine. This implies that this novel extract may definitely be utilized, for instance, as a cellulitis preventative. The in vivo test using the most promising extract (hydroalcoholic) revealed that they may be regarded as safe for topical application with regard to irritating effects since no skin irritation was seen. Additionally, Rodrigues et al. used this coffee waste in other cosmetic formulations, including a body lotion and a hand cream. The in vitro results were perfect since neither keratinocytes nor fibroblasts showed any signs of toxicity. Additionally, the in vivo tests showed that both formulations were secure. Sensorial tests were sent out to participants and indicated the level of overall satisfaction with both goods. The same research team recently created a face formulation with coffee silverskin and an enhanced version of the cream with hyaluronic acid [47]. Over the course of 28 days, participants (n = 20 for each formulation) applied facial formulations twice daily. Using approved instruments (Corneometer and Cutometer), the effect on skin moisture and viscoelastic characteristics was examined. At time 0 and after 28 days, the depth, roughness, cavity volume and Visioface pictures of the wrinkles were analyzed. Volunteers were questioned about perceived effectiveness. The results showed that after eight hours, roughly 20% of the coffee silverskin extract penetrated the skin of the pig's ear. In none of the formulations was there any evidence of cytotoxicity. Both formulations showed significant improvements in skin hydration and viscoelastic characteristics, with no differences between them. For both formulations, there were no variations in the amount of cavities, roughness or wrinkle depth. This makes coffee silverskin a useful component for cosmetic treatments meant to boost the moisture and firmness of the skin. Almeida et al. assessed the topical applications of *C. Sativa* leaf ethanol/water extracts used in a skin care product [48, 49]. A strong 280 nm absorption band indicates promise for topical use of chestnut leaf extract in the prevention of UV radiation-induced skin damage. Finally, the polyphenol family of antioxidants includes phytonutrients called oligomeric proanthocyanidins, as has been previously documented. Following their inclusion in particular topical cream or lotion formulations, Hughes-Formella et al. showed that oligomeric proanthocyanidins from grapes (extracted from Vitis vinifera seeds) have anti-inflammatory and skin hydration activities in the skin of human volunteers [50].

9.2.3 Food industry

However, the influence of bioactive compounds on microbial cells can be clarified through the attack of the phospholipid bilayer of the cell membrane, disruption of enzyme systems, compromise of the genetic material of bacteria and oxidation of unsaturated fatty acids leading to the formation of fatty acid hydroperoxides. The precise mechanisms of antimicrobial action of bioactive compounds have not been fully described [51]. Several reports have established a connection between plant-based bioactive chemicals and the suppression of bacteria, moulds and fungi. Bioactive compounds have been shown to have antibacterial effects against a variety of bacteria, including Aerobacter hydrophila, Bacillus sp., Clostridium jejuni, Clostridium perfringens, Escherichia coli, Listeria monocytogenes, Pseudomonas sp., Salmonella sp., Shigella sp. and Staphylococcus aureus [51]. According to reports, bioactive substances can inhibit the growth of moulds such as Aspergillus flavus, Aspergillus niger and Aspergillus parasiticus. Additionally, it is known that some fungi, including Aspergillus niger, Aspergillus flavus, Aspergillus parasiticus and spergillus and Penicillium species, are inhibited by bioactive substances [51]. In general, bioactive substances prevent bacteria, moulds and fungi from growing and acting in ways that are harmful to human health and food quality. As a result, bioactive substances made from waste can be used in food items to increase their shelf life and reduce the amount of nutrients lost while being transported, processed, stored and cooked. The three main groups—chlorophyll, flavonoids and carotenoids—among the numerous bioactive substances obtained from plants that contribute to colour are described in this section. Due to its impact on the sensory quality of food, colour is a crucial aspect of food items. Customers' preferences have recently shifted from artificial colours to natural colours as a result of the negative health effects associated with artificial colours [52]. In several investigations, naturally coloured chemicals have been isolated and extracted from plants in an effort to be used as a food colouring. However, it should be emphasized that, compared to artificial colouring agents, natural colouring pigments are more expensive and less stable, and further study is needed to lower their price and increase their stability.

When weak acids, oxygen or light are present, the two major kinds of chlorophyll can break down into other types. Numerous plants have large amounts of chlorophyll, which is crucial to photosynthesis. It emits a green tint as it preferentially absorbs light from the visible red and blue spectrums [53]. Chlorophyll has been extracted and separated for use in foods using a variety of solvents and techniques. When used as a colouring additive in food goods, chlorophyll has several drawbacks. It is not water-soluble, the exact amount of colour it contains is unknown and it is prone to instability depending on the pH of the food item to which it is added. It is also less stable and more costly than artificial colourings. Therefore, to increase its stability as a food colouring component, chlorophyll has to undergo chemical alteration where the magnesium centre is swapped out for a copper ion [53]. Additionally, it is possible to saponify oleoresin to make it water-soluble. This water-soluble component is known as chlorophyllin. It is noteworthy that different nations employ different amounts of chlorophyll in food items. For instance, while it is prohibited in the US, it is permissible in the EU and other places [54].

The most prevalent and extensively dispersed class of plant phenolic chemicals, flavonoids, are a key element in plant colours. Anthocyanins, anthoxanthins and betalains are examples of flavonoid pigments. There are 19 different forms of anthocyanins, but the six most common are pelargonidin, cyanidin, peonidin, delphinidin, petunidin and malvidin. The hues of red, purple and blue are shown by anthocyanins. The more hydroxyl groups there are on their B-ring, the bluer the hue is. pH can also have an impact on colour. Anthocyanins accentuate the red hue in an acidic environment, but they may also become green in an alkaline environment [55].

The cream or white colour of cauliflower, white potatoes and turnips is due to anthoxanthins, which are made up of flavones, flavonols and flavanones. Anthoxanthins can be whiter in acidic environments, but in alkaline water, they could take on an unfavourable yellow hue. They even turn blue-black or red-brown when heated excessively or when there is iron or copper present [55, 56]. The reddish-purple betacyanins and the water-soluble yellow betaxanthins are the two categories of pigments that make up betalains. While betaxanthins are conjugates of amines or amino acids and betalamic acid, betacyanidins are conjugates of cyclo-DOPA and betalamic acid. Betalains may transform the purple-red into a brighter red hue in acidic circumstances, but in an alkaline medium, the red colour can turn yellow. Carotenoids are lipid-soluble pigments with strong red, yellow, or orange hues [54]. Lycopenes, xanthophylls and carotenes are examples of carotenoids. Carrots and winter squash are reddish-orange in colour because of carotenes. Alpha, beta and gamma carotenes make up carotenes; beta-carotenes are the most well-known and have been used to colour dairy products, which can have a high fat content. As a result, cheese and margarine frequently contain beta-carotene. Orange is the colour of lycopene. However, due to its high cost as a pigment and high susceptibility to oxidative degradation, it is scarcely ever utilized as a colourant [54]. The xanthophylls are what give pigment a bright yellow hue. Because carotenoids are heat- and light-sensitive, adding them to food can change their amount. Meat and meat products are prone to lipid oxidation because they contain significant quantities of lipids, ranging from 4.5% to 11%. It should be noted that lipid levels vary depending on the type of meat or the animal parts utilized. In order to prevent microbial growth and slow down the enzyme activities, that is, to increase the shelf life and/or quality of meat and meat products, bioactive substances originating from plant materials as well as waste have been used in meat and meat products [57]. Meat and its products may be exposed to light and air while being stored or processed, in addition to being impacted by bacteria. By chelating iron, the most active catalyst for oxidative rancidity in meat, the addition of bioactive substances with powerful antioxidants can reduce lipid oxidation [58]. Furthermore, bioactive substances can snare superoxide, hydroxyl and peroxyl radicals as well as inhibit and end free radical chain reactions, which stops the oxidation of meat and meat product lipids. Additionally, the addition of bioactive substances can reduce the development of bacteria by targeting the phospholipid bilayer of the cell membrane or upsetting the balance of the enzyme systems [57]. Additionally, bioactive substances (catechins) have been discovered to enhance and protect the colour of meat products, such as beef patties packed under aerobic circumstances during refrigeration for nine days or minced beef patties packed under aerobic conditions during refrigeration for seven days [59].

Waste produced by the food and agricultural industries contains a variety of bioactive compounds and is a fantastic source that might be used in meat and meat products to enhance their quality and increase shelf life. For use in meat and meat products, waste extracts have been used in several research. In order to enhance quality, reduce lipid oxidation and limit microbial development in sausages, bacon, meat patties and meat juice, for instance, extracts from grape seeds, cherry and blackcurrant leaves, Ginkgo and olive leaves have been successfully used [57].

9.2.4 Energy industry

Biofuels and other forms of Renew. Energ. have become crucial due to the increase in the human population. The food and agroforestry industries create five billion tonnes of biomass waste. There is great promise for producing bioactive substances that can be used as biofuels to cut down on biomass waste. In biorefineries, ethanol and vegetable oil are frequently generated as biofuels [60]. A pulse electric field treatment was used by Gorte et al. (2020) to extract lipids from live oleaginous yeast cells [61]. Although the extraction method is costly and ineffective, single-cell oils or microbial oils that are extracted from yeasts, fungi, microalgae and bacteria can be used as alternative fuels. These are still the most pressing issues that need to be resolved. Potential Renew. Energ. source *C. Camphora* has been examined for its mass production as well as its volatile components (camphor, eucalyptol, limonene and -pinene) [62]. Bioenergy and biofuels like ethanol, methanol and butanol can also be produced from sugar-based waste (sugar cane, sugar beets), animal waste (cow, swine, poultry), food industry waste, starch-based wastes (corn), lignocellulosic waste (switchgrass, micanthus, corn stover, corn fibre) and glycerine [63]. Microorganisms are the power source in microbial fuel cells (MFCs). Some of the secondary metabolites, such as epigallocatechin-3-gallate, gallic acid, gallocatechin and anthocyanin, have been demonstrated to have electron-shunting properties, leading to increased power density. Power generation in MFCs is enhanced by the incorporation of fungal and algal metabolites [64]. Dimers are created by the processing of -pinene condensation and are a great choice for a renewable, high-energy-density jet fuel that can also be used as diesel. An economical replacement for fossil fuels, the effects of a dual biofuel mix with various proportions of turpentine oil and jatropha biodiesel were examined in a single-cylinder diesel engine. Nitrous oxide, hydrocarbon and carbon monoxide emissions all decreased by 13.04%, 17.5% and 4.21%, respectively, but CO_2 emissions rose by 11.04% [65].

9.2.5 Bioremediation sector

Coagulants, biofilms, bioactive extracts and so on are only some of the many forms in which bioactive chemicals have been put to use in the bioremediation industry. Botryococcus braunii's α-linolenic, oleic and palmitic acids and Stoechospermum marginatum's diethyl phthalate and many more are only a few examples of the bioactive compounds isolated from different sources that have demonstrated promising efficacy against harmful algal blooms (HAB) [66]. Similar to this, Sargassum's bioaccumulation of nutrients and metals (Cd, Cu, Zn, Pb, Cr) has three advantages: it reduces eutrophication and coastal metal pollution, sequesters metals and produces useful bioactive compounds that can be used in the fertilizer, cosmetics and pharmaceutical industries [67]. Due to the presence of substances like monocerin, a fungal endophyte known as Drechslera (strain 678) has been shown by d'Errico et al. (2020) to be capable of performing double duty as a biopesticide and bioremediation of the soil contaminant methyl tert-butyl ether, which is typically used as a gasoline additive [68].

Furthermore, tannins have found widespread application in sewage treatment facilities. Suspended particles have been removed and flocculated using tannin-based coagulants. Acacia mearnsii condensed tannins and tannic acid have been utilized to treat wastewater contaminated with cationic and anionic dyes [69]. The application of tannin cryogels and wattle tannins in the removal of heavy metals and methylene blue, respectively, from polluted water, has also been investigated by Das et al. (2020) [70].

There is a lot of promise for bioremediation using biofilms produced by intertidal and marine microorganisms. It is encouraging for the management and clean-up of oil spills that Mugge et al. (2021) evaluated the changes in bacterial populations and biofilm compositions in surface and deep-sea water when exposed to crude oil or chemical dispersants [71]. Bacillus subtilis CN2 lipopeptides have shown intriguing capabilities regarding the breakdown of polycyclic aromatic hydrocarbons and the recovery of motor oil from polluted soil. Consequently, bioactive substances show promise in fields such as wastewater treatment and hydrocarbon breakdown [72].

9.2.6 Other industries

Polymers and biomaterials, textile dyes and processing, leather manufacturing, fragrance and cosmetics are just some of the many chemical sectors that might benefit from bioactive chemicals. For example, the oil and soap industries have long relied on bioactive compounds like oils and fatty acids, but the toxicity and unsustainable nature of traditional chemicals, processes or end-products has recently accelerated the use of bio-based substitutes in novel materials, catalysts and certain raw materials [73].

For prospective use in biomedical and cosmetic applications, Spiridon et al. (2020) created a biomaterial composed of cellulose, collagen and polyurethane that allows for the slow release of antioxidants such as tannin and lipoic acid. Aloe vera agro-wastes were also used to encapsulate antioxidants by integrating them into polyethylene oxide electro spun nanofibers in a similar fashion [74]. By combining biopolymers (starch, chitosan, gluten) with bioactive ingredients, these technologies have made it possible to create biodegradable and even edible food-packaging films with improved antibacterial, antioxidant and mechanical capabilities (essential oils, polyphenols, carotenoids) [75]. Utilizing macromolecules derived from tannins, other intriguing features have been attained for textiles, including UV light protection, antimicrobial capabilities and flame retardancy. Dyes have also been created, including the run dye from the roots of Rubia tinctorum L. and naturally derived anthraquinone dyes from the stem tissues of Miscanthus Sinensis Andersson [76].

The perfume and cosmetic industries favour a variety of essential oils, including lavender, carvone, linalool, limonene, citronellol and eucalyptus. They have positive benefits on the skin and serve as active components and preservatives in addition to aroma [77]. Due to their anti-ageing, anti-acne, antimicrobial, skin glow-enhancing, moisture-retaining, UV protection and anti-allergic properties, seaweed and microalgae have been reported as excellent sources for cosmeceuticals. This is because they contain phlorotannins, polysaccharides (laminarin, carrageenan, etc.), astaxanthin and several bioactive peptides [78].

Due to the numerous uses, which vary from the production of food oils and hair care products to soaps and grease, the industrial extraction of oils (palm, coconut and castor oils) is another significant area of the chemical industry [79]. Additionally, a number of Malaysian petrochemical companies are investigating the potential use of palm oil and glycerol derived from vegetable oil as feedstocks in the production of lubricants, thereby moving resources from fossil fuels to bioactive chemicals. In a similar vein, castor oils are used to create biodegradable polyesters, lubricants and paints [79, 80].

Tannins, on the other hand, are highly prevalent and significant industry bioactive substances that have been utilized in the wood (adhesive and preservation) and leather processing industries for centuries. There have also been several new attempts at utilizing tannins as a sustainable alternative in 3D printing. With proper steps adopted for effective extraction and maintenance of biodiversity, bioactive chemicals hold great promise for establishing a long-term chemical business.

9.3 CHALLENGES AND FUTURE TRENDS

The major challenges to the commercial use of bioactive compounds from food waste include the extraction and separation of bioactive compounds from the food waste matrix. Conventional extraction methods are helpful, but they take a long time and are inefficient. Nonconventional extraction methods have been created to get around the aforementioned restrictions. However, they also have certain drawbacks. The issues addressed regarding UAE, MAE and SFE are the susceptibility of thermosensitive compounds, non-uniformity of large-scale extraction, high operating costs and CO_2 consumption resulting in high-value compounds. Since astaxanthin and phycobiliproteins in microalgae require costly extraction and purification techniques, they are among the most valuable chemicals in the field of SFE. As has been mentioned, there are a number of other restrictions that come with using each extraction technique.

Variable, stability and activity loss of bioactive chemicals, particularly in foods, is a significant barrier to their broad usage because the majority of studies that confirm their beneficial characteristics are conducted in carefully controlled environments. While researching the nutritional and therapeutic advantages of bioactive substances, variation across people is also an important issue that must be taken into account. The effects of these substances on a population may vary due to individual differences in absorption and metabolism, as well as those associated with differences in age, gender and lifestyle. From an ecological standpoint, addressing the rising need for bioactive chemicals would place a great deal of strain on biodiversity, land and marine resources, which might endanger the existence of extremely rare species.

Problems with bioanalytical methods of characterization, such as interference, the need for clean-up during sample preparation, limited sensitivity, inaccurate results and unreliable techniques, are often disregarded. When downstream processes make up 50–80% of the production value, other factors such as bioavailability, bioaccessibility, safe and "green" manufacturing procedures, safety and toxicity must be taken into account [81].

Demand for healthier, natural, immune-boosting and bio-fortified foods, novel antibiotics and pharmaceuticals, bio-based raw materials for various processes and biomaterials has increased as a result of the recent pandemic and the emergence of non-communicable diseases like cancer, obesity, diabetes, etc., over the last few decades. According to a Grand View Research, Inc. (2016) estimate, the market for bioactive components will reach US$ 51.71 billion worldwide by 2024, with functional foods and drinks accounting for 25% of that market and derived from plants and marine species [80]. In order to keep up with consumer demand, scientists must investigate more efficient and environmentally friendly strategies for screening, extracting, characterizing, processing and commercializing high-quality bioactive chemicals. Fu et al. (2019) proposed switching from traditional chromatography methods to multi-targeted ways to simultaneously screen several bioactive chemicals using biosensor and microfluidic chip-based technologies [82]. The discovery of these compounds is constrained when screening is limited to in vitro tests because, in certain situations, the mechanism of action of bioactive compounds is better known via in vivo screening approaches [83]. It could be possible to screen such molecules using innovative pathways by efficiently developing in vivo tests, particularly for antibiotic efficacy. To reduce costs and enable scalability, it is necessary to further investigate techniques to enhance selectivity and yield of extraction, such as modelling solvent-compound interactions and affinities and optimizing physical parameters in the case of nonconventional methods.

In order to combat the exploitation of scarce and finite marine resources for commercial production, metabolic engineering is a viable method to enhance the microbial synthesis of bioactive chemicals like terpenoids, omega-3 PUFAs and so on. Other methods of locating the best sources from agricultural waste and by-products of the food processing industry can aid in adopting sustainable waste recycling and high-value product recovery techniques, supporting a circular economy.

When delivering bioactive compounds into functional food products, nanoencapsulation techniques offer a strong chance of maintaining bioavailability, improving stability and enabling the regulated release of bioactive compounds. Several strategies include encapsulation inside spontaneously assembled structures and the use of biopolymer sheets in food packaging. Making them economically viable alternatives to traditional treatments is the difficult part.

9.4 CONCLUSION

Due to the environmental consequences caused by its disposal, food waste is increasingly seen as an issue that has to be treated, minimized and prevented. One-third of the world's edible component of food produced for human use is lost or wasted, as reported by the FAO. The outlook for developing technologies that enable the recovery, recycling and sustainable manufacture of high-added-value ingredients is difficult. These residual materials, in particular, might be useful for the cosmetics industry since bioactive chemicals have real-world applications. Due to their diverse compositions, food by-products present a chance to acquire fresh active ingredients for cosmetics and pharmaceuticals, making use of the benefits of the vast quantities that are often thrown away at competitive rates and offering noteworthy possible uses for the waste. Even though their chemical compositions indicate their significant potential as a source of natural substances, such as antioxidants, fatty acids, minerals or caffeine, which may be employed as functional components, the food by-products previously described are still underestimated. Even though there is a growing body of research studies on clinical anti-ageing, for instance, more studies are still required to show the effectiveness of these various components in topical formulations.

The energy, chemical and food-packaging industries are just a few examples of those that might benefit from bioactive compounds derived from food waste. In the recent years, which are covered in this chapter, there has been clear evidence in the published literature of a rise in the application of bioactive compounds in the abovementioned industrial sectors. However, a significant amount of research effort must still be put into determining the best extraction method, which will depend on a variety of aspects, such as the sample, the amount of energy and money that will be required, the impact on the environment and the extracted material's quality.

ACKNOWLEDGEMENT

The study is supported by the Indian National Academy of Engineering (INAE/121/AKF/22), Gurgaon, India. The authors are solely responsible for the opinions, results and conclusions expressed in this study; INAE's viewpoints are not necessarily reflected in any of these aspects.

REFERENCES

1. P. Duarah, D. Haldar, M.K. Purkait, Technological advancement in the synthesis and applications of lignin-based nanoparticles derived from agro-industrial waste residues: A review, *Int. J. Biol. Macromol.* 163 (2020) 1828–1843.
2. S. Sinha, P. Tripathi, Trends and challenges in valorisation of food waste in developing economies: A case study of India, Case Studies in Chemical and Environmental *Engineering* 4 (2021) 100162.
3. S. Ben-Othman, I. Jõudu, R. Bhat, Bioactives from agri-food wastes: Present insights and future challenges, *Molecules* 25(3) (2020) 510.
4. D. Haldar, P. Duarah, M.K. Purkait, Chapter 16 - Progress in the synthesis and applications of polymeric nanomaterials derived from waste lignocellulosic biomass, in: D. Giannakoudakis, L. Meili, I. Anastopoulos (Eds.), *Advanced Materials for Sustainable Environmental Remediation, Elsevier2022*, pp. 419–433.
5. R. Theagarajan, L. Malur Narayanaswamy, S. Dutta, J.A. Moses, A. Chinnaswamy, Valorisation of grape pomace (cv. Muscat) for development of functional cookies, *Int. J. Food Sci.* 54(4) (2019) 1299–1305.
6. B. Debnath, D. Haldar, M.K. Purkait, Environmental remediation by tea waste and its derivative products: A review on present status and technological advancements, *Chemosphere* 300 (2022) 134480.
7. B. Debnath, D. Haldar, M.K. Purkait, A critical review on the techniques used for the synthesis and applications of crystalline cellulose derived from agricultural wastes and forest residues, *Carbohydr. Polym.* 273 (2021) 118537.
8. A. Mohanty, M. Mankoti, P.R. Rout, S.S. Meena, S. Dewan, B. Kalia, S. Varjani, J.W.C. Wong, J.R. Banu, Sustainable utilization of food waste for bioenergy production: A step towards circular bioeconomy, *Int. J. Food Microbiol.* 365 (2022) 109538.
9. C. Espro, E. Paone, F. Mauriello, R. Gotti, E. Uliassi, M.L. Bolognesi, D. Rodríguez-Padrón, R. Luque, Sustainable production of pharmaceutical, nutraceutical and bioactive compounds from biomass and waste, *Chem Soc Rev* 50(20) (2021) 11191–11207.
10. B. Debnath, D. Haldar, M.K. Purkait, Potential and sustainable utilization of tea waste: A review on present status and future trends, *Journal of Environmental Chemical Engineering* 9(5) (2021) 106179.
11. T.N. Baite, B. Mandal, M.K. Purkait, Ultrasound assisted extraction of gallic acid from Ficus auriculata leaves using green solvent, *Food Bioprod. Process.* 128 (2021) 1–11.
12. V.L. Dhadge, M. Changmai, M.K. Purkait, Purification of catechins from Camellia sinensis using membrane cell, *Food Bioprod. Process.* 117 (2019) 203–212.
13. P. Duarah, D. Haldar, A.K. Patel, C.-D. Dong, R.R. Singhania, M.K. Purkait, A review on global perspectives of sustainable development in bioenergy generation, *Bioresour. Technol.* 348 (2022) 126791.
14. P. Duarah, A. Bhattacharjee, P. Mondal, M.K. Purkait, Green Synthesized Carbon and Metallic Nanomaterials for Biofuel Production: Effect of Operating Parameters, in: M. Srivastava, M.A. Malik, P.K. Mishra (Eds.), *Green Nano Solution for Bioenergy Production Enhancement*, Springer Nature, 2022, pp. 105–126.

15. P. Mondal, A. Anweshan, M.K. Purkait, Green synthesis and environmental application of iron-based nanomaterials and nanocomposite: A review, *Chemosphere* 259 (2020) 127509.
16. V. Musolino, M. Gliozzi, E. Bombardelli, S. Nucera, C. Carresi, J. Maiuolo, R. Mollace, S. Paone, F. Bosco, F. Scarano, M. Scicchitano, R. Macrì, S. Ruga, M.C. Zito, E. Palma, S. Gratteri, M. Ragusa, M. Volterrani, M. Fini, V. Mollace, The synergistic effect of Citrus bergamia and Cynara cardunculus extracts on vascular inflammation and oxidative stress in non-alcoholic fatty liver disease, *J. Tradit. Complement. Med.* 10(3) (2020) 268–274.
17. B. Choi, E. Koh, Spent coffee as a rich source of antioxidative compounds, *Food Sci. Biotechnol.* 26(4) (2017) 921–927.
18. Refined Spruce Resin to Treat Chronic Wounds: Rebirth of an Old Folkloristic Therapy, *Adv. Wound Care* 5(5) (2016) 198–207.
19. EU, *Bio-based Products –from Idea to Market*, 2019.
20. Y. Shi, P.C.J. Kamer, D.J. Cole-Hamilton, Synthesis of pharmaceutical drugs from cardanol derived from cashew nut shell liquid, *Green Chem.* 21(5) (2019) 1043–1053.
21. M. Cerone, E. Uliassi, F. Prati, G.U. Ebiloma, L. Lemgruber, C. Bergamini, D.G. Watson, T. de A. M. Ferreira, G.S.H. Roth Cardoso, L.A. Soares Romeiro, H.P. de Koning, M.L. Bolognesi, Discovery of sustainable drugs for neglected tropical diseases: cashew nut shell liquid (cnsl)-based hybrids target mitochondrial function and atp production in Trypanosoma brucei, *ChemMedChem* 14(6) (2019) 621–635.
22. M. Zorić, M. Banožić, K. Aladić, S. Vladimir-Knežević, S. Jokić, Supercritical CO2 extracts in cosmetic industry: Current status and future perspectives, *Sustain. Chem. Pharm.* 27 (2022) 100688.
23. A.S. Ribeiro, M. Estanqueiro, M.B. Oliveira, J.M. Sousa Lobo, Main benefits and applicability of plant extracts in skin care products, *Cosmetics* 2(2) (2015) 48–65.
24. O. Zillich, U. Schweiggert-Weisz, P. Eisner, M. Kerscher, Polyphenols as active ingredients for cosmetic products, *Int. J. Cosmet. Sci.* 37(5) (2015) 455–464.
25. B.D. Craft, A.L. Kerrihard, R. Amarowicz, R.B. Pegg, Phenol-Based Antioxidants and the In Vitro Methods Used for Their Assessment, *Compr. Rev. Food Sci. Food Saf.* 11(2) (2012) 148–173.
26. W. Zhu, J. Gao, The Use of Botanical Extracts as Topical Skin-Lightening Agents for the Improvement of Skin Pigmentation Disorders, *J. Investig. Dermatol. Symp. Proc.* 13(1) (2008) 20–24.
27. L.J. Rios, E. Bas, C.M. Recio, Effects of natural products on contact dermatitis, *Current Medicinal Chemistry - Anti-Inflammatory & Anti-Allergy Agents* 4(1) (2005) 65–80.
28. J.A. Nichols, S.K. Katiyar, Skin photoprotection by natural polyphenols: Anti-inflammatory, antioxidant and DNA repair mechanisms, *Arch. Dermatol.* 302(2) (2010) 71–83.
29. S. Bastianetto, Y. Dumont, A. Duranton, F. Vercauteren, L. Breton, R. Quirion, Protective action of resveratrol in human skin: Possible involvement of specific receptor binding sites, *PloS One* 5(9) (2010) e12935.
30. S. Giardina, A. Michelotti, G. Zavattini, S. Finzi, C. Ghisalberti, F. Marzatico, [Efficacy study in vitro: Assessment of the properties of resveratrol and resveratrol + N-acetyl-cysteine on proliferation and inhibition of collagen activity], *Minerva Ginecologica* 62(3) (2010) 195–201.
31. N.Y. Schurer, P.M. Elias, The biochemistry and function of stratum corneum lipids, *Adv Lipid Res.* 24 (1991) 27–56.
32. P. Viola, M. Viola, Virgin olive oil as a fundamental nutritional component and skin protector, *Clin. Dermatol.* 27(2) (2009) 159–165.
33. F. Rodrigues, F.B. Pimentel, M.B.P. Oliveira, Olive by-products: Challenge application in cosmetic industry, *Ind Crops Prod.* 70 (2015) 116–124.
34. F. Rodrigues, I. Almeida, B. Sarmento, M.H. Amaral, M.B.P. Oliveira, Study of the isoflavone content of different extracts of Medicago spp. as potential active ingredient, *Ind Crops Prod.* 57 (2014) 110–115.
35. V. Micol, N. Caturla, L. Pérez-Fons, V. Más, L. Pérez, A. Estepa, The olive leaf extract exhibits antiviral activity against viral haemorrhagic septicaemia rhabdovirus (VHSV), *Antiviral Research* 66(2) (2005) 129–136.
36. S. Stavroulias, C. Panayiotou, Determination of optimum conditions for the extraction of squalene from olive pomace with supercritical CO2, *Chem Biochem Eng Q* 19(4) (2005) 373–381.
37. A. Herman, A.P. Herman, Caffeine's mechanisms of action and its cosmetic use, *Skin Pharmacol Physiol* 26(1) (2013) 8–14.
38. L. Bresciani, L. Calani, R. Bruni, F. Brighenti, D. Del Rio, Phenolic composition, caffeine content and antioxidant capacity of coffee silverskin, *Food Res. Int.* 61 (2014) 196–201.
39. M. Del Carmen Velazquez Pereda, G. De Campos Dieamant, S. Eberlin, C. Nogueira, D. Colombi, L.C. Di Stasi, M.L. De Souza Queiroz, Effect of green Coffea arabica L. seed oil on extracellular matrix components and water-channel expression in in vitro and ex vivo human skin models, *J. Cosmet. Dermatol.* 8(1) (2009) 56–62.

40. N. Dragicevic, H.I. Maibach, *Percutaneous Penetration Enhancers Physical Methods in Penetration Enhancement*, Springer, 2017.
41. A.V. Rawlings, P.J. Matts, Stratum corneum moisturization at the molecular level: An update in relation to the dry skin cycle, *J. Invest. Dermatol.* 124(6) (2005) 1099–1110.
42. S. Stevenson, J. Thornton, Effect of estrogens on skin aging and the potential role of SERMs, *Clin Interv Aging* 2(3) (2007) 283–97.
43. F. Rodrigues, A.F. Vinha, M.A. Nunes, M.B.P. Oliveira, *Potential Application of Bioactive Compounds from Agroindustrial Waste in the Cosmetic Industry, Utilisation of Bioactive Compounds from Agricultural and Food Waste*, CRC Press, 2017, pp. 358–382.
44. T. Tanaka, M. Shnimizu, H. Moriwaki, Cancer Chemoprevention by Carotenoids, *Molecules* 17(3) (2012) 3202–3242.
45. A. Prema, U. Janakiraman, T. Manivasagam, A. Justin Thenmozhi, Neuroprotective effect of lycopene against MPTP induced experimental Parkinson's disease in mice, *Neurosci. Lett.* 599 (2015) 12–19.
46. R.C. Alves, S. Casal, M.R. Alves, M.B. Oliveira, Discrimination between arabica and robusta coffee species on the basis of their tocopherol profiles, *Food Chem.* 114(1) (2009) 295–299.
47. F. Rodrigues, R. Matias, M. Ferreira, M.H. Amaral, M.B.P.P. Oliveira, In vitro and in vivo comparative study of cosmetic ingredients Coffee silverskin and hyaluronic acid, *Exp. Dermatol.* 25(7) (2016) 572–574.
48. I.F. Almeida, P.C. Costa, M.F. Bahia, Evaluation of functional stability and batch-to-batch reproducibility of a Castanea sativa leaf extract with antioxidant activity, *AAPS PharmSciTech* 11(1) (2010) 120–5.
49. I.F. Almeida, J. Maleckova, R. Saffi, H. Monteiro, F. Góios, M.H. Amaral, P.C. Costa, J. Garrido, P. Silva, N. Pestana, M.F. Bahia, Characterization of an antioxidant surfactant-free topical formulation containing Castanea sativa leaf extract, *Drug Dev. Ind. Pharm.* 41(1) (2015) 148–155.
50. B. Hughes-Formella, O. Wunderlich, R. Williams, Anti-inflammatory and skin-hydrating properties of a dietary supplement and topical formulations containing oligomeric proanthocyanidins, *Skin Pharmacol Physiol* 20(1) (2007) 43–9.
51. M. Tajkarimi, S.A. Ibrahim, D. Cliver, Antimicrobial herb and spice compounds in food, *Food control* 21(9) (2010) 1199–1218.
52. A. Aberoumand, A review article on edible pigments properties and sources as natural biocolorants in foodstuff and food industry, *World J Dairy Food Sci* 6(1) (2011) 71–78.
53. A. Hosikian, S. Lim, R. Halim, M.K. Danquah, Chlorophyll extraction from microalgae: A review on the process engineering aspects, *Int. J. Chem. Eng. Res.* 2010 (2010).
54. A. Mortensen, Carotenoids and other pigments as natural colorants, *Pure Appl. Chem.* 78(8) (2006) 1477–1491.
55. A.C. Brown, *Understanding Food: Principles and Preparation*, Cengage Learning, 2018.
56. B. Amy, Understanding Food & Principles and Preparation, US Am. *Thomson Wadsworth* 3 (2008) 27.
57. Q.V. Vuong, M.l.A. Atherton, *Utilisation of Bioactive Compounds Derived from Waste in the Food Industry, Utilisation of Bioactive Compounds from Agricultural and Food Waste*, CRC Press, 2017, pp. 342–357.
58. S. Tang, D. Sheehan, D.J. Buckley, P.A. Morrissey, J.P. Kerry, Anti-oxidant activity of added tea catechins on lipid oxidation of raw minced red meat, poultry and fish muscle, *Int. J. Food Sci.* 36(6) (2001) 685–692.
59. S. Tang, S. Ou, X. Huang, W. Li, J. Kerry, D. Buckley, Effects of added tea catechins on colour stability and lipid oxidation in minced beef patties held under aerobic and modified atmospheric packaging conditions, *J. Food Eng.* 77(2) (2006) 248–253.
60. P. Ferreira-Santos, E. Zanuso, Z. Genisheva, C.M. Rocha, J.A. Teixeira, Green and sustainable valorization of bioactive phenolic compounds from pinus by-products, *Molecules* 25(12) (2020) 2931.
61. O. Gorte, N. Nazarova, I. Papachristou, R. Wüstner, K. Leber, C. Syldatk, K. Ochsenreither, W. Frey, A. Silve, Pulsed electric field treatment promotes lipid extraction on fresh oleaginous yeast Saitozyma podzolica DSM 27192, *Front. Bioeng. Biotechnol.* 8 (2020) 575379.
62. J. Zhang, C. Wu, L. Gao, G. Du, X. Qin, Astragaloside IV derived from Astragalus membranaceus: A research review on the pharmacological effects, *Adv. Pharmacol.* 87 (2020) 89–112.
63. P.K. Swain, Utilisation of agriculture waste products for production of bio-fuels: A novel study, *Mater. Today: Proc.* 4(11) (2017) 11959–11967.
64. D. Nath, M.M. Ghangrekar, Plant secondary metabolites induced electron flux in microbial fuel cell: Investigation from laboratory-to-field scale, *Sci. Rep.* 10(1) (2020) 17185.
65. P. Dubey, R. Gupta, Influences of dual bio-fuel (Jatropha biodiesel and turpentine oil) on single cylinder variable compression ratio diesel engine, *Renew. Energ.* 115 (2018) 1294–1302.

66. S.E.A. Zerrifi, R. Mugani, E.M. Redouane, F. El Khalloufi, A. Campos, V. Vasconcelos, B. Oudra, Harmful Cyanobacterial Blooms (HCBs): Innovative green bioremediation process based on anti-cyanobacteria bioactive natural products, *Arch. Microbiol.* 203(1) (2021) 31–44.
67. S. Saldarriaga-Hernandez, G. Hernandez-Vargas, H.M.N. Iqbal, D. Barceló, R. Parra-Saldívar, Bioremediation potential of Sargassum sp. biomass to tackle pollution in coastal ecosystems: Circular economy approach, *Sci. Total Environ.* 715 (2020) 136978.
68. G. d'Errico, V. Aloj, G.R. Flematti, K. Sivasithamparam, C.M. Worth, N. Lombardi, A. Ritieni, R. Marra, M. Lorito, F. Vinale, Metabolites of a Drechslera sp. endophyte with potential as biocontrol and bioremediation agent, *Nat. Prod. Res.* 35(22) (2021) 4508–4516.
69. K. Grenda, J. Arnold, J.A. Gamelas, M.G. Rasteiro, Up-scaling of tannin-based coagulants for wastewater treatment: Performance in a water treatment plant, *Environ. Sci. Pollut. Res.* 27(2) (2020) 1202–1213.
70. A.K. Das, M.N. Islam, M.O. Faruk, M. Ashaduzzaman, R. Dungani, Review on tannins: Extraction processes, applications and possibilities, *S. Afr. J. Bot.* 135 (2020) 58–70.
71. R.L. Mugge, J.L. Salerno, L.J. Hamdan, Microbial functional responses in marine biofilms exposed to deepwater horizon spill contaminants, *Front. Microbiol.* 12 (2021) 636054.
72. F.A. Bezza, E.M.N. Chirwa, Production and applications of lipopeptide biosurfactant for bioremediation and oil recovery by Bacillus subtilis CN2, *Biochemical Engineering Journal* 101 (2015) 168–178.
73. S. Basak, A.S.M. Raja, S. Saxena, P.G. Patil, Tannin based polyphenolic bio-macromolecules: Creating a new era towards sustainable flame retardancy of polymers, *Polymer Degradation and Stability* 189 (2021) 109603.
74. I. Solaberrieta, A. Jiménez, I. Cacciotti, M.C. Garrigós, Encapsulation of bioactive compounds from aloe vera agrowastes in electrospun poly (ethylene oxide) nanofibers, *Polymers* 12(6) (2020) 1323.
75. G.F. Nogueira, R.A.d. Oliveira, J.I. Velasco, F.M. Fakhouri, Methods of incorporating plant-derived bioactive compounds into films made with agro-based polymers for application as food packaging: A brief review, *Polymers* 12(11) (2020) 2518.
76. I.A.P. Pinzon, R.A. Razal, R.C. Mendoza, R.B. Carpio, R.C.P. Eusebio, Parametric study on microwave-assisted extraction of runo (Miscanthus sinensis Andersson) dye and its application to paper and cotton fabric, *Biotechnol. Rep.* 28 (2020) e00556.
77. J.B. Sharmeen, F.M. Mahomoodally, G. Zengin, F. Maggi, Essential oils as natural sources of fragrance compounds for cosmetics and cosmeceuticals, *Molecules* 26(3) (2021) 666.
78. V. Jesumani, H. Du, M. Aslam, P. Pei, N. Huang, Potential use of seaweed bioactive compounds in skincare—A review, *Marine Drugs* 17(12) (2019) 688.
79. V.R. Patel, G.G. Dumancas, L.C.K. Viswanath, R. Maples, B.J.J. Subong, Castor oil: Properties, uses, and optimization of processing parameters in commercial production, Lipid insights 9 (2016) *LPI.* S40233.
80. S. Pai, A. Hebbar, S. Selvaraj, A critical look at challenges and future scopes of bioactive compounds and their incorporations in the food, energy, and pharmaceutical sector, *Environ. Sci. Pollut. Res.* 29(24) (2022) 35518–35541.
81. S. Cuellar-Bermudez, I. Aguilar-Hernandez, D. Cardenas-Chavez, N. Ornelas-Soto, M. Romero-Ogawa, R. Parra-Saldivar, Extraction and purification of high-value metabolites from microalgae: Essential lipids, astaxanthin and phycobiliproteins. *Microb Biotechnol* 8 (2): 190–209, 2015.
82. Y. Fu, J. Luo, J. Qin, M. Yang, Screening techniques for the identification of bioactive compounds in natural products, *J. Pharm. Biomed. Anal.* 168 (2019) 189–200.
83. C.S. Ahamefule, B.C. Ezeuduji, J.C. Ogbonna, A.N. Moneke, A.C. Ike, B. Wang, C. Jin, W. Fang, Marine bioactive compounds against Aspergillus fumigatus: Challenges and future prospects, *Antibiotics* 9(11) (2020) 813.

10 Food waste management and valorization policies of various countries

10.1 INTRODUCTION

Scientists regard food waste (FW) to be a critical problem that has to be better understood in order to create systems for effective food management. A United Nations research from 2015 estimates that by 2030, there will be 8.5 billion people on the planet. Even though we need to innovate to boost land productivity and establish new agricultural techniques, we also need to reduce FW, which accounts for one-third of all food produced [1]. Food management is known, to some degree and attempts are being made to innovate in this field through the development of procedures and the progress of technical innovations. Despite this, the amount of wasted food is growing at an alarming rate [2]. The current analysis indicates FW in the supply chain and during preparation and consumption. Managing wasted food might alleviate hunger [3]. To meet both the near-term and the long-term demands for food, novel approaches food supply chain management are required. In light of this, the innovation of food management and waste reduction is essential to the process of identifying problems and possibilities [4]. The global issue of wasted food is becoming an increasingly urgent one. Researchers have either conducted an in-depth analysis of the existing research or conducted a literature study in order to propose a variety of strategies, some of which are education-based, while others are structural and technology-based. Both Centobelli et al. (2016) and Shashi et al. (2018) have conducted reviews on the topics of FW and the cold food supply chain, respectively [5, 6].

Management of FW encourages a circular economy and has the potential to improve social, political and environmental conditions. Technology, according to the authors, may improve the sustainability of production and the use of waste as a feeder to satisfy dietary, individual and societal needs along the food chain [7]. Anaerobic digestion and composting are two examples of technologies that may process leftover food to provide inputs (like organic fertilizer) for food production. This strategy also promotes the circular economy. To improve the efficacy and efficiency of these devices, further effort must be made [8, 9]. Although the major focus of FW management approaches is waste reduction, the possibility of new waste valorization solutions is also a feasible and maybe lucrative choice. These solutions must have long-term effects on the environment and society in order to be sustainable [10]. On the other side, creative approaches to food that take FW into account across the whole food life cycle, including initial production, processing, packing, manufacturing, distribution, retail and consumption, can also promote food chain sustainability.

United Nations (UN) Sustainable Development Goals (SDGs) highlight FW and loss as impediments to sustainable consumption. Goal 12.3 aims to decrease FW at retail and consumer levels and along production and supply chains, including post-harvest losses, by 2030. The European Commission (the Commission) has established a reduction objective of 50% for wasted edible food by the year 2020 at both the regional and national levels, as well as a reduction target of 20% for resource inputs in Europe's food chain by the same year. In 2015, United States Department of Agriculture (USDA) and United States Environmental Protection Agency (EPA) established the first-ever domestic aim to decrease the amount of food loss and waste by half by the year 2030. In 2016, France passed a new rule that prohibits shops in the country from wasting or destroying food

DOI: 10.1201/9781003315469-10

that has not been sold and requires them to donate it to charitable organizations or to be used as animal feed [11]. Similarly, other countries have also taken several measures to limit FW.

It is important to have a good understanding of the different ways to cut down on FW. This chapter discusses how different countries manage food, including their policies for FW management.

10.2 INTERNATIONAL GOALS AND AGREEMENTS

The problem of food loss and waste (FLW) is now one of the most important issues to consider while developing sustainable food systems. Wasted resources from FLW have an adverse effect on nutrition and food security and contribute considerably to greenhouse gas (GHG) emissions, environmental damage, deterioration of natural ecosystems and loss of biodiversity. The aim of responsible consumption and production is covered by Sustainable Development Goal number 12, which is depicted in Figure 10.1(a) and (b).

There are a total of 17 SDGs, and reducing FLW is the specific aim of SDG 12.3. United Nations Food and Agriculture Organization (FAO) and United Nations Environment Programme (UNEP), in their roles as custodians of this aim, measure and monitor progress on initiatives to minimize FLW in comparison to the Global Food Loss and Waste Index.

10.3 POLICIES OF VARIOUS COUNTRIES

The identification of the problem and the definition of the targets and goals is carried out through the use of SMART (Specific, Measurable, Achievable, Relevant, Time-Bound) objectives, as close monitoring, performance evaluation and follow-up activities are critical components of the successful reduction of FW.

Key performance indicators (KPIs) need to be defined and measured so that preventative measures may be tracked, followed up and prioritized. These metrics will keep tabs on how well each initiative is progressing towards its objectives, assist in making tough choices, like weighing the environmental advantages of action against the expense of putting it into practise. The KPIs are analyzed to locate areas that need some work and devise a strategy for further action, assuring the KPIs' continued viability over the long run. This assessment framework utilizes a three-stage methodology to plan, implement and evaluate various activities and strategies to reduce FW. However, the indicators in the assessment framework should be adapted to the nature and aims of the activity, as FW reduction initiatives might vary considerably. That's why it's important to compare just those activities that use the same KPIs. In order to create and evaluate effective actions and strategies to reduce FW, it is crucial that the right information is made available to calculate the SMART goals and KPIs of the assessment framework [12]. Different nations have implemented various approaches to the problem of managing FW. By mandating rules for FW prevention, segregation and correct disposal, these initiatives encourage interventions at various stages of the process. The subsections that follow detail a few instances.

10.3.1 European Union

There were around 638 Mt of food commodities accessible for human consumption in the European Union 7 (EU7) in 2018 and about 129 Mt (fresh weight) of FW was produced along the whole food supply chain, as determined by a Mass Flow Analysis by Caldeira et al. (2019) [12]. Therefore, 20% of all food produced is lost or wasted. This value is far greater than the amount of FW generation predicted by the FUSIONS, which is often utilized by policymakers to develop waste management strategies. When broken down by category (as shown in Table 10.1), the most wasted foods are fish (51%), vegetables (46%), fruits (41%), oil crops (36%) and meat (23%). Despite having the highest rates of FW (51% and 31%, respectively), the fish and egg food categories, which represent the smallest segments of the food supply chain, also create the lowest quantities of FW in absolute terms [13].

FIGURE 10.1 (a) The global goals for sustainable development; (b) Targets of sustainable development goal 12.

TABLE 10.1
The percentage of each food group's value of waste produced in the EU-28

Food waste per food supply chains stage	Food waste/total food available
Meat	23%
Fish	51%
Dairy	5%
Eggs	29%
Cereals	20%
Fruits	41%
Vegetables	46%
Potatoes	22%
Sugarbeets	4%
Oil crops	36%

Data source: Caldeira et al. (2019) [13].

The consuming phase accounts for 46% of all FW, followed by the primary production phase (25%) and the processing and manufacturing phases (24%). Approximately 5% of food is wasted throughout the distribution and retail stages.

Fruit and vegetables make up just 21% of the readily available food. Yet, they are responsible for 76% and 41% of the FW produced during primary consumption and production. Due to their large inedible proportion at the point of purchase and high perishability in comparison to other food groups, these food categories account for sizeable portions of the FW produced during the consuming stage.

Oil crops account for 33% of the entire amount of wasted food that is produced as a by-product of food processing and manufacturing. This is mainly attributable to the production of olive oil (e.g., waste pomace). Since the processing of fruit and fish creates a substantial quantity of FW that is typically not valorized, significant quantities are also contributed at this stage by fish (10%) and fruit (20%). On the other hand, a substantial portion of the inedible by-products that are generated during the processing of various categories of food are valorized by other sectors. As a result, they are not considered to be examples of FW. For instance, during processing meat, the bones, blood, inedible organs and skin are repurposed as fertilizer, feedstuffs, binders, clothing, pharmaceuticals and so on. Additionally, residues from the processing of cereals, brewer's spent grain from the brewing of beer, oilcake from the production of vegetable oil and residues from the potato processing industry are frequently used as animal feed [14, 15].

Reducing the amount of food that is wasted is essential to attaining sustainability. The Commission has made a commitment to reducing the amount of food that is wasted per capita at retail and consumer levels by the year 2030 as mentioned earlier and recently approved EU initiatives on issues related to FW. Using data from planned Member State reporting in 2022, the Commission intends to propose legally binding objectives to decrease FW throughout the EU and to incorporate the prevention of food loss and waste into existing EU programmes. Initially created in the 1970s to help prioritize waste management measures, the waste hierarchy has now been modified to apply to perishable food items. The preferred strategies are arranged in a pyramid (as shown in the Figure 10.2), with prevention measures at the top, followed by pathways for reusing excess food fit for human consumption, for reusing food no longer intended for human consumption as animal feed,

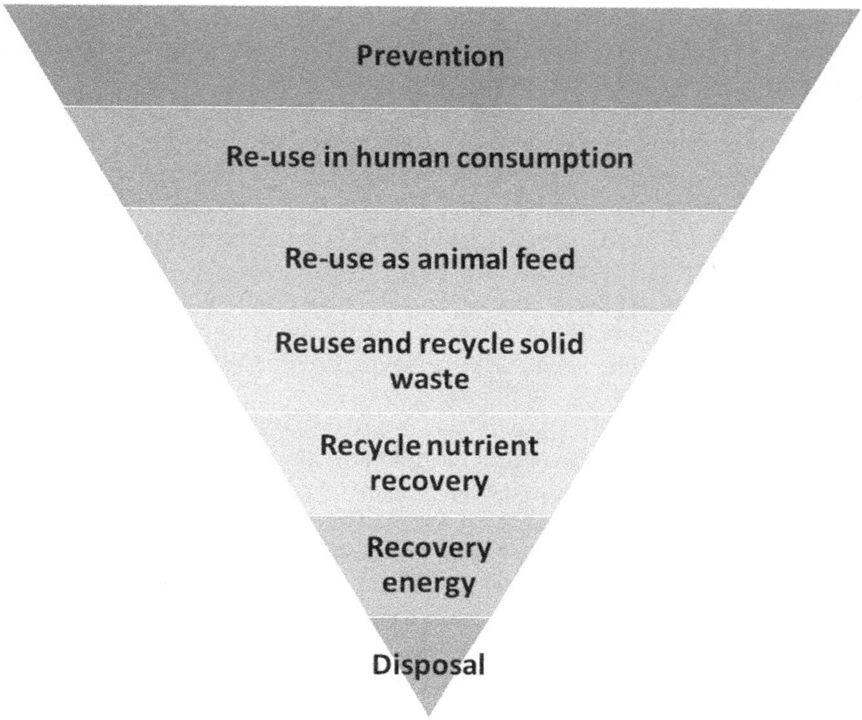

FIGURE 10.2 Prioritization of food waste (FW) prevention techniques according to a hierarchy.

for recycling materials into high-added-value products (without complete degradation), for nutrient recycling, for recovering energy and for disposing of FW as the least preferred choice.

The economic viability of the initiatives further down the pyramid might be impacted by the decision to reduce FW (for instance, there will be less need for investing in recycling and valorization technologies or for human resources to distribute extra food). FW reduction methods should be developed with the FW hierarchy in mind. Strategies like this rely on enactments whose results need to be assessed in terms of quality, efficacy, efficiency, longevity, replicability, scalability and cross-sectoral collaboration [12].

Directive No. 2018/851(EU) of the European Parliament and of the European Council on May 30, 2018, amending Regulation (EC) No. 2008/98/EC is the primary piece of EU law governing municipal solid waste (MSW) at the present time. According to the "waste hierarchy," which categorizes waste management into five stages—prevention, reuse, recycling, incineration and disposal—the Directive of the European Parliament (Dir) reaffirmed the necessity of these practices [16].

First, in order to encourage waste reduction and recycling strategies, member states are required to draught and approve waste management plans as well as waste prevention programmes (Article 29 Dir 2008/98/EC). With this in mind, the EU has reaffirmed the objectives established by Dir 2008/98/EC (Article 11), which promoted 50% preparation for reuse and recycling of MSW by 2020. Additional loftier objectives were authorized by Dir 2018/851. Enhancing the separation of the waste collected is necessary to raise recycling rates. Effective collection programmes will promote recycling habits and improve the standard of secondary raw materials. Dir 2008/98/EC urges "by 2015 separate collection should be established up for at least: paper, metal, plastic and glass" in addition to these measures.

The EU also aims to minimize the amount of packaging waste produced by addressing the decrease in the amount of packaging goods put on the market and, on the other hand, by ensuring better management of packaging products at the end of their useful lives through recycling and recovery techniques. Dir (EU) 2018/852, which modifies Dir 94/62/EC, establishes this. EU supports Extended Producer Responsibility (EPR) programmes, among other things. In fact, industries are urged to reduce waste creation provided they bear financial responsibility for the processing and disposal of their goods. This will set in motion a positive feedback loop that can help promote things like the development of reusable packaging and innovative reuse techniques (Article 5 (1) of Dir 2018/ 852), the dissemination of recycled-content packaging and the production of recyclable packaging (using materials that conform to multiple recycling processes or bio-based materials) (Article 6 (6) of Dir 2018/ 852). EU offers numerous strategies that member states might use, including the use of deposit-return programmes, quantitative or qualitative goals, economic incentives and minimum percentages of recyclable products (for each stream) introduced to the market annually are only a few of the topics covered [16].

It is obvious that waste management based on segregated collection is required to fulfil those aims. The reduction of waste that is landfilled and the secure placement of such waste in sites are also addressed by EU action. For this reason, Council Dir 1999/31/EC's landfill limits were increased by Dir 2018/850. Reducing waste disposal in landfill to a minimum is essential to ensure a transition towards prevention, reuse and recycling. Particularly, the types of waste include plastic, glass, paper, metals and biodegradable municipal garbage that might be better managed through alternative recovery techniques. Article 5(3)f mandates that member states intervene in order to limit the landfill for all wastes that have already been gathered for reuse or recycling due to this. With regard to municipal biodegradable garbage that has not been treated, this purpose is particularly crucial. If left unchecked, this last one has been shown to have significant detrimental environmental effects. By 2035, member states must reduce the quantity of MSW they dump to less than 10% of the entire amount of municipal solid waste they create, according to Article 5(3a) of the Waste Convention [16].

In conclusion, member states have improved their waste management techniques towards circular economy during the past 20 years. The primary practices and requirements to decrease waste generation were regulated by Dir 2008/98/EC. Additionally, Dir 1999/31/ and Dir 94/62/EC, which both work to remove garbage from landfills, encourage separate collection for waste recycling. Additionally, the 2018 revisions to the EU Directives emphasize the goals of waste reduction and increased recycling. As a consequence, between 2006 and 2016, the amount of MSW that was landfilled in the EU reduced from 220 kg per person to 118 kg per person, while waste incineration increased from 103 kg to 133 kg per person and trash recycling increased from 119 kg to 141 kg per person. Despite those outcomes, waste management still needs development in order to end the material loops. The next part will give four case studies that were chosen to examine the various strategies around the EU connected to the switch from waste management to the circular economy and their results [16].

10.3.2 United States of America

From 1970 to 2017, the annual domestic plant FLW output in the United States rose from 3.34×10^{10} kg to 6.18×10^{10} kg. This doubling may be attributed to increasing rates of urbanization and gross domestic product (GDP) as well as a more prosperous population and higher average salaries at home. Vegetable FLW was more strongly correlated with these socioeconomic characteristics than grain FLW was, whereas fruit FLW was not substantially influenced. Half of all FLW was comprised of vegetables, with production increasing from 1.73×10^{10} kg/year in 1970 to 3.39×10^{10} kg/year in 2017. The yearly quantity of FLW produced by vegetables was between five and two times larger than the annual amount of FLW produced by fruits and grains. This disparity across dietary groups can be explained by the fact that Americans consume a greater quantity of vegetables in comparison to their intake of grains and fruits. Previous research found that the community that had the highest consumption of veggies also had the highest rate of vegetable waste, which was 4.7 times higher than the group that had the lowest consumption of vegetables [17]. The urbanization rate in the United States grew from 73.6% in 1970 to 82.7% in 2017, which means that 114 million more individuals now have access to diets that are predominantly comprised of vegetables and are thus healthier. In addition, social marketing efforts encourage the use of vegetables to better the health of families [18]. Inappropriate food storage, which fails to consume food, is another possible contributor to the increased amount of wasted vegetable food [19, 20]. High-income areas have a higher incidence of vegetable FW because affluent people are more likely to reject or ignore items with flaws (e.g., misshapen, unattractive or surface-damaged products). Businesses and farmers must throw away perfectly edible food because it doesn't meet the standard aesthetic requirements [21]. Another significant aspect that contributes to FW outside of aesthetics, is the sensory qualities of vegetables. For instance, the bitterness and pungency of vegetables might affect customer preferences. Due to the variability in sensory aversion, it is challenging to forecast customer behaviour, and this uncertainty in retail marketing results in vegetable waste [22].

Reducing FW needs multi-level solutions that span across a "solution hierarchy," which encompasses governance, stakeholder collaboration and public awareness, as depicted in Figure 10.3. Governments have the ability to initiate and carry out a wide variety of projects by means of regulation and educational campaigns [23]. Government initiatives to lessen FLW fall into three categories: prevention, recovery and recycling. Continuing to allow more variation through broader

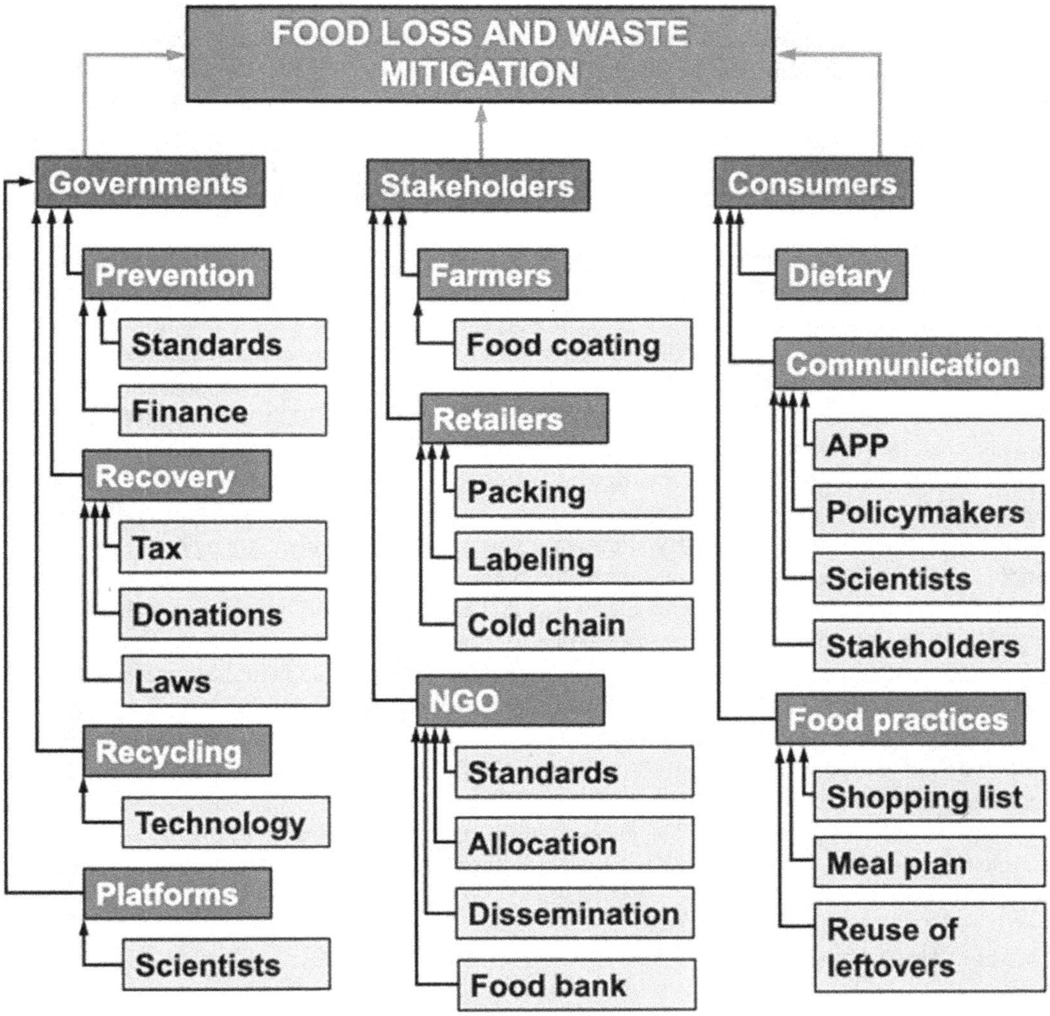

FIGURE 10.3 Role of government, stakeholder and consumers in food loss and waste mitigation [Reproduced with permission from Sun et al. (2022) [26]].

cosmetic standards for agricultural products, implementing standardized date labelling systems and imposing financial penalties for the disposal of FW are all examples of measures that could be taken to reduce the problem's origins and thus increase the effectiveness of prevention efforts [24, 25]. For instance, customers frequently get the terms "sell-by," "use-by," "best-by," and "best before" mixed up, which results in around 7% of FLW during the consuming stage. When date labels are standardized, there is the potential for a decrease in FW of 5.82×10^8 kg per year [26]. Recovery means feeding those who are hungry by offering or increasing financial incentives to promote farm-level food recovery and commercial food contributions, harmonizing health department rules on donation liability laws and promoting and developing potential food donors [27]. Donations from businesses such as manufacturers, merchants or restaurants have the potential to divert 1.10×10^9 kg of FW each and every year [26]. In order to gather food in a timely and effective manner that is accessible for contributions, logistics and transportation are required. For instance, 6.43×10^8 kg of FW may be prevented and 1.07×10^9 meals can be recovered if the infrastructure for small-scale transportation is improved, long-haul transportation capacities are increased or other ways that enable donations from more enterprises are implemented [26]. Composting, recycling and animal feeding are all examples of recycling. Feeding animals might reduce annual FW by 6.04×10^7 kg. Additionally, governments might establish prohibitions or penalties for disposing of FW in landfills, give rewards for reusing FW and enable kerbside pickup of biodegradable food scraps for the convenience of customers [25]. The Nashville Food Waste Initiative (NFWI), launched in 2015 by the Natural Resources Defense Council (NRDC) in the United States, is a representative programme that promotes ideas to address FLW from the perspectives of prevention, recovery and recycling. The main accomplishments include (i) creating proposals for a citywide FW strategy for the metro government, (ii) including lessons on FW in the curriculum of public schools and (iii) leading the neighbourhood in gathering and evaluating FW indicators for the city. Governments should also set up communication channels to make sure data on FW is reported and shared [28]. This information might be used to create food policies that address FW and enhance food consumption plans. Before implementing successful interventions, governments should build policy dissemination platforms to increase the public's knowledge of the implications of edible FW generated across a variety of food kinds and production stages [27]. The Chinese government, for instance, adopted the Anti-Food Waste Law in April 2021 and began the "Clean Plate Campaign" in 2020 to reduce FW. The rules and laws had far-reaching implications for other countries efforts to reduce FW [23].

The potential energy and water savings by recycling plant-based FW into the agricultural and philanthropic sectors of the food supply chain are considerable (e.g., energy capture). Energy and materials may be recovered from FW through a variety of management strategies. These include landfilling, burning, composting and anaerobic digestion. Utilizing the Waste Reduction Model (WARM) developed by the EPA, we determined that composting and burning vegetable and fruit wastes with energy capture may prevent the majority of carbon dioxide emissions relative to landfilling treatment. Composting could be the most effective and scalable method of waste treatment, taking into account the fact that burning waste is inconvenient for households and communities. Composting at home on a small scale is said to be able to provide a net financial advantage of 4.87 million dollars and get rid of at least 9.36×10^7 kg of FW annually. Software that allows for two-way communication with clients, like The Nashville Waste and Recycling App, can also help enhance the FW management programme. The purpose of the app is to give locals all the data they need to recycle their FLW, from rules to how to categorize their FW to where they may bring it [26].

10.3.3 Asian countries

10.3.3.1 Japan

In order to improve resource efficiency and productivity, waste management and recycling regulations in Japan throughout the 1990s placed a strong emphasis on the idea of the Sound Material-Cycle

Society. This strategy incorporates specific aspects of sustainable consumption and production (SCP), such as lifestyle modifications, consumer behaviour adjustments and green buying. Japan's waste management and recycling policy switched from focusing on cleanliness and disposal to concentrating on resource usage when their environmental agency was elevated to a Ministry (the Ministry of the Environment, MOE) in 2000. The "Fundamental Law for Establishing a Sound Material-Cycle Society (2000)" incorporates a wide range of policies, both old and modern. Similar to the national goal of creating a "sound material-cycle society," the Fundamental Plan for Establishing a Sound Material-Cycle Society (2003) sets numerical targets, assigns specific roles to various stakeholders and offers guidelines to motivate individual efforts to align with this broader vision [11, 29]. Japan was one of the first countries to undertake a policy effort centred on managing FW, making it one of the pioneering nations in the fight against FLW. Despite this, FLW continues to be significant problem in the nation because of the low level of food self-sufficiency and the lack of appropriate landfill sites for garbage disposal. An array of institutional initiatives have been proposed since the "Food Waste Recycling Law" (also known as the "Law for Promotion of Recycling and Related Activities for Treatment of Cyclical Food Resources") was approved in 2001. For instance, the national government has not only published national studies on the subject but has also conducted surveys on FW production among different food-producing enterprises, wholesalers, retailers, eateries and householders. Similar to this, several studies, including a variety of academic research articles on the subject, have been conducted in recent years concentrating on national FW concerns. Nagano (2014) gave a succinct overview of the state of the food recycling industry in Japan. Shoji (2014) offered an overview of the existing situation and forecast for the development of FW recycling among food-related industry operators and the Ministry of the Environment's initiatives to encourage the use of waste biomass [30]. Ishikawa (2014) released a report on initiatives made by a working committee to look at corporate practices to minimize FW [31]. A view for municipal solid waste management initiatives intended to encourage FW reduction and recycling was presented by Sakai et al. (2014) [32]. Marra (2013) investigated national waste stream logistics and policy in connection to the social and cultural history of Japan [33]. Policy advancements in Japan and the United Kingdom concerning FW have been surveyed by the OECD [34] and evaluated by Takata et al. (2012), who evaluated the environmental and economic efficiency of food recycling facilities included in the recycling loop [35]. However, a thorough investigation of FLW throughout Japan's food distribution system has not yet been undertaken, leading to a muddled and incomplete image of problem regions and potential solutions.

In a nutshell, Japan passed the Food Waste Recycling Law in 2001, which was later revised in 2007 and again in 2015. The purpose of the law is to recycle FW produced by companies and enterprises associated with the food industry. As a result of this, the recycling goals for each sector of commerce and industry have been set at the following percentages by March 2020: 95% for food producers, 70% for wholesalers, 55% for retailers and 50% for restaurants [35]. In addition, large volume generators of FW (those who produce more than 100 tonnes of FW annually) are required to submit data related to FW on an annual basis. This data includes, among other things, information on the total amount of FW produced as well as the total amount recycled [35]. In order to facilitate effective waste transportation and collection services, the Japanese government has also developed a certification programme for enterprises called "Recycling Loops," under which certified firms are excluded from rules on the unloading and loading of FW. Additionally, firms involved in the food industry have established a working group to examine business practices by executing pilot projects to assess expiry dates and labelling practices [34]. Other bottom-up initiatives, including the construction of food banks by the Japanese people, complement the efforts made by the Japanese government. These programmes give consumers the chance to share undesired food that would otherwise be thrown away and they create food banks.

10.3.3.2 Hong Kong

Hong Kong is a cosmopolitan metropolis that provides convenient access to a wide variety of culinary goods from around the world. The local FW production rate is around 3,648 t/day, with one-third coming from the commercial and industrial sectors and two-thirds from homes. As a result of Hong Kong's rapid rise to a higher standard of living in recent years, the quantity rose dramatically. Up to 38% of municipal solid trash that was disposed of in landfills in 2019 is composed of FW [36]. FW in Hong Kong is expected to be reduced from 3,600 t/d in 2014 to 2160 t/d in 2022 as a result of community mobilization, promotion of FW separation, recycling and treatment of separated FW, as well as treatment of non-separated FW prior to its final disposal [37]. The pace at which municipal solid trash is disposed of in Hong Kong is targeted for a 40% reduction by the year 2022 and "A Food Waste and Yard Waste Plan for Hong Kong 2014–2022" aims to reduce the quantity of FW that is sent to landfills by a further 40% by the same year. In order to achieve this, it has started campaigns like the "Food Wise Hong Kong Campaign," which disseminates best practices in the industrial and commercial sectors to collaborate with governmental agencies, educational institutions and non-profit organizations to prevent and reduce FW at the source. The "Green Lunch Charter" is the product of collaboration between the general public, academic institutions and the Education Bureau of Hong Kong. This charter urges schools to limit the amount of FW they produce and to utilize disposable lunch boxes [36].

There are many initiatives taken by the government which includes (1) waste management at the source: this is the most important first step to prevent creating FW from the very beginning, and it may be accomplished through the Food Waste Hong Kong Campaign, with applicable programmes and activities; (2) food donation: through the same campaign, commercial enterprises are encouraged to donate food to charitable groups. Additionally, Environmental Product Declarations (EPD) provides financing to nongovernmental organizations (NGOs) that start food recovery programmes under the Environment and Conservation Fund (ECF), lowering the quantity of food wasted. (3) reduction programmes: the ECF has also sponsored "food waste reduction programmes" (on the local level) to better recycle kitchen waste by sorting them at the source. FW composting is one component of community-sponsored trash reduction efforts. An investigation of high-rise and high-density urban centres, like those found across Hong Kong, has revealed that economic incentives are the most effective way to encourage the recycling of waste. The scheme has a strong and positive link with the amount of recycled waste collected from each household [38].

Additionally, since July 2018, the government has been providing a "food waste collection and recycling service," wherein daily collected FW is converted into electricity and compost (by-product) at O Park 1. In 2021, the pilot programme will enter its second phase, collecting FW on a wider scale and eventually including FW from homes. FW created in households and in the commercial and industrial sectors must be handled differently since each set of rules is designed to maximize the recycling potential of the corresponding types of waste. The government also gives grants to businesses at Eco-Park that recycle various materials, including food scraps. (4) Waste-to-energy/resources: to improve the total FW treatment capabilities, anaerobic co-digestion of FW and sewage sludge will be developed and optimized at the O Parks [36].

10.3.3.3 Malaysia

FW in Malaysia is often caused by agricultural techniques and various activities carried out by humans. The agricultural industry of Malaysia made a contribution of 8.9% to the country's GDP in 2015. The oil palm industry was responsible for 46.9% of this GDP, while other agricultural goods accounted for 17.7%. Palm oil had the largest agricultural product production, with 19.96 million metric tonnes produced. Only 10% of the entire oil palm produced was recovered as palm oil, with around 90% of it being wasted [39]. Eighty-three million metric tonnes of oil palm biomass, including fronds, stems, empty fruit bunches, oil palm fibres and shells, were harvested in 2012. Moreover, the pineapple canning process typically produces enormous volumes of trash. While just 0.45 million metric tonnes of pineapples were harvested, an estimated 30–50% of that amount was thrown away as wastage [40].

In Malaysia, the majority of FW is disposed of as MSW, and the garbage is not correctly classified. In line with the Malaysia Solid Waste and Public Cleansing Management Act 2007, Malaysian FW is also included in the category of MSW. Waste accounted for 44.5% of all solid waste in

Malaysia in 2012, per the country's solid waste composition. Malaysians generated 38,000 tonnes of solid waste daily in 2016. Around 39% of this solid waste came from FW. Surprisingly, 3,000 tonnes of FW may have been prevented. A combination of factors, including the country's rapidly growing population (31.7 million in 2016) and its high standard of living, has led to an alarmingly high amount of wasted food. However, it is noteworthy that between the years 2012 and 2016, the percentage of the solid waste comprised of FW in Malaysia reduced from 44.5% to 39% as a direct result of the adoption of more effective FW management in Malaysia [41].

Two of Malaysia's most important waste management tools are the landfill and the incinerator. While landfilling might save money, the vast majority of landfill sites in Malaysia are open dumps, which can have serious environmental consequences [42]. The landfill is no longer a desirable alternative as a result of the rising population, absurdly high property prices, a lack of available land and increased garbage disposal. Since it doesn't result in unpleasant odours or rodent issues and takes up less space than a landfill site, incinerating waste has been employed in several locations in Malaysia [43]. However, incineration comes at a hefty financial and energy expenditure. Second-generation waste management techniques have recently been developed, turning FW into goods with additional value, like chemicals, flavours and biofuels. Since pineapple is one of the key crops that creates substantial waste volumes, bioconversion of pineapple waste to value-added goods has been studied in Malaysia by a number of researchers.

The Solid Waste and Public Cleansing Management Act has been in effect in Malaysia since 2011 and features severe penalties for the unauthorized disposal of trash as well as strict source-separation requirements [44]. It has also mandated the 2 + 1 collection system, wherein residual garbage (including FW) is collected twice weekly, while recyclable and bulky waste is collected once weekly.

10.3.3.4 Singapore

FW is a major problem in densely populated Singapore. As reported by the Ministry of Environment and Water Resources (MEWR), only 1 in 10 tonnes of the 703,200 tonnes of FW produced in Singapore in 2012 gets recycled into animal feeds, organic fertilizer and bioenergy. While Singapore Green Plan 2012 set a 30% recycling rate goal, the actual percentage in 2012 fell short of that mark. In 2002, when the FW recycling rate was the second lowest (6%, 2001), MEWR set the objective at 30%. It was predicted that recycling rates might be increased by a factor of five, leading to a 60% recycling rate by the end of the decade [45].

The recycling rate for FW in Singapore is calculated by dividing the total amount of FW recycled by the total amount of FW recycled plus the total amount of FW disposed of. Daily truck weigh-ins at incinerator facilities and yearly samples of solid waste composition are used to calculate the quantity of FW discharged. In contrast, phone surveys of all possible FW recyclers are used to estimate the amount of FW that gets recycled. Although energy may be produced via incineration, the primary goal is to reduce MSW volumes by as much as 90%. Compared to recycling, incineration's energy recovery isn't nearly as impressive; hence it's sometimes seen as a disposal option rather than recycling. MSW management techniques are a subset of solid waste management. Since 1972, Singapore has used a centralized system for dealing with its solid waste, consisting mostly of landfills and incinerators [46]. A single anaerobic digestion (AD) facility for organic waste (including FW) was in operation at Tuas, Singapore, from 2008 to 2011. Because it is methodical, efficient and easy, the vast majority of Singaporeans have adjusted to the centralized system. Simultaneously, vermicomposting and decentralized aerobic composting (AC) had a relatively small impact on Singapore's waste management.

Most nations, including the more industrialized ones, use landfills as their primary waste management option. Singapore's FW problems were previously unsolvable until 1979 when incineration technology was first implemented. Many environmental concerns were brought on by the practice of dumping FW in landfills. In particular, secondary contamination of groundwater, soil and the air is considerably facilitated by the high microbial decomposition activity facilitated by the high moisture content of FW, which is released as leachate and methane in an anaerobic environment following landfilling. Thus, it is preferable to discover a treatment option for Singapore, which has a severe shortage of available land. As a last resort, incineration is used to handle the ever-increasing amounts of MSW and FW. In modern times, the sole landfill in Singapore, the offshore Semakau landfill, is used for the disposal of burned ash and non-incinerable MSW. It is anticipated that landfill waste will continue to rise until 2030. Among the top five incinerable waste kinds, FW is one of the main causes of this rising pattern [47]. In order to accomplish the additional goals outlined in SGP2012, such as increasing the lifespan of Semakau Landfill to 50 years (2000–2050) and working towards "zero landfill" requirements, an intensive effort to recycle waste is being prioritized [45].

After the Lorong Halus Dumping Ground was closed down in 1999, all waste that needed to be disposed of was sent to incineration plants instead of being dumped. After then, incineration will serve as the primary method of waste disposal in Singapore. Following incineration, FW can have a drop in volume of up to 99.9%. Due to the considerable amount of moisture that it contains, FW was thought to be a tough input during the planning stages of the first incineration facility. The amount of water included in municipal solid waste brings down its total calorific value. As a direct consequence of this, its running costs can go up. In the end, FW is still counted as a substantial input due to the fact that it makes up a considerable amount of MSW and is difficult to separate [48].

With 30 years of incineration under its belt, Singapore is better able to alleviate the effects of its chronic land shortage. However, it's possible that constructing new incineration facilities would be detrimental to the economy and the environment. For the sake of comparison, the Tuas South Incineration Plant processes 3,000 tonnes of municipal solid waste per day and costs roughly S$900 million. Therefore, AD has been developed as a third method of recycling FW [45].

AD is a process in which biogas is produced as a by-product of the organic molecules being digested by anaerobic bacteria [9]. Complex organic matter is broken down by the many microbes and bacteria involved in the process into simpler compounds including methane, carbon dioxide, hydrogen sulphide, ammonium and water. The decomposition of organic matter involves four distinct steps: (1) hydrolysis, (2) acidogenesis, (3) acetogenesis and (4) methanogenesis. AD is the preferred method of treating FW from an ecological standpoint, despite the fact that it does not reduce volume. In 2013, the organic fraction of MSW was treated by more than 560 AD plants with a combined capacity of more than 7.3 TWh [49]. Since bioenergy is one of AD's most important by-products, the process is often used for FW as a kind of recycling. Singapore's annual electrical consumption rose from 37.7 TWh in 2009 to 42.6 TWh in 2012. The local power rate has increased from S$0.205/kWh in 2009 to S$0.279/kWh in 2012 as a result of the increase in worldwide fuel prices. The majority of Singapore's electrical needs are met by using imported natural gas. AD produces biogas with a high energy content; thus, it may be used as a renewable alternative to natural gas.

The rate of recycling of Singapore's MSW increased from 9% to 12% in 2008. The initiation of operations at a centralized AD facility is the cause of the improvement. The increase in the recycling rate is expected to reach 16% by the year 2010. During these three years, the facility served as the community's primary recycler of MSW, transforming the majority of this waste into bioenergy as well as organic fertilizer. In light of this recent advancement, two life cycle assessment (LCA) studies were carried out to compare the environmental effect of AD and incineration in the setting of Singapore. The results of both types of research indicate that aerobic decomposition has a smaller impact on the environment than incineration, particularly in terms of the production of GHGs [49].

The National Environmental Agency (NEA) and the Agri-Food and Veterinary Authority (AVA) in Singapore are leading efforts to decrease FW by educating the public about the perils of wasteful eating habits like last-minute grocery shopping and eating out. These initiatives also include providing tips on meal preparation, food storage and creative ways to repurpose leftovers [45].

10.3.3.5 India

The development of biofuels is the primary emphasis of India's current practices for the management of agro-food waste. It is estimated that only 23% of the country's energy needs are satisfied by domestically produced crude oil; the other 76% is dependent almost entirely on imports of fossil fuels. The mission of the Ministry of New and Renewable Energy (MNRE) is to guarantee that the minimum levels of biofuels are easily accessible in the market to fulfil the demand at any given moment. This is in accordance with the requirement that gasoline and diesel include 5% ethanol by volume. After accounting for competing applications, such as animal bedding, calf feed, organic fertilizer and fuel for cooking and heating, approximately 34% of the total amount of food residue is accessible as excess trash. It is estimated that the bioenergy potential of this excess garbage is 4.15 exajoules (EJ) per year, which is equivalent to approximately 17% of India's primary energy usage. As a direct result of this, ongoing efforts have been directed towards the development of technology for the enhanced production of biofuels utilizing renewable feedstocks. There have been efforts by the Indian Government's Department of Biotechnology (DBT) to establish facilities like California's Joint BioEnergy Institute and Canada's BioFuelsNet, which are devoted to the research and development of biofuels and bioenergy. Following strict requirements and extensive preparation, DBT established the world's first multi-feedstock, second-generation (2G) bioethanol demonstration facility in Uttarakhand in 2016 [4]. Future goals include converting waste into 4-(hydroxymethyl) phenoxyacetic acid (HMP), 2,5-furandicarboxylic acid (FDCA) and 100% polyethylene terephthalate/polyethylene furanoate (PET/PEF), in addition to producing bioethanol on a big scale. On the other hand, municipal solid waste that is produced in India is often converted into energy via the use of traditional technologies such as gas-to-energy from landfills, refuse-derived fuel (RDF) and biogas [50]. These have undergone extensive changes and modifications in India. According to the Planning Commission Report from 2014, urban Indian cities generate 62 million tonnes of MSW, which can produce 439 thousand kilowatts (kW) of power from combustible components and RDF and 72 thousand kW of power from landfill gas. Together, these sources can produce 72 thousand kW of electricity [4].

To uncover and develop them into more inventive goods, the chemical and material potential of FW has to be further investigated. FW streams include a variety of functionalized molecules, including proteins, carbohydrates and biopolymers, which may be collected, concentrated and utilized to create precursors with added value for use in the chemical, cosmetic, pharmaceutical and food sectors [51]. Therefore, efforts have been made to convert these waste streams into high-value chemical intermediates and new bio-based goods. Some of these new bio-based products include enzymes, oligosaccharides, pectin and biosurfactants. Other examples are furfural and lactic acids.

The MSW (Management and Handling) Rule of India, initially enacted in 2000 and revised in 2013 and 2015, requires that waste be separated into three categories before being sent to garbage collectors: biodegradable, nonbiodegradable and hazardous [41]. In addition, businesses that establish collection centres at farms or retail food locations are eligible to receive financial aid from the government. In addition, mega food parks have been developed as a means of linking the agricultural and food sectors together in order to increase productivity while simultaneously decreasing waste.

10.3.3.6 Thailand

First-generation waste valorization strategies, such as repurposing scraps as animal feed or biofuel, are the norm in Thailand, where conventional waste management is practised. Ruminant animals in Thailand are fed mostly on crop wastes such as corn stover, rice straw, cassava chip and sugarcane bagasse. Swine and poultry feed often consists of rice bran, broken rice and an oilseed meal (such as soybean meal).

With an abundance of resources to fuel its facilities, Thailand offers promising prospects for biogas generation [52]. An average biogas plant can convert organic waste (such as that generated by a slaughterhouse or a fresh produce market) into biogas at a rate of 1,987.50 $m^3.day^{-1}$. Molasses and cassava are the primary sources of the cellulosic bioethanol produced in Thailand, alongside biogas. Twenty-one bioethanol facilities in Thailand generated 1.3 billion litres of ethanol in 2016, operating at 81% of their capacity. Due to its uniform composition, FW generated by the agro-food sector is an ideal feedstock for these facilities. The agricultural industry of Thailand relies mostly on rice, cassava and sugarcane [53]. Since rice is the primary source of carbohydrates for Thais and sugarcane is often grown under contract for sugar mills, both rice and sugarcane output are typically stable. The FAO found that in South East Asia, including Thailand, rice accounted for 72% of all cereals wasted [54]. Waste rice can produce as much as 3.4 kg CO_2 eq/kg during decomposition. This might cause the carbon intensity in Asia to increase. As such, it is a pretty pragmatic approach to use these plentiful bioresources for transformation into high-value-added goods.

FW management policies in Thailand have also been issued. In 2015, for instance, the Thai government ran a national campaign called "The National Save Food" in conjunction with FAO to address the issue of FLW. Using measures to separate FW at the source and enforcing the 3Rs Act and the National 3Rs Strategy, it seeks to reduce FW by 5%, 30% and 50% by 2016, 2021 and 2026, respectively [41].

The government of Thailand is making an effort to address the impending issue of FLW. The "Eleventh National Economic and Social Development Plan (2012–2016)" from the Ministry of Agriculture and Cooperatives highlighted the significance of reducing FW and maximizing food production. The plan's primary objective was to improve the food production system, which would involve the improvement of production technology and logistical management. On March 5, 2016, Thailand's Ministry of Natural Resources and Environment highlighted the issue of garbage in its National Agenda. The development of the trash management programme cost the government 6,325 million baht (about US$ 179.05 million). The key challenge for this initiative is to reduce the production of waste and to find an appropriate technology for waste disposal [41].

10.3.3.7 China

Each year, China throws away between 80 and 100 million metric tonnes of edible food. China has thus initiated national-level campaigns like the "Plan of Action for Implementation of Resource Saving and Loss Reduction in the Grain Production Sector" and the "13th Five-Year Plan (2016–2020)," both of which include detailed technical standards and codes for designing grain loss reduction and other strategies for enhancing the efficacy of the country's food distribution networks. The "Food Security Law," which was passed in 2009 and governed the safety concerns of FW treatment and the "Grain Law," which contains provisions to encourage grain conservation and fight FW, are two examples of policies targeted at the treatment of FW [55].

In China, the responsibility for the reduction, reuse and recycling of FW is split among around 12 different governmental departments. The functions of the four ministries that are indicated in Table 10.2 interact with one another and overlap in certain cases. They are all on the same administrative level.

In this context, China's many government departments and agencies operate largely autonomously, despite some overlap and mutual influence between their mandates. Since every FW manager has priorities and spheres of influence, adopting FW policies and standards can be difficult. To handle FW concerns from a legislative perspective, all relevant parties, including federal, state and municipal governments, as well as relevant committees and businesses, must collaborate to create appropriate rules and regulations [2, 56].

10.4 MAJOR CHALLENGES

Nonetheless, in spite of the increased public attention and awareness in recent years, a review of the relevant international literature discovered a paucity of data on food waste as well as loss, with projections that varied widely. Additionally, there were uncertainties in the estimated waste due to a lack of a standardized measurement protocol. As a direct consequence, the nature and scope of FLW across the supply chain continues to be poorly understood. There is a growing realization of the need to reconcile the various definitions of FW in addition to clarifying uncertainties with regard to waste quantities and generation patterns. In recent years, there has been a significant

TABLE 10.2
Different ministries of the People's Republic of China and their responsibilities in food waste management

Ministry	Responsibilities
The Ministry of Housing and Urban-Rural Development of the People's Republic of China (MOHURD)	Solid waste treatment and disposal in China.
The National Development and Reform Commission of the People's Republic of China (NDRC)	Investing and allocating resources, together with coming up with and enacting strategies in yearly, medium-term and long-term development plans.
The Ministry of Environmental Protection (MEP)	Comprehensive management, monitoring and coordination of environmental pollution prevention and control.
The Ministry of Agriculture (MOA)	Control and management of the use of FW as an animal feedstock and agricultural fertilizer.

increase in the amount of food that has been thrown away. In 2013, the World Resources Institute (WRI) started leading a multi-stakeholder initiative to produce a Food Loss and Waste Standard. This standard is intended to reduce the amount of food that is lost or wasted [11]. These uncertainties also highlight the difficulties in determining where and how to decrease waste throughout the whole food production and consumption life cycle. Understanding waste and loss from production and supply networks at the local and global level requires evaluating FW concerns from a life cycle perspective. Although only a few studies have used the LCA approach to the problem of FW, the vast majority of the research has concentrated on the disposal and treatment of FW farther down the food chain.

10.5 CONCLUSIONS

The tremendous amount of food that is thrown out around the world represents a valuable bioresource that can be put to use in the manufacture of high-quality goods while also helping to alleviate the significant challenge posed by the disposal of FW. In the majority of countries, a new paradigm in the environmentally friendly and sustainable production of industrially important products can be observed. This is particularly the case as a result of the consistent efforts of researchers and the increased interest shown by governments in such procedures. In addition to the traditional methods of producing bioenergy, researchers are investigating novel methods of product valorization. In conclusion, even if the implementation of these procedures in large-scale industrial settings has not yet taken place, the present trends are congruent with the objectives of sustainable development.

ACKNOWLEDGEMENT

The study is supported by the Indian National Academy of Engineering (INAE/121/AKF/22), Gurgaon, India. The authors are solely responsible for the opinions, results and conclusions expressed in this study; INAE's viewpoints are not necessarily reflected in any of these aspects.

REFERENCES

1. M. Al-Obadi, H. Ayad, S. Pokharel, M.A. Ayari, Perspectives on food waste management: Prevention and social innovations, *Sustain. Prod. Consum.* 31 (2022) 190–208.
2. N.B.D. Thi, G. Kumar, C.-Y. Lin, An overview of food waste management in developing countries: Current status and future perspective, *J. Environ. Manage.* 157 (2015) 220–229.

3. Food, A.O.o.t.U. *Nations, Global Food Losses and Food Waste—Extent, Causes and Prevention*, Food and Agricultural Organisation of the United Nations, 2011.
4. S. Otles, S. Despoudi, C. Bucatariu, C. Kartal, *Food Waste Management, Valorization, and Sustainability in the Food Industry, Food Waste Recovery*, Elsevier, 2015, pp. 3–23.
5. P. Centobelli, R. Cerchione, E. Esposito, M. Raffa, The revolution of crowdfunding in social knowledge economy: Literature review and identification of business models, *Adv. Sci. Lett.* 22(5–6) (2016) 1666–1669.
6. S. Shashi, R. Cerchione, R. Singh, P. Centobelli, A. Shabani, Food cold chain management: From a structured literature review to a conceptual framework and research agenda, *Int. J. Logist. Manag.* (2018).
7. W.B. Traill, M. Meulenberg, Innovation in the food industry, *Agribusiness: an International Journal* 18(1) (2002) 1–21.
8. C.W. Babbitt, *Foundations of Sustainable Food Waste Solutions: Innovation, Evaluation, and Standardization*, Springer, 2017, pp. 1255–1256.
9. P. Duarah, D. Haldar, A.K. Patel, C.-D. Dong, R.R. Singhania, M.K. Purkait, A review on global perspectives of sustainable development in bioenergy generation, *Bioresour. Technol.* 348 (2022) 126791.
10. M. Mourad, Recycling, recovering and preventing "food waste": Competing solutions for food systems sustainability in the United States and France, *J. Clean. Prod.* 126 (2016) 461–477.
11. C. Liu, Y. Hotta, A. Santo, M. Hengesbaugh, A. Watabe, Y. Totoki, D. Allen, M. Bengtsson, Food waste in Japan: Trends, current practices and key challenges, *J. Clean. Prod.* 133 (2016) 557–564.
12. C. Caldeira, V. De Laurentiis, S. Sala, *Assessment of Food Waste Prevention Actions, Development of an Evaluation Framework to Assess the Performance of Food Waste Prevention Actions* (2019). file:///C:/Users/ENV/Downloads/caldeira_et_al_2019_del2_online.pdf
13. C. Caldeira, V. De Laurentiis, S. Corrado, F. van Holsteijn, S. Sala, Quantification of food waste per product group along the food supply chain in the European Union: A mass flow analysis, *Resour. Conserv. Recycl.* 149 (2019) 479–488.
14. B. Debnath, D. Haldar, M.K. Purkait, Potential and sustainable utilization of tea waste: A review on present status and future trends, *J. Environ. Chem. Eng.* 9(5) (2021) 106179.
15. A.M. Shabbirahmed, D. Haldar, P. Dey, A.K. Patel, R.R. Singhania, C.-D. Dong, M.K. Purkait, Sugarcane bagasse into value-added products: A review, *Environ. Sci. Pollut. Res.* (2022).
16. E. Chioatto, P. Sospiro, Transition from waste management to circular economy: The European Union roadmap, *Environ. Develop. Sustain.* (2022). https://doi.org/10.1007/s10668-021-02050-3
17. Z. Conrad, Daily cost of consumer food wasted, inedible, and consumed in the United States, 2001–2016, *Nutr. J.* 19(1) (2020) 1–9.
18. C.M. Pollard, M.R. Miller, A.M. Daly, K.E. Crouchley, K.J. O'Donoghue, A.J. Lang, C.W. Binns, Increasing fruit and vegetable consumption: Success of the Western Australian Go for 2&5® campaign, *Public Health Nutr.* 11(3) (2008) 314–320.
19. G. Waitt, C. Phillips, Food waste and domestic refrigeration: A visceral and material approach, *Social & Cultural Geography* 17(3) (2016) 359–379.
20. R.A. Neff, M.L. Spiker, P.L. Truant, Wasted food: US consumers' reported awareness, attitudes, and behaviors, *PloS One* 10(6) (2015) e0127881.
21. S.T. Hingston, T.J. Noseworthy, On the epidemic of food waste: Idealized prototypes and the aversion to misshapen fruits and vegetables, *Food Qual Prefer* 86 (2020) 103999.
22. A.A. Poelman, C.M. Delahunty, C. de Graaf, Vegetables and other core food groups: A comparison of key flavour and texture properties, *Food Qual Prefer* 56 (2017) 1–7.
23. L. Xue, X. Liu, S. Lu, G. Cheng, Y. Hu, J. Liu, Z. Dou, S. Cheng, G. Liu, China's food loss and waste embodies increasing environmental impacts, *Nat. Food.* 2(7) (2021) 519–528.
24. B. Lipinski, *The Complex Picture of On-farm Loss*, Crawford Fund, 2016. https://doi.org/10.22004/ag.econ.257220
25. D. Gunders, J. Bloom, *Wasted: How America is Losing up to 40 Percent of Its Food from Farm to Fork to Landfill*, NRDC, 2017.
26. H. Sun, Y. Sun, M. Jin, S.A. Ripp, G.S. Sayler, J. Zhuang, Domestic plant food loss and waste in the United States: Environmental footprints and mitigation strategies, *Waste Manag.* 150 (2022) 202–207.
27. M.K. Muth, C. Birney, A. Cuéllar, S.M. Finn, M. Freeman, J.N. Galloway, I. Gee, J. Gephart, K. Jones, L. Low, A systems approach to assessing environmental and economic effects of food loss and waste interventions in the United States, *Sci. Total Environ.* 685 (2019) 1240–1254.
28. J. Zhuang, F.E. Löffler, G.S. Sayler, Creating a research enterprise framework for transdisciplinary networking to address the food–energy–water nexus, *Engineering.* 11 (2022) 95–100.

29. A.V. Shekdar, Sustainable solid waste management: An integrated approach for Asian countries, *Waste Manag.* 29(4) (2009) 1438–1448.
30. M. Shoji, The state and problem of recycling of food waste, *Mater Cycles Waste Manag Res.* 25(1) (2014) 13–19.
31. T. Ishikawa, Reporting on efforts by working group for business practice improvement regarding food waste reduction, *Mater. Cycles Waste Manag. Res.* 25(1) (2014) 43–54.
32. S. Sakai, J. Yano, Outlook for municipal solid waste management strategies for promoting reduction and recycling of food waste, *Mater Cycles Waste Manag Res.* 25(1) (2013) 55–68.
33. F. Marra, *Fighting Food Loss and Food Waste in Japan*, Unpublished Master's Thesis, Leiden University. Retrieved from http://www. fao. org/fileadmin/user_upload/save-food/PDF/FFLFW_in_J apan. pdf (accessed 8 October 2018) (2013).
34. A. Parry, P. Bleazard, K. Okawa, *Preventing Food Waste: Case Studies of Japan and the United Kingdom*, OECD, 2015. https://doi.org/10.1787/18156797
35. M. Takata, K. Fukushima, N. Kino-Kimata, N. Nagao, C. Niwa, T. Toda, The effects of recycling loops in food waste management in Japan: Based on the environmental and economic evaluation of food recycling, *Sci. Total Environ.* 432 (2012) 309–317.
36. M.H. Wong, Integrated sustainable waste management in densely populated cities: The case of Hong Kong, *Sustain Horizons* 2 (2022) 100014.
37. H. ENB, *A food waste & yard waste plan for Hong Kong 2014–2022*, 2014.
38. Y. Yau, Domestic waste recycling, collective action and economic incentive: The case in Hong Kong, *Waste Manag.* 30(12) (2010) 2440–2447.
39. M.F. Awalludin, O. Sulaiman, R. Hashim, W.N.A.W. Nadhari, An overview of the oil palm industry in Malaysia and its waste utilization through thermochemical conversion, specifically via liquefaction, *Renew. Sustain. Energy Rev.* 50 (2015) 1469–1484.
40. O.K. Lun, T.B. Wai, L.S. Ling, Pineapple cannery waste as a potential substrate for microbial biotransformation to produce vanillic acid and vanillin, *Int. Food Res. J.* 21(3) (2014) 953.
41. K.L. Ong, G. Kaur, N. Pensupa, K. Uisan, C.S.K. Lin, Trends in food waste valorization for the production of chemicals, materials and fuels: Case study South and Southeast Asia, *Bioresour. Technol.* 248 (2018) 100–112.
42. L. Abd Manaf, M.A.A. Samah, N.I.M. Zukki, Municipal solid waste management in Malaysia: Practices and challenges, *Waste Manag.* 29(11) (2009) 2902–2906.
43. H.A. Rahman, Incinerator in Malaysia: Really needs?, 1 (2013) 678–681. http://psasir.upm.edu.my/id/eprint/29781/
44. I. Lardinois, C. Furedy, *Separation at Source*, WASTE, 1999.
45. B.J.H. Ng, Y. Mao, C.-L. Chen, R. Rajagopal, J.-Y. Wang, Municipal food waste management in Singapore: Practices, challenges and recommendations, *J. Mater. Cycles Waste Manag.* 19(1) (2017) 560–569.
46. R. Bai, M. Sutanto, The practice and challenges of solid waste management in Singapore, *Waste Manag.* 22(5) (2002) 557–567.
47. H.H. Khoo, L.L. Tan, R.B. Tan, Projecting the environmental profile of Singapore's landfill activities: Comparisons of present and future scenarios based on LCA, *Waste Manag.* 32(5) (2012) 890–900.
48. Y.S. Tan, T.J. Lee, K. Tan, *Clean, Green and Blue, Clean, Green and Blue*, ISEAS Publishing, 2008.
49. J.W. Levis, M.A. Barlaz, N.J. Themelis, P. Ulloa, Assessment of the state of food waste treatment in the United States and Canada, *Waste Manag.* 30(8–9) (2010) 1486–1494.
50. N. Gupta, K.K. Yadav, V. Kumar, A review on current status of municipal solid waste management in India, *J. Environ. Sci.* 37 (2015) 206–217.
51. R.A.D. Arancon, C.S.K. Lin, K.M. Chan, T.H. Kwan, R. Luque, *Advances on Waste Valorization: New Horizons for a More Sustainable Society*, Waste Management and Valorization, Apple Academic Press, 2017, pp. 23–66.
52. P. Aggarangsi, N. Tippayawong, J. Moran, P. Rerkkriangkrai, Overview of livestock biogas technology development and implementation in Thailand, *Energy Sustain. Develop.* 17(4) (2013) 371–377.
53. T. Silalertruksa, S.H. Gheewala, Security of feedstocks supply for future bio-ethanol production in Thailand, *Energy Policy* 38(11) (2010) 7476–7486.
54. F.W. Footprint, *Food Wastage Footprint: Impacts on Natural Resources: Summary Report*, Food & Agriculture Org, 2013.
55. G. Liu, *Food Losses and Food Waste in China: A First Estimate*, OECD, 2014. https://doi.org/10.1787/18156797
56. Y. Li, Y. Jin, A. Borrion, H. Li, Current status of food waste generation and management in China, *Bioresour. Technol.* 273 (2019) 654–665.

Index

I

K

L

M

N

O

P

S

T

U

V

X

For Product Safety Concerns and Information please contact our EU representative GPSR@taylorandfrancis.com
Taylor & Francis Verlag GmbH, Kaufingerstraße 24, 80331 München, Germany

www.ingramcontent.com/pod-product-compliance
Lightning Source LLC
Chambersburg PA
CBHW081446280126
38381CB00044B/2729
9781032325255